THE THEORY AND PRACTICE OF HYDRODYNAMICS AND VIBRATION

ADVANCED SERIES ON OCEAN ENGINEERING

Series Editor-in-Chief
Philip L-F Liu (*Cornell University*)

Published

Vol. 13 Water Wave Propagation Over Uneven Bottoms
Part I — Linear Wave Propagation
by Maarten W Dingemans (Delft Hydraulics)
Part II — Non-linear Wave Propagation
by Maarten W Dingemans (Delft Hydraulics)

Vol. 14 Coastal Stabilization
by Richard Silvester and John R C Hsu (The Univ. of Western Australia)

Vol. 15 Random Seas and Design of Maritime Structures (2nd Edition)
by Yoshimi Goda (Yokohama National Univ.)

Vol. 16 Introduction to Coastal Engineering and Management
by J William Kamphuis (Queen's Univ.)

Vol. 17 The Mechanics of Scour in the Marine Environment
by B. Mutlu Sumer and Jørgen Fredsøe (Technical Univ. of Denmark, Denmark)

Vol. 18 Beach Nourishment: Theory and Practice
by Robert G. Dean (Univ. of Florida)

Vol. 19 Saving America's Beaches: The Causes of and Solutions to Beach Erosion
by Scott L. Douglass (Univ. of South Alabama)

Vol. 20 The Theory and Practice of Hydrodynamics and Vibration
by Subrata K. Chakrabarti (Offshore Structure Analysis, Inc., Illinois, USA)

Vol. 21 Waves and Wave Forces on Coastal and Ocean Structures
by Robert T. Hudspeth (Oregon State Univ., USA)

Vol. 22 The Dynamics of Marine Craft: Maneuvering and Seakeeping
by Edward M. Lewandowski (Computer Sciences Corporation, USA)

Vol. 23 Theory and Applications of Ocean Surface Waves
Part 1: Linear Aspects
Part 2: Nonlinear Aspects
by Chiang C. Mei (Massachusetts Inst. of Technology, USA),
Michael Stiassnie (Technion–Israel Inst. of Technology, Israel) and
Dick K. P. Yue (Massachusetts Inst. of Technology, USA)

Vol. 24 Introduction to Nearshore Hydrodynamics
by Ib A. Svendsen (Univ. of Delaware, USA)

Vol. 25 Dynamics of Coastal Systems
by Job Dronkers (Rijkswaterstaat, The Netherlands)

Vol. 26 Hydrodynamics Around Cylindrical Structures (Revised Edition)
by B. Mutlu Sumer and Jørgen Fredsøe (Technical Univ. of Denmark, Denmark)

Vol. 27 Nonlinear Waves and Offshore Structures
by Cheung Hun Kim (Texas A&M Univ., USA)

Advanced Series on Ocean Engineering — Volume 20

THE THEORY AND PRACTICE OF HYDRODYNAMICS AND VIBRATION

Subrata K. Chakrabarti
Offshore Structure Analysis, Inc.
Illinois, USA

World Scientific

NEW JERSEY • LONDON • SINGAPORE • BEIJING • SHANGHAI • HONG KONG • TAIPEI • CHENNAI

Published by

World Scientific Publishing Co. Pte. Ltd.
5 Toh Tuck Link, Singapore 596224
USA office: 27 Warren Street, Suite 401-402, Hackensack, NJ 07601
UK office: 57 Shelton Street, Covent Garden, London WC2H 9HE

British Library Cataloguing-in-Publication Data
A catalogue record for this book is available from the British Library.

First published 2002
Reprinted 2008

THE THEORY AND PRACTICE OF HYDRODYNAMICS AND VIBRATION

Copyright © 2002 by World Scientific Publishing Co. Pte. Ltd.

All rights reserved. This book, or parts thereof, may not be reproduced in any form or by any means, electronic or mechanical, including photocopying, recording or any information storage and retrieval system now known or to be invented, without written permission from the Publisher.

For photocopying of material in this volume, please pay a copying fee through the Copyright Clearance Center, Inc., 222 Rosewood Drive, Danvers, MA 01923, USA. In this case permission to photocopy is not required from the publisher.

ISBN-13 978-981-02-4921-2
ISBN-10 981-02-4921-7
ISBN-13 978-981-02-4922-9 (pbk)
ISBN-10 981-02-4922-5 (pbk)

Printed by Fulsland Offset Printing (S) Pte Ltd, Singapore

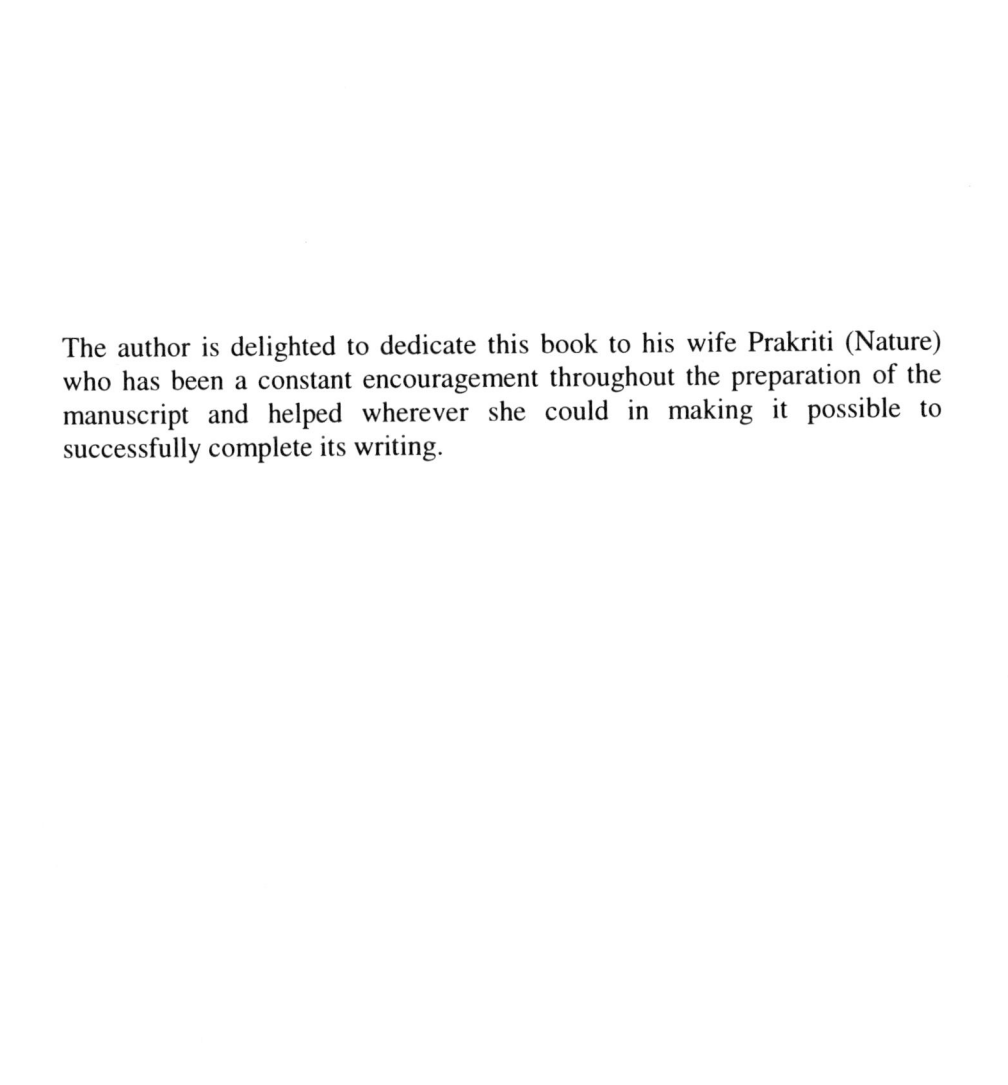

The author is delighted to dedicate this book to his wife Prakriti (Nature) who has been a constant encouragement throughout the preparation of the manuscript and helped wherever she could in making it possible to successfully complete its writing.

ACKNOWLEDGEMENTS

Many individuals were kind to spend their valuable times in reviewing this book. Keith Melin, a friend from the Chicago Bridge and Iron Co., Plainfield, IL. critically reviewed the entire book and provided valuable comments that improved its readability and quality. Dennis Cotter, a past colleague, also reviewed several chapters and checked the contents. Dr. Julio Meneghini reviewed Chapter 8 and improved its technical quality. Chapter 9 was reviewed by Dr. Erling Huse, who made several corrections of an earlier draft. Prabal Chakrabarti edited the preface and first chapter and improved their contents. The author is indebted to all these individuals. There is no doubt that the quality of the book improved because of these contributions.

PREFACE

This book is written in a textbook format such that students of hydrodynamics and vibration may receive a comprehensive understanding of the subject of vibration as it is generated by and related to hydrodynamics. The main objectives of this book are:

- To provide a broad review aimed at a clear understanding of the fundamental concepts underlying the associated fields of hydrodynamics and structural vibrations,
- To describe the basic theories of hydrodynamics and vibration,
- To focus on applications of structural vibrations to problems encountered in structural engineering, with particular emphasis on the response of structures to fluid flow,
- To provide an understanding of the modeling techniques used in the broad field of hydrodynamics and vibration,
- To illustrate practical problems with basic themes in structural vibration related particularly to fluid structure interaction.

Each of the subjects of hydrodynamics and vibration is by itself a broad topic and it is not possible to do a fair treatment of these subjects in their entirety in one single book. No attempt is made to cover them extensively. Instead the fundamentals of these subjects are covered such that their combined application in a structure is understood. The basic objective is to describe the application of vibration to the problems encountered in hydrodynamics. The first part of the book covers the basic development of hydrodynamics and vibration analysis. In the following few chapters, particular reference is made to their interactive effects. The final few chapters of the book are devoted to application of these principles to structures where their combined effects are important. Thus the main purpose of the book is to cover the application of these two subjects of hydrodynamic and vibration in structures. The book, however, is self-contained and the basic principles that are needed to understand their application to structures are included.

The book covers the basics of hydrodynamics and vibration of structures subjected to environmental loads with easy and simple explanations. The vibration control is briefly covered. However, it does not cover shock or acoustics. Particular emphasis is placed in the application of the theory to practical problems. Examples are provided to show how the theory outlined in the book is applied in the design of structures. Examples are borrowed mainly from the novel structures of interest today including offshore structures and components.

The book is written as a textbook with the intention that it may be used as a teaching guide in the universities. The level of the book is expected to be suited for Advanced Engineering Undergraduate (senior and junior level) and First Year Graduate interested in vibrating structures and their design aspects. It should serve as a handy reference book for the design engineers and consultant involved with the design of structures subjected to dynamics and vibration. Such a subject does not appear to be covered in one single book.

CONTENTS

Chapter 1
Introduction to the Book — 1
1.1 What is Hydrodynamics? — 1
1.2 What is Vibration? — 2
1.3 Interrelationship of Hydrodynamics and Vibration — 4
1.4 Overview of Contents — 5
1.5 References — 7

Chapter 2
Basic Theory of Hydrodynamics — 9
2.1 Introduction — 9
2.2 Equations of Motion — 10
    2.2.1 Fluid Flow and Continuity — 10
    2.2.2 Rotational Flow — 11
    2.2.3 Potential Function — 12
    2.2.4 Laplace Equation — 12
2.3 Bernoulli's Equations — 13
2.4 Laminar and Turbulent Flow — 14
    2.4.1 Steady Drag and Lift Force — 18
2.5 Vorticity Parameter, Reduced Velocity and Strouhal Number — 24
2.6 Moving Structures in Fluid — 26
2.7 Free Surface Flows — 34
    2.7.1 Kinematical Condition at the Floor — 36
    2.7.2 Kinematical Condition at the Free Surface — 37
    2.7.3 Dynamic Condition at the Free Surface — 38
    2.7.4 Linear Wave Theory — 39
    2.7.5 Oscillatory Drag and Inertia Forces — 42
    2.7.6 Summary of Hydrodynamic Forces — 45
    2.7.7 Combined Steady and Oscillating Flows — 46
2.8 Exercises — 49
2.9 References — 50

Chapter 3
Basic Theory of Vibration — 51
3.1 Introduction — 51
3.2 Periodic Vibration — 52
3.3 Linear Single Degree of Freedom System — 54
    3.3.1 Transient Response — 56
    3.3.2 Forced Linearly Damped System — 63
    3.3.3 Energy Dissipation — 68
    3.3.4 Mechanical Oscillation — 69
3.4 Nonlinear Single Degree of Freedom System — 73
    3.4.1 Phase-Plane Diagram — 75
    3.4.2 Nonlinearly Damped Free Vibration — 76
    3.4.3 Nonlinearly Damped Forced Vibration — 80
    3.4.4 General Nonlinear Damping — 81
    3.4.5 Nonlinear Restoring Force — 82
    3.4.6 Duffing's Equation — 86
    3.4.7 Nonlinear Damping and Restoring Force — 90
3.5 Vibration of a Rigid Body — 95
3.6 Linear Multi-Degree Freedom System — 100
    3.6.1 Two Degree-of-Freedom System — 100
    3.6.2 Multi-Mass System — 106
    3.6.3 Multiple Degree-of-Freedom System — 107
3.7 Nonlinear Multiple Degree of Freedom Systems — 108
3.8 Continuous Systems — 110
    3.8.1 Modes of Vibration — 111
    3.8.2 Vibration of an Elastic Structure — 112
3.9 Self-Excited Vibration — 115
3.10 Exercises — 116
3.11 References — 118

Chapter 4
Experiments in Hydrodynamics and Vibration — 119
4.1 Introduction — 119
4.2 Planning a Model Test — 120
    4.2.1 Benefits of Model Testing — 121
    4.2.2 Requirements of a Facility — 121

Contents xiii

4.3	Description of Test Facilities	122
	4.3.1 Current Generating Facility	122
	4.3.2 Towing Tanks	127
	4.3.3 Planar Motion Mechanism	129
	4.3.4 Wave Generating Facility	132
	4.3.5 Random Wave Generation	134
	4.3.6 Wind Tunnels	135
4.4	Modeling Laws	137
	4.4.1 Dimensional Method	137
	4.4.2 Froude Similitude	139
	4.4.3 Reynolds Similitude	140
	4.4.4 Strouhal Similitude	143
	4.4.5 Cauchy Similitude	144
	4.4.6 Dynamic Similarity	146
4.5	Hydrodynamic Model	148
	4.5.1 Modeling of Rigid Structure	148
	4.5.2 Modeling of Elastic Structure	149
	4.5.3 Modeling of Mooring Line	150
	4.5.4 Modeling of Riser	153
4.6	Vibration Instruments	158
	4.6.1 Accelerometers	160
	4.6.2 Motion Sensors	161
	4.6.3 Load Measuring Device	163
	4.6.4 Design of a Load Cell	165
	4.6.5 Design of a Propeller Dynamometer	169
	4.6.6 Waterproofing of Transducers	170
	4.6.7 Excitation Level for Transducers	171
4.7	In-Situ Vibration Measurements	171
4.8	Exercises	173
4.9	References	173

Chapter 5
Statistical Theory in Vibration — 175
5.1	Introduction	175
5.2	Random Variables and Processes	176
	5.2.1 Probability Distribution Function	178

	5.2.2 Probability Density Function	179
	5.2.3 Gaussian Distribution	181
	5.2.4 Rayleigh Distribution	183
	5.2.5 Weibull Distribution	184
	5.2.6 Gumbel Distribution	185
5.3	Power Spectrum and Statistics	186
5.4	Extreme Values and Envelope Functions	189
	5.4.1 Extreme Value Function	189
	5.4.2 Design Extreme Value	190
5.5	Response Spectra	193
	5.5.1 Transfer Function	194
	5.5.2 Cross Spectral Analysis	198
	5.5.3 Error Analysis	199
5.6	Exercises	202
5.7	References	206

Chapter 6
Random Vibration   207

6.1	Introduction	207
6.2	Random Single Degree of Freedom Linear System	207
	6.2.1 Non-Harmonic Vibration	208
	6.2.2 Random Vibration	211
6.3	Response of SDOF System to Random Input	212
	6.3.1 Frequency Domain Analysis	212
	6.3.1.1 Frequency domain linear system	213
	6.3.1.2 Frequency domain nonlinear system	214
	6.3.2 Time Domain Analysis	216
	6.3.2.1 Time domain linear system	217
	6.3.2.2 Time domain nonlinear system	218
	6.3.3 Random Decrement Technique	222
6.4	Random Response of Multi-Degree of Freedom System	225
	6.4.1 Multi-Moduled Time-Domain Solution	225
6.5	Response of Nonlinear Systems to Random Input	231
	6.5.1 Stochastic Linearization of Force	231
	6.5.2 Stochastic Linearization of Response	233
6.6	Numerical Methods in Stochastic Dynamics	235

Contents xv

	6.6.1 Eigenvalues and Eigenvectors	235
	6.6.2 Structure Finite Element Method	237
6.7	Response Statistics in Random Excitation	239
6.8	Exercises	245
6.9	References	247

Chapter 7
Fluid Structure Interaction — 249

7.1	Introduction	249
7.2	Structures in Steady Flow	250
	7.2.1 Vertical Cylinder in Uniform Flow	254
	7.2.2 Vertical Cylinder in Shear Flow	257
	7.2.3 Horizontal Cylinder in Uniform Flow	261
	7.2.4 Horizontal Cylinder in Shear Flow	263
	7.2.5 Blockage Factor in Steady Flow	269
7.3	Structures in Waves	271
	7.3.1 Linear Diffraction Theory	272
	7.3.2 Two-Body Interaction Problem	276
	7.3.3 Fluid-Structure Finite Element Method	278
	7.3.4 Stability of Motion	279
7.4	Structure Damping	284
	7.4.1 Damping of a TLP in Heave	285
	7.4.2 Damping in Ship Roll	286
	7.4.2.1 Skin friction	287
	7.4.2.2 Eddy making damping	290
	7.4.2.3 Lift damping	294
	7.4.2.4 Wave damping	295
	7.4.2.5 Bilge-keel damping	295
	7.4.2.6 Damping results	297
	7.4.3 Roll Suppression	302
7.5	Tensioned Riser Analysis	303
7.6	Flexible Line Analysis	308
7.7	Coupled Dynamic Analysis	311
7.8	Exercises	313
7.9	References	314

Chapter 8
Fluid Induced Vibration — 317
- 8.1 Introduction — 317
- 8.2 Flow-Induced Vortices — 319
  - 8.2.1 Formation of Vortices — 319
  - 8.2.2 Reynolds and Strouhal Dependence — 321
  - 8.2.3 Vortex Formation in Steady Uniform Flow — 321
  - 8.2.4 Vortex Formation in Shear Flow — 323
  - 8.2.5 Vortex Formation in Oscillatory Flow — 325
- 8.3 Theory of Vortex-Induced Vibration — 327
  - 8.3.1 Empirical VIV Models — 330
  - 8.3.2 Transverse Oscillator Model — 331
  - 8.3.3 Analytic VIV Model — 335
- 8.4 Eigenfrequencies of Structures — 344
- 8.5 Vortex-Induced Vibration of Risers, Cables and Pipelines — 346
  - 8.5.1 VIV of Risers — 346
  - 8.5.2 VIV of Cables — 348
  - 8.5.3 VIV of Subsea Pipeline — 355
- 8.6 Interference Effect in Vortex-Induced Vibration — 357
  - 8.6.1 VIV between Two Closely Spaced Cylinders — 358
  - 8.6.2 VIV among Multiple Cylinders — 361
- 8.7 VIV Spoilers — 362
  - 8.7.1 Streamline Geometry — 363
  - 8.7.2 Super Smooth Surface — 364
  - 8.7.3 External Damper — 365
  - 8.7.4 Pneumatic Damper — 367
  - 8.7.5 Damping Plate — 367
  - 8.7.6 Helical Strakes — 369
  - 8.7.7 Wavy Edge — 371
  - 8.7.8 External Disturbance — 372
  - 8.7.9 Porosity — 374
  - 8.7.10 Tension Control — 376
- 8.8 Exercises — 377
- 8.9 References — 377

Chapter 9
Practical and Design Case Studies 383
9.1 Introduction 383
9.2 Damping Experiment of a Tethered TLP Caisson 383
    9.2.1 Caisson with Leaf Springs 384
    9.2.2 Caisson with Coil Springs 387
9.3 Field Experiment of Vibration of a Riser 390
    9.3.1 Collision Within Riser Group 390
    9.3.2 Towing of Riser 393
9.4 Impact Loading on a Fixed Platform 397
    9.4.1 Current Effect on Jacket 398
    9.4.2 Jacket Structure in Regular Waves 400
    9.4.3 Pile Structure in Random Waves 402
9.5 Sloshing in a Hydroelastic Vertical Caisson 405
    9.5.1 Simplified Sloshing Theory 407
    9.5.2 Elastic Model Design 415
    9.5.3 Elastic Model Tests 416
9.6 Articulated Tower Instability in Waves 422
    9.6.1 Theory of Inclined Tower 422
    9.6.2 Model Test of Inclined Tower 424
    9.6.3 Analysis of SLIM Response 425
9.7 Hydroelasticity of a Multi-Moduled Structure 428
    9.7.1 Mobile Offshore Base 429
    9.7.2 Analysis of a MOB 430
9.8 Dynamics of a Flexible Jack-Up Unit 436
    9.8.1 Hydrodynamic Jack-Up Analysis 437
    9.8.2 Jack-Up Structural Analysis 438
    9.8.3 Wave Response Method 439
    9.8.4 Response in Regular Waves 441
9.9 Impact of a Submarine with a Jacket Structure 444
    9.9.1 Safety Check on Residual Strength 445
    9.9.2 Structural Damping 446
9.10 References 449

List of Symbols	**453**
List of Acronyms	**457**
Conversion Factors	**457**
Author Index	**459**
Subject Index	**463**

Chapter 1

Introduction to the Book

The purpose of this book is to address separately, and then jointly, the subjects of hydrodynamics and vibration of structures. The book connects these two independent topics and shows the effect each has on the other. It concentrates on the vibration aspects of structures caused by fluid loading. The basics of both subjects are covered and the interrelationship of these two fundamental aspects of structures immersed in fluids is illustrated. The book includes both the theory and the application of the two concepts as they apply to incompressible fluids.

Hydrodynamics and vibration may be considered to be a cause and an effect. Generally the hydrodynamic interaction between a fluid system and a structure initiates a motion or vibration in the system encompassing the structure. Conversely, the vibration of a structure in the fluid, in turn, generates a change in the fluid motion in the vicinity of the structure.

1.1 What is Hydrodynamics?

Hydrodynamics is the science of expressing the dynamic movement of fluids in mathematical formulas. The discipline provides an analytical description of fluid dynamics following certain basic laws of fluid and some engineering principles. Since it is often difficult to represent a physical phenomenon with an equivalent mathematical analogy, it is critical that the hydrodynamic description adequately defines the physical concept. In developing the mathematical description, certain physical laws are satisfied giving rise to the basic equation of the fluid particle motion. This requires certain simplification and assumptions that are justified by the physical constraints. Similarly, certain boundary conditions must be met for a given hydrodynamic problem. This gives rise to what is commonly known as a boundary value problem consisting of a fundamental differential equation and a set of boundary conditions.

Take, for example, a structure that is placed in a fluid field, which is in motion. Mathematically, one needs first to describe the uninterrupted motion of the fluid. Even then, certain boundary conditions are imposed arising from the boundaries defining the containment of the fluid. For example, in an open channel flow, the bottom boundary is usually a solid surface, such as the river floor where the fluid must satisfy the bottom boundary condition. Similarly, the interface between the fluid free surface and the air will impose a second boundary condition known as the free surface boundary condition. When a structure is placed in the fluid, then the fluid flow in the vicinity of the structure will be distorted, which should be taken into account by another boundary condition that is determined by the shape of the submerged structure. This is commonly referred to as the body surface boundary condition. The fluid in this case should satisfy a far-field condition as well, at which point it should restore to its original form.

Once identified, the boundary conditions are then satisfied by posing the boundary value problem, and the solution is determined either analytically or numerically, depending on the problem's complexity. Essentially, this solution describes the fluid flow around the structure in terms of mathematical functions. This expression reveals the effect of the fluid upon the structure, including the distribution of fluid pressures or forces. If the structure is allowed to move or is flexible enough to distort due to fluid flow around it, then the displacement of the structure may be determined by considering the hydrostatic and dynamic properties of the structure.

The theory of hydrodynamics has been thoroughly described by Lamb (1945) and Milne-Thompson (1960), which should be excellent reference books for the readers. The books cover many aspects of theoretical hydrodynamics and develop the theory and formulation for fluid flows past submerged structures.

1.2 What is Vibration?

The subject of vibration deals with the dynamic behavior of structures when excitation forces are imposed upon them. The excitation can occur in the form of steady or oscillatory forces. The oscillatory forces themselves may be of single or multiple frequencies. These forces are caused by some external means such as the environment surrounding the structure or even internal means arising from the motion of the structure itself. In mechanical systems such as rotating machinery, the main cause of vibration arises from an imbalance in the rotation of the machine components. This may cause excessive wear and tear and even result in failure of the machine parts. Another cause of vibration is the interaction of a structure subjected to fluid flow. In this case, the vibration may cause excessive motions and stresses in the members of the structure and thus become the critical cause for the capsizing or failure of the structure.

Free vibration takes place when a system is displaced from its equilibrium position and released. In the absence of any external force, the system attempts to return to its equilibrium position and thus undergoes oscillation about this position. In all cases the structure or system has a natural frequency inherent to the system. The natural frequency is based on the mass of the structure and the restoring force in the system. While most systems are concerned with a single degree of oscillation with one natural frequency, it is possible to have multiple degrees of oscillation each having its own natural frequency from the free motion. Generally, the oscillatory motions from external excitation are periodic, having a single frequency. However, vibrations may be random, having a band of frequencies. In the former case, the excitation is a harmonic function while in the latter case, the forces are generally random. When the forcing frequency coincides with the natural frequency of the system, the system experiences a condition called resonance. In hydrodynamics, resonance is one of the most important phenomena and calls for careful examination.

In this regard, the natural frequency and damping in a system are important computation. In most cases the natural frequency is easy to determine. The restoring force may be represented as a spring in the system. The spring may be an external device, e.g., attached springs, or mooring lines or it may be internal, e.g., buoyancy or righting moment of a floating structure. All systems include an inherent damping function. The damping arises from the dissipation of energy by resistance, e.g., friction or interaction of the structure with the fluid. While the magnitude of damping may vary depending on the system under consideration, no physical structural system exists without some damping. Theoretically, the response of a system "blows up" or is greatly magnified in the absence of damping; however, the response is finite (although perhaps unacceptably large in magnitude) for all physical systems due to the presence of this damping.

The vibration problem associated with many types of structures has been extensively covered in the Handbook of Vibration [Harris and Crede (1976)], which is an excellent reference book for the reader. This book may be consulted to learn more about vibration of machines and structures and their applications.

1.3 Interrelationship of Hydrodynamics and Vibration

As stated earlier, this book aims to address the aspects of hydrodynamics causing vibration and motions in a structure. The structure resides in a fluid flow, which exerts an external force (or moment) on the structure. The structure displaces from its equilibrium position under the action of this force and tends to return to its original equilibrium position due to the restoring forces (or moments) acting on it. This introduces an oscillation of the structure in the fluid. The oscillation frequency may correspond to the frequency of the fluid motion or else one of its harmonic or subharmonic frequencies. The initial disturbance of the structure introduces a motion at the natural frequency of the structure as well. The structure continues to oscillate as long as the external force from the fluid is present. In the absence of the fluid force, the structure comes to rest at its equilibrium position as long as it is stable. For certain conditions, instability may result, which may lead the structure to a new equilibrium position.

The interrelationship of these two subjects is complex and multifaceted. The fluid flow introduces forces on the structure, and the structure and its restraints react to the forces exerted by the fluid flow around or through it. If the structure is floating and restrained by external means, then the restraining members experience reactive forces from the motion of the structure. On the other hand, if the structure is flexible, then the pressure distribution on the component of the structure introduces deflections. If the flow is time dependent, then this motion or deflection becomes time dependent, having generally the same frequencies as the fluid flow. The structure itself has its own modes of frequencies. If the fluid flow frequencies happen to coincide with any of the modal frequencies of the structure, then the structure is excited by these frequencies, giving rise to the natural frequency of vibration. This book will address the aspects of fluid flow and how it imposes forces on structures under various conditions. In addition, it will show how these forces give rise to vibration in the structure.

1.4 Overview of Contents

The basic theories of hydrodynamic and vibration are given respectively in the first two chapters. Most of the remaining chapters make use of these theories in dealing with their inter-relations. The book specifically addresses how the theory in these two areas can be applied to practical structures residing in fluids.

The basic theory of hydrodynamics is the subject of Chapter 2. The description of common quantities, such as potential function and the basic governing equations of fluid flow, including the continuity equation and Bernoulli's equation are given. The forces on a structure due to steady fluid flow and time dependent flow are derived. The effect of the free surface flow on a submerged structure is developed.

The basic theory of the vibration problems is introduced in Chapter 3. The chapter covers the modes of vibration and different aspects of vibration of a rigid body. The free and forced vibrations of structures are derived. Here the single and multiple degrees of vibration of rigid bodies subjected to simple harmonic motion are analyzed. The multi-degree system includes multiple degrees of freedom of a single body as well as motions of interconnected multiple bodies. Handling of certain nonlinearities, such as nonlinear restoring forces and nonlinear damping in the system are addressed and specific examples of such types of nonlinearities are illustrated. Examples are given showing how the theory may be applied in a practical problem.

Chapter 4 describes the modeling laws and model testing for systems where hydrodynamic forces are coupled with vibration. This section describes the methods of testing vibratory systems in prototype and model scale. Facilities that are used in performing these tests are described. The instruments and testing methods generally used for such systems are addressed. Specific examples of testing moving structures in fluid and some associated problems encountered are illustrated.

Many of the vibration problems encountered in the fluid field are random in nature. Chapter 5 deals with the probabilistic nature of the random signals. It reviews the statistical quantities of importance that are necessary to describe a random signal. The probable maxima of a random signal are best described by the probability theory. The probability distribution and density functions are defined, and certain common probability distributions necessary in the vibration theory are introduced. A brief description of the power spectrum is given. The method of computing the response spectrum for a linear and nonlinear vibrating system is provided. An example shows how these quantities are estimated, and discusses the practical applications of these estimates.

Chapter 6 develops the theory of random vibration based on the material presented in Chapter 5. Many structures placed in fluids experience random motion resulting from the fluid forces. Theory of random input and random output that can describe such a system is addressed in this chapter. The statistical and spectral theory applied in latter chapters is discussed here. The method of developing the solution in the frequency domain as well as the time domain is shown. Several examples are discussed in which the vibration of structures in fluid was experienced.

The flow of fluid past a structure not only affects the structure, but also changes the fluid field round it. The fluid structure interaction problems are described in Chapter 7. This chapter is divided into two parts. The first part considers the structure placed in steady flow while the second part describes the effect of time dependent flow on a submerged structure. In the latter case, the flow is considered oscillatory as well as random, thus both harmonic and random vibration of structures in fluid are considered. Damping present in the system is quantified, and the effect of damping in limiting the motion of structures is shown. Flexible structures undergo vibration, which includes deformation of the structure as well. In particular, the equations of motion of a tensioned riser placed vertically in the fluid field are considered. The practical aspects of the various terms in this equation are discussed, and the methods to solve for the vibration of the flexible body in fluid are provided. The coupling effect of a large body attached to smaller flexible members may be important. In particular, examples of floating structures moored in the ocean with mooring lines are given, and the dynamics of mooring lines themselves are discussed.

Vortex shedding is an important fluid effect that has very important practical consequences in the design of structures placed in flowing fluids. The effect of vortices and the Vortex-Induced Vibration (VIV) is the subject of Chapter 8. In this chapter, the mechanism of the formation of vortices past a fixed or moving structure in fluid is revealed. Convenient guidelines that can be used to determine vortex formation in terms of nondimensional variables are highlighted. The frequency of vortex induced vibration and the effect it will have on the vibration of a structure is addressed. The lock-in mechanism in which the frequencies of the fluid flow and structure vibrations coincide is explored. The VIV of cables, risers and pipelines in steady and oscillatory flows is described, and the corresponding eigenfrequency problem is evaluated. Finally, the mitigation of VIV problem and its various means are investigated.

The theories outlined in Chapters 1 through 8 have many applications in the real world. Chapter 9 compiles several practical prototype problems of fluid-structure interaction. These practical examples apply the theory outlined in the previous chapters. Many of these examples are actual field studies or laboratory experiments. The important lessons learned or the pitfalls and success of these studies are illustrated here.

The interrelationship among the various topics is schematically shown in Fig. 1.1. The chapters, in which these subjects are covered, are also indicated in parenthesis.

1.5 References

1. Harris, C.M., and Crede, C.E. (Editors), Shock and Vibration Handbook, McGraw-Hill, 1976.

2. Lamb, H., Hydrodynamics, 6th Edition, Cambridge University Press, New York, 1945.

3. Milne-Thompson, L.M., Theoretical Hydrodynamics, Macmillan Company, 1960.

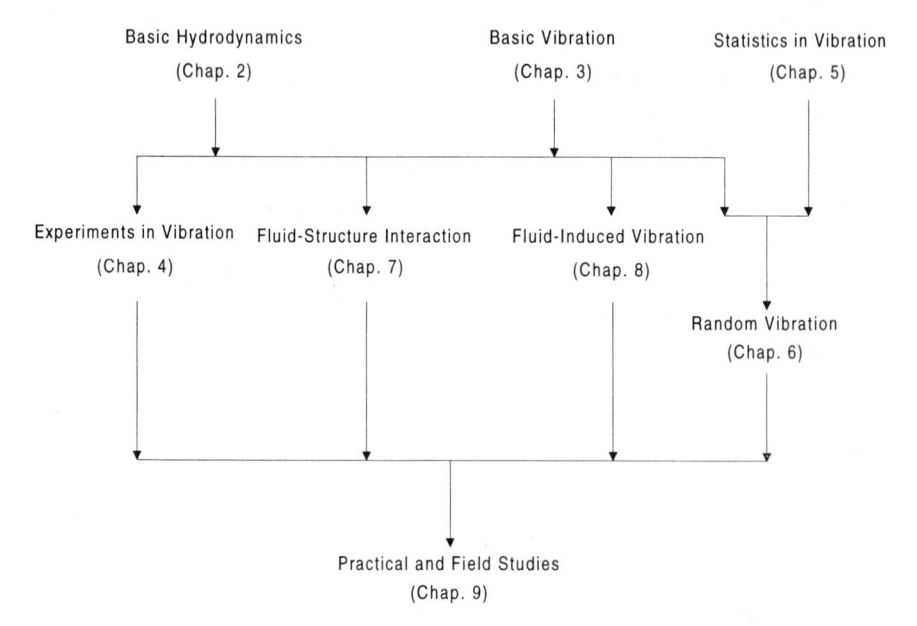

Fig. 1.1 Interrelationships among Various Topics in the Book

Chapter 2

Basic Theory of Hydrodynamics

2.1 Introduction

Theoretical hydrodynamics is based on mathematical theories that describe the flow of fluids in prescribed confinements. These confinements are called boundaries in fluid mechanics, and the mathematical theories that are developed to describe the fluid flow take into consideration the appropriate boundary conditions. Theoretical hydrodynamics is based on the concept of describing the motion of an elementary mass called a particle of fluid. This particle does not have a well-defined existence. It is a matter of convenience to consider the fluid continuum to be composed of these fluid particles. In this respect it is similar in concept to the material particle in the mechanics of solid bodies. This particle is infinitely small and is assumed to be homogeneous and isotropic. The description of hydrodynamics assumes that the fluid particle has a density representative of the entire fluid mass and this density is invariable (i.e., incompressible fluid). While some treatment of aerodynamics has been applied in this book, the mathematical treatment presented here is strictly applicable to the hydrodynamics problems.

The mathematics within hydrodynamics attempts to describe the motion of this fluid particle within the fluid mass by its kinematics and dynamics. The kinematics considers the particle velocities and accelerations, while the dynamics describe the pressure exerted on and by the fluid particle. These kinematics and dynamic properties of the fluid particle are obtained along a line or throughout a volume of the fluid. Therefore, a general differential equation is needed that governs the motion of the fluid particle.

It should be noted here that the investigation of fluid motion follows two independent but complementary disciplines. They are (1) theoretical development that attempts to explain the physics of the fluid motion using simplified logical assumptions, and (2) experimental or full-scale observation that establishes the fluid field and verifies the theory while pointing out its deficiencies or limitations. These two areas go hand in hand to explain the physical phenomenon and to provide the tools to the practicing engineers in order to design structures within the fluid field with confidence. Therefore, both subjects have been covered in this book. It should be stressed that the purpose of this chapter is to introduce the basic hydrodynamic theory that is needed to study the vibration of structures placed in a fluid medium. The experimental methods have been described in Chapter 4.

The mathematical treatment makes use of vector algebra whenever convenient. In some cases the same equation is shown as scalar and vector quantities. Matrices are used when multiple directions are used. Alternatively, tensor notation is also applied. Complex algebra is introduced as an alternative to demonstrate that the mathematical development is much more straightforward in deriving the solution. These are mixed, and the relationships among them are demonstrated so that the readers can follow the different forms and develop skills to use different treatments as appropriate and convenient.

2.2 Equations of Motion

In order to analyze the interaction of fluids and structures placed in the fluid flow, governing laws for fluid flow in the absence of the structure should first be established. These governing laws are based on certain simplifying assumptions that describe a physical phenomenon. Simplifications are assumed in order that the system may be described mathematically in simple terms. These simplifications are justifiable in most practical problems in order that useful results may be obtained without significant distortion. The results are applicable in the design of practical structures placed in fluids. These are described in an equation describing the motion of particles of fluid as they move through the field as a function of time and space.

2.2.1 Fluid Flow and Continuity

In developing hydrodynamic theory, irrespective of the kind of fluid, the flow is assumed to be continuous throughout the region under consideration.

Chapter 2 Basic Theory of Hydrodynamics

The continuity of flow states that the mass of the fluid is conserved, hence no addition or depletion of fluid mass is possible. In this context it is also assumed that fluid is incompressible so that the fluid density is constant. Then the mass conservation may be equivalently stated as the conservation of volume. It is possible to derive the equation of continuity by considering a cube of fluid and the velocity field in and out of this cube. In terms of fluid velocities, the continuity equation in a three-dimensional flow is expressed as

$$\frac{\partial u}{\partial x} + \frac{\partial v}{\partial y} + \frac{\partial w}{\partial z} = 0 \quad (2.1)$$

in which x, y, and z are the rectangular Cartesian coordinate system and u, v, and w are the components of the fluid velocity vector $\vec{V}$ along the x, y and z directions respectively. Note that x is the direction of flow and y is vertical while z is transverse normal to the x-y plane. This choice is made so that in two-dimensional flow the coordinate system reduces to an x-y plane. For a two or one-dimensional flow, the additional terms in z and in y and z, respectively will drop out of Eq. 2.1. If we introduce the operator ∇ in which ∇ is written in terms of the unit vectors $\vec{i}$, $\vec{j}$, and $\vec{k}$ along x, y, and z respectively as

$$\nabla = \frac{\partial}{\partial x}\vec{i} + \frac{\partial}{\partial y}\vec{j} + \frac{\partial}{\partial z}\vec{k} \quad (2.2)$$

the vector representation of the continuity equation becomes

$$\nabla \cdot \vec{V} = 0 \quad (2.3)$$

in which the dot product of the two quantities is implied. The velocity vector $\vec{V}$ is written in terms of the unit vectors as

$$\vec{V} = u\vec{i} + v\vec{j} + w\vec{k} \quad (2.4)$$

2.2.2 Rotational Flow

A flow is considered rotational if each of the fluid particles in the flow undergoes a rotation about its axis in addition to a pure translation and pure strain (or deformation). The amount of rotation is defined by a rotation vector or an angular velocity vector $\vec{\omega}$ such that

$$\vec{\omega} = \frac{1}{2}\nabla \times \vec{V} \tag{2.5}$$

Therefore it is clear that the flow may be considered to be irrotational if the rotational vector is identically zero. Note that for the flow to be irrotational, it does not necessarily have to be non-viscous. Moreover, the flow may still be inviscid in the greater region with the exception of the region near $\vec{\omega} = 0$.

2.2.3 Potential Function

The concept of irrotational flow is very important in hydrodynamics. Many practical fluid flow problems in hydrodynamics may be solved as irrotational flow. For an irrotational flow, the velocity field may be defined in terms of a scalar quantity called potential function, Φ, as follows:

$$u = \frac{\partial \Phi}{\partial x}, \quad v = \frac{\partial \Phi}{\partial y}, \quad w = \frac{\partial \Phi}{\partial z} \tag{2.6}$$

This relationship may be expressed in vector form as

$$\vec{V} = \nabla \Phi \tag{2.7}$$

The mathematical treatment of the fluid structure interaction problem has analytical advantages, if it is assumed that the flow is irrotational. It is obvious from Eq. 2.6 that the fluid particle kinematics is known once the velocity potential is solved.

2.2.4 Laplace Equation

If the expression for the velocity vector $\vec{V}$ in Eq. 2.7 for irrotational flow is substituted in Eq. 2.5, then it is found that the vorticity $\vec{\omega}$ is identically zero (being the cross product $\nabla \times \nabla$). Thus the existence of the velocity potential Φ implies that the flow is irrotational. On the other hand, if the flow is

Chapter 2 Basic Theory of Hydrodynamics 13

assumed irrotational then one can conclude that a velocity potential may be found to describe the flow.

Now the irrotational flow, in addition, should be continuous, which implies that Eq. 2.1 should also be satisfied. If Eq. 2.6 is substituted in Eq. 2.1, then one obtains the following differential equation:

$$\frac{\partial^2 \Phi}{\partial x^2} + \frac{\partial^2 \Phi}{\partial y^2} + \frac{\partial^2 \Phi}{\partial z^2} = 0 \tag{2.8}$$

This equation may also be written in terms of the operator as follows:

$$\nabla^2 \Phi = 0 \tag{2.9}$$

This is commonly known as the Laplace equation and is one of the most powerful relationships that is applied in many hydrodynamic problems. It has been shown that for larger submerged structures and especially those having streamlined boundaries the fluid-structure interaction with the assumption of irrotational flow provides realistic results.

In some cases, it may be convenient to express this equation in cylindrical polar coordinates r, θ and z. Thus for a horizontal cylinder, substituting $x = r \cos \theta$ and $y = r \sin \theta$ for the coordinates x and y, and substituting in Eq. 2.8, one gets

$$\frac{\partial^2 \Phi}{\partial r^2} + \frac{1}{r^2}\frac{\partial^2 \Phi}{\partial \theta^2} + \frac{\partial^2 \Phi}{\partial z^2} = 0 \tag{2.10}$$

Since many submerged structures that exist have cylindrical members, this form of the equation provides more convenient mathematical representation for these cases.

2.3 Bernoulli's Equations

Once the solution for Eq. 2.9 or 2.10 is obtained for the velocity potential, the kinematics of the fluid particle is known. For example, the three Cartesian components of particle velocities are obtained from Eq. 2.6. In order to apply the effect of the fluid on a structure, the dynamics of the fluid particle must also be known. Bernoulli's equation provides the relationship for the fluid pressure on the structure based on the fluid potential as a function of time. For this purpose, one examines the Navier-Stokes equation and applies the condition of irrotationality and inviscid flow. This gives the equation of the unsteady Bernoulli's equation

$$\rho\frac{\partial \Phi}{\partial t} + p + \rho g y + \frac{1}{2}\rho\left[\left(\frac{\partial \Phi}{\partial x}\right)^2 + \left(\frac{\partial \Phi}{\partial y}\right)^2 + \left(\frac{\partial \Phi}{\partial z}\right)^2\right] = C(t) \tag{2.11}$$

where p = dynamic pressure, ρ = fluid density, and $C(t)$ is a spatial constant which is a function of time. If the flow is steady, then the first term in Eq. 2.11 becomes zero and the constant becomes time-invariant. Then one obtains the familiar three-dimensional form of Bernoulli's equation.

$$p = -\rho g y - \frac{1}{2}\rho\left[\left(\frac{\partial \Phi}{\partial x}\right)^2 + \left(\frac{\partial \Phi}{\partial y}\right)^2 + \left(\frac{\partial \Phi}{\partial z}\right)^2\right] + Q \tag{2.12}$$

where Q is the Bernoulli constant. This is the complete expression in three dimensions.

If a two-dimensional description for a planar motion is desired, then the last term on the left-hand side of Eq. 2.11 becomes zero, leaving

$$\rho\frac{\partial \Phi}{\partial t} + p + \rho g y + \frac{1}{2}\rho\left[\left(\frac{\partial \Phi}{\partial x}\right)^2 + \left(\frac{\partial \Phi}{\partial y}\right)^2\right] = Q \tag{2.13}$$

If, in addition, only the linearized theory is considered then the higher order terms beyond the first order may be considered insignificant and eliminated. In this case, the velocity-squared terms are deemed negligible and dropped from the left-hand side. The Bernoulli equation by linear theory is reduced to

$$\rho\frac{\partial \Phi}{\partial t} + p + \rho g y = Q \tag{2.14}$$

2.4 Laminar and Turbulent Flow

If in a flowing field the path of a fluid particle is traced, the emerging line tracing the fluid particle movement with tangents representing the direction of the fluid velocity is called a streamline. In a steady flow the streamlines are constant and do not change with time. In other words, if one takes a snapshot of a steady uniform flow, the picture will remain the same at all times. This instant picture showing all of the streamlines in the flow is called the flow pattern. If, however, the fluid velocity at a given point changes with time as well, then the flow pattern will be different at each instant of time, and the flow is considered unsteady.

Consider a two-dimensional steady uniform flow past a horizontal cylinder as shown in Fig. 2.1. The direction of flow is from the left to the right. The cylinder is held in a fixed position and is not subjected to any external forces other than the fluid flow. The streamline flow is symmetric about the horizontal diameter AC of the cylinder in the direction of the flow. The central streamline from the left meets the cylinder at the point A. At this point the fluid particle comes to rest. Point A where the fluid particle velocity is zero is called a stagnation point. Once the streamline reaches the point A, it divides going around the cylinder above and below yet staying attached to the cylinder until it reaches point C. At this point the flow becomes parallel to the incoming flow, resulting in a streamline that is aligned with the earlier streamline. Point C is a second stagnation point. The other streamlines parallel to the middle streamline bend around the cylinder with the fluid particles moving from left to right, becoming straight again away from the cylinder.

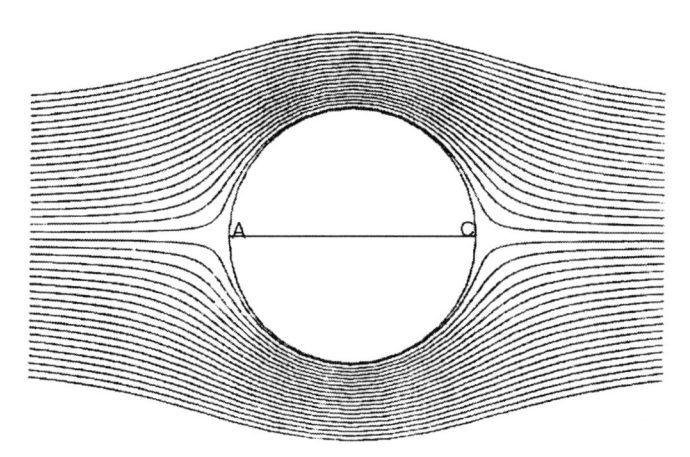

Fig. 2.1: Steady Laminar Flow Past a Horizontal Cylinder

This type of flow has been confirmed from observation at low particle velocity or, equivalently, low Reynolds number (which will be introduced shortly), and is considered a potential flow. For such flow, the differential equation called the Laplace equation derived earlier is applicable. The flow in which the streamlines remain attached to the cylinder as the fluid particles move around the cylinder is called a laminar flow.

The above description is suitable for an ideal fluid. For a real fluid such as water, there is internal friction, which causes the fluid particle in contact with the cylinder to remain at rest. This defines a boundary layer that forms close to the cylinder where the tangential component of the flow velocity goes from zero to the mean stream velocity over a very short distance. For the laminar flow problems, this boundary layer is quite narrow and is ignored in practice, as its effect is generally insignificant.

For a bluff body (such as a cylinder) in a free stream, the pressure of a fluid particle approaching the leading edge of the body increases in magnitude compared to the free stream pressure. This introduces the boundary layer on and about both sides of the body. As the flow velocity increases, it is expected that the flow will separate from the cylinder. As the fluid particle approaches the widest part of the body, the boundary layer separates from the body surface at the downstream side of the bluff body, particularly at a high Reynolds number. The point at which this separation takes place will depend on the magnitude of the velocity. The greater the velocity, the sooner the flow will separate. Upon separation, the flow velocity will no longer remain tangent to the surface of the cylinder. This introduces a pair of shear layers on the aft or downstream side of the body. Since the innermost region of the shear layer moves slower than the outermost region in contact with the free stream, the flow curls up. The upward roll of the shear layer introduces discrete swirling vortices.

During this separation, the fluid particle undergoes a transition near and beyond the separation point. The (downstream) region in the area of the cylinder beyond the separation point (the region between these two shear layers in the flow) will experience reduced flow activity, characterized by low pressure, and is called a wake (Fig. 2.2). The wake is a region of low pressure compared to the pressure on the "upstream" side.

Chapter 2 Basic Theory of Hydrodynamics 17

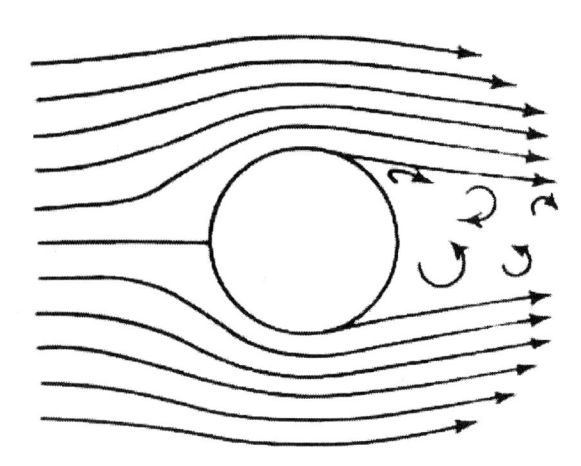

Fig. 2.2: Steady Turbulent Flow Past a Horizontal Cylinder

In this case the flow in the greater region away from the cylinder will still remain laminar (potential) flow. However, in the immediate area in the anterior region of the cylinder, where the separation has taken place, the flow becomes turbulent. Based on the flow velocity, this flow field behind the neighborhood of the cylinder is very complex and is difficult to describe in simple mathematical terms.

These vortices are found to form alternately on either side aft of the bluff body, and the separation of these vortices from the body, referred to as shedding, is associated with frequencies with which the vortices are shed. These frequencies are known as the vortex shedding frequencies. The shed vortices move with the flow away from the body and eventually disintegrate. Thus a regular pattern of vortices develop behind the bluff body creating an asymmetric pressure distribution around the body. The net effect is the introduction of a drag force and a fluctuating lift force on the body. The frequency of the lift force is dependent on the frequency of vortex shedding and the number of shed vortices present in the fluid field. If the body is free to move, the force in the transverse direction introduces an oscillatory motion of the body from the flow-induced pressure asymmetry which is called vortex induced vibration (VIV). During VIV, the frequency of vibration of the cylinder often coincides with the vortex shedding frequency of the fluid.

A motorist driving behind a large truck will experience similar flow fields. If the motorist is far behind the truck, the effect of the flow field around the truck is not felt by the motorist. However, as he gets closer to the truck, the turbulent flow created by the truck will be felt. A small car will shake in this region. The frequency of this vibration will depend on the speed and size of the truck. As the motorist gets even closer, the flow becomes much calmer, indicating that he is in the wake region of the truck. This region is a fuel-efficient zone for the car because it is an area of reduced pressure, or head-wind; however, it is not recommended as a good driving practice. It is sometimes attempted in a car or bicycle racing, called drafting.

2.4.1 Steady Drag and Lift Force

Consider a two-dimensional cylinder placed across the flow field. The flow is uniform with a constant velocity from the left to the right of the cylinder (Fig. 2.1). The flow in the region outside the vicinity of the cylinder is considered laminar. The flow will change its direction upon encountering the cylinder, flowing around it, causing a flow separation. The extent of this separation is based on the incoming velocity (Fig. 2.2). The cylinder will experience a pressure from the flow, which acts normal to the cylinder surface. There will also be a tangential pressure from the friction between the fluid and cylinder surface, which will be considered separately later.

Now if the flow remains virtually attached to the cylinder with minimum separation as it flows around the cylinder, the normal pressure distribution is expected to be nearly symmetric. For very low flow velocity, even though the drag coefficient is found to be large, the low flow velocity makes the inline force negligibly small.

As the flow velocity increases a wake region is formed behind the cylinder. This will cause an asymmetry in the pressure distribution in the inline direction, causing a net inline horizontal force. This force is referred to as the drag force.

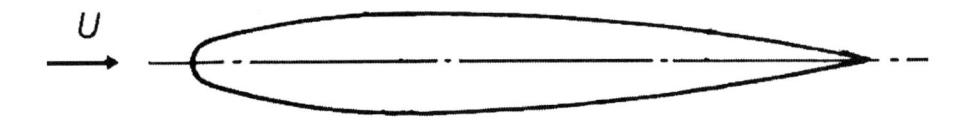

Fig. 2.3: Streamline Shape of an Airfoil

In order to reduce the effect of flow separation, the shape of the member in the flow may be made more streamlined. The airfoil used in the wings of an aircraft or fins and rudders of a submarine has a streamline shape. Figure 2.3 shows the typical cross section of a rudder. These shapes have been determined experimentally and their geometry is available as standard shapes in the literature. The flow around such shapes is similar to the one shown in Fig. 2.1 with a possible trailing vortex at the tip. In these cases the flow in most part remains attached and if the body is symmetric about the inline direction, the net inline force will be quite small.

If a two-dimensional structure is placed in a steady uniform flow and the flow separates from the structure, then the force experienced by the structure will depend on the fluid density ρ, the steady uniform flow velocity U, and the frontal area A of the structure encountering the flow. The force is found to vary with the square of the flow velocity:

$$f = \frac{1}{2}\rho C_D A U^2 \tag{2.15}$$

where C_D is a constant known as the drag coefficient, which for steady flow has been shown to be a function of the Reynolds number Re, where

$$Re = \frac{UD}{\nu} \tag{2.16}$$

and where D = structure characteristic dimension across the flow and ν = kinematic viscosity of fluid. For a circular cylinder across the flow, D is the diameter of the cylinder.

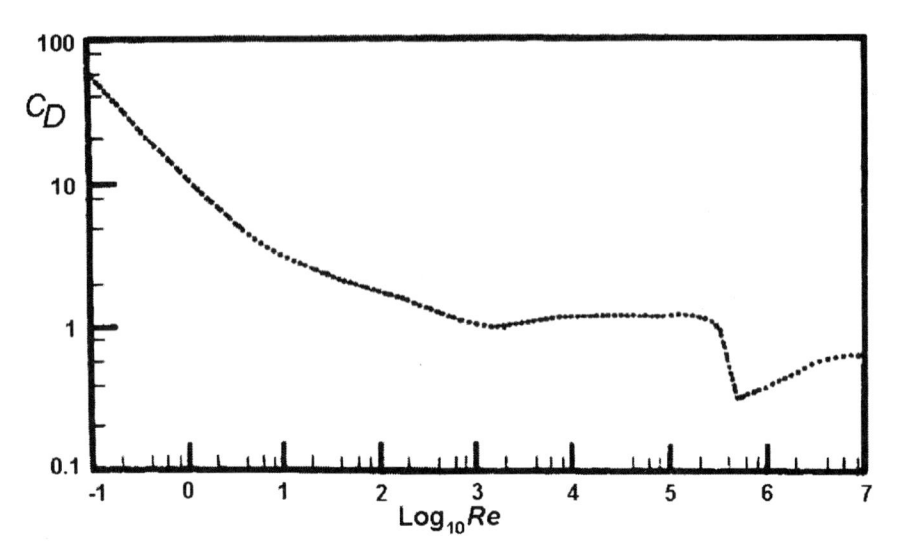

Fig. 2.4: Drag Coefficient for a Smooth Circular Cylinder in Steady Flow

The drag coefficient for a smooth circular cylinder has been obtained through laboratory testing in a steady flow and is shown in Fig. 2.4. For a Reynolds number less than 2×10^5, the flow is considered to be in the subcritical range. For Re less than 50 the flow is strictly laminar and steady. The drag coefficient in this range decreases linearly as a function of Reynolds number. The laminar flow is maintained up to a Reynolds number of about 200, beyond which the flow starts to become turbulent. The flow becomes fully turbulent for $Re > 5000$. The range of Reynolds number between 2×10^5 and 5×10^5 is termed critical where flow is in the transition mode. This area actually causes the wake to be narrower with a corresponding decrease in the pressure gradient, causing a sharp drop in the actual value of the drag coefficient. This is known as drag crisis. As the Reynolds number increases, the supercritical range is reached where the flow is strictly turbulent. In this range the drag coefficient slowly increases again. Beyond Re of 3×10^6 the turbulent flow is called post-supercritical. Here the drag coefficient approaches a steady value because less dramatic changes occur in the boundary layer at higher velocities.

This figure clearly demonstrates the difficulty associated with small scale testing of structures in which the drag force is important. The data from the model test is difficult to scale up due to difference in the values of the drag coefficients between the model and the full scale due to disparity in the value of *Re*. This area will be discussed further in Chapter 4.

In practice, the surfaces of many structures in operational mode are not smooth. The roughness of the structure surface may be contributed from several sources. The appendages attached to a structural component may introduce irregularities causing flow tripping. An example of this type of structural component exists in ships and submarines. Sources of roughness present on a structural member are different kinds of marine growth, which attach to the surface of the structure near the ocean surface. Examples of such structures are floating buoys, bridge piers, cylindrical piles, and members of offshore rigs. It should also be noted that this type of roughness on the surface often increases the effective overall size of the structure. This means that there is an increase in drag force on the structure simply from the larger projected area to the flow. Moreover, if the surface of the structure is not smooth, the roughness changes the flow separation point and wake behind the structure, resulting in a change in the drag coefficient as well. In order to quantify the roughness, a roughness coefficient *K/D* is defined in terms of a roughness parameter *K*, and a transverse dimension *D* (e.g., diameter of the structure). The roughness parameter is a measure of the mean diameter of the roughness particles attached on the surface.

Figure 2.5 illustrates the general relationship from past studies (most of these with air as the flowing fluid) between drag coefficient C_D and Reynolds number for flow past a stationary cylinder. This figure shows that, at a Reynolds number of about 2×10^5 (the "critical" Reynolds number range), the drag coefficient (especially for the smooth cylinder) diminishes dramatically, a phenomenon known as the "drag crisis". This corresponds to a change in the boundary layers from laminar to turbulent. Since turbulent boundary layers produce flow reattachment (after separation) much more easily than laminar boundary layers, the pressure on the downstream side of the cylinder is reduced. For rough cylinders, the drag crisis is much less pronounced. With sufficient roughness on the body surface the drag crisis can be eliminated. Surface roughness plays an important role, not just in the drag coefficient, but in the VIV response as well.

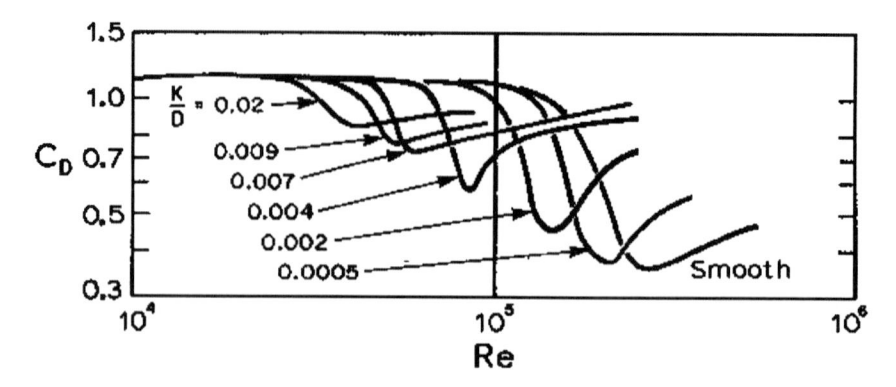

22 — Theory and Practice of Vibration and Hydrodynamics

Fig. 2.5: Drag Coefficient for a Rough Circular Cylinder in Steady Flow

The main reason for the increase in the drag coefficient in the presence of roughness is that the roughness on the surface breaks up (i.e., trips) the flow earlier in the velocity field creating a larger wake. Thus, the flow on a rough surface separates from the surface ahead of the point of separation for a smooth cylinder. The greater the roughness, the earlier the separation of flow. This causes a larger pressure gradient between the upstream and downstream faces of the structure. The effect is a larger drag coefficient and, hence, a larger drag force for the same Reynolds number. The drag coefficients for a rough cylinder for a roughness coefficient value of up to 0.02 are shown in Fig. 2.5 as a function of Re and the roughness coefficient, K/D. The increased roughness moves the transition area (drag crisis) to lower values of Re. This is clear from the plots in Fig. 2.5. Outside of this region, the value of C_D is higher with the higher values of K/D. Note that at very low Reynolds number where the flow is strictly laminar, the cylinder roughness does not appreciably alter the drag coefficient on the cylinder. This is simply because there is minimal flow separation in this range.

It has already been noted that the wake formed behind the body and any associated vortex shedding from the boundary of the body surface is not symmetric with respect to the direction of flow. This is mainly because the formation (and shedding) of vortices on either side of the flow direction does not take place at the same time. In fact, it has been found experimentally that the shedding of vortices alternates on the two sides (Fig. 2.2). Thus the pressure distribution around the structure is not symmetric

Chapter 2 Basic Theory of Hydrodynamics 23

about the flow direction. Consequently, an additional force is generated transverse to the flow. This force is generated from the asymmetric pressure distribution due to the wake and the uneven formation of vortices behind the body. This force is referred to as the transverse or lift force, and is written in a form similar to the inline drag force as

$$f_L = \frac{1}{2}\rho C_L A U^2 \qquad (2.17)$$

Since the vortex shedding is alternating between the two sides, the net force in the transverse direction will change direction every time the vortices are shed. Thus the lift force will be oscillating even for a steady flow at the frequency of vortex shedding. The formation of vortices and the associated wake field are somewhat irregular with respect to time. In other words, the change in the wake field with time even in steady flow conditions is not highly predictable. Therefore, the force generated in the transverse direction is irregular and the lift coefficients are not well defined.

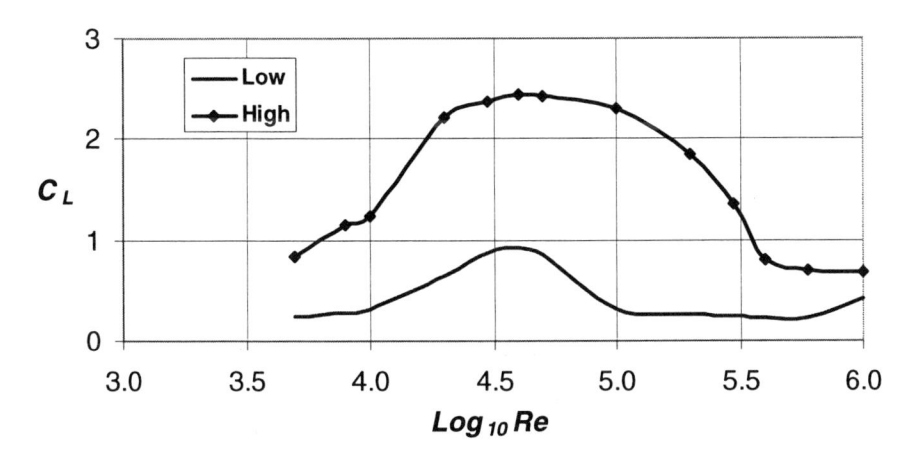

Fig. 2.6: Lift Coefficient for a Smooth Circular Cylinder in Steady Flow

Because the lift force is irregular, the coefficients are expressed either as rms values over one measured cycle or maximum values corresponding to the maximum lift force. Numerous experiments have been conducted with circular cylinder transverse to steady flow to determine the values of lift coefficients. There is considerable scatter in the experimental data. Instead of attempting to show these individual experiments, the range of the lift coefficients are shown versus Reynolds number in Fig. 2.6. The lift coefficients obtained from these experiments involve maximum, mean and rms values. The low and high curves in Fig. 2.6 cover all these values ranging from a low of 0.25 to a high of about 2.5. The highest values appear in the Reynolds number range of 2×10^4 and 3×10^5. Most of this scatter may be attributed to the free stream turbulence in the flow, flow over the ends of the cylinder, lack of rigidity in the mounting system, and other physical sources.

2.5 Vorticity Parameter, Reduced Velocity and Strouhal Number

The vorticity or the vorticity vector was previously introduced in section 2.2.2. If the vorticity vector is zero at certain region of a fluid field, then the fluid region will be irrotational, as discussed earlier. Motion is initially irrotational if it starts from rest. In real fluid the vorticity is almost always present, particularly near a structure as the fluid flows past the structure. Generally for a bluff body, the near wake region is an important region where the vortex at the body is produced, travels away from its surface and finally disintegrates. The assumption of potential flow in this region may be questioned. In the greater region of the fluid flow, however, the potential flow is often an acceptable assumption, particularly if the structure is large. For a small structure the near-wake vorticity is important since it represents a significant part of the flow region around the structure.

The transverse vibration of a body resulting from the wake field depends on a parameter called the reduced velocity. Consider the lateral vibration of a spring-loaded cylinder (Fig. 2.7). The dimension of the cylinder normal to the direction of free stream is given by D. If U_∞ is the free stream velocity and f is the frequency of normal vibration, then U_∞/f is the length of the path for one cycle of vibration. The nondimensional reduced velocity V_r is defined as (= path length per cycle / body length)

$$V_r = \frac{U_\infty}{fD} \qquad (2.18)$$

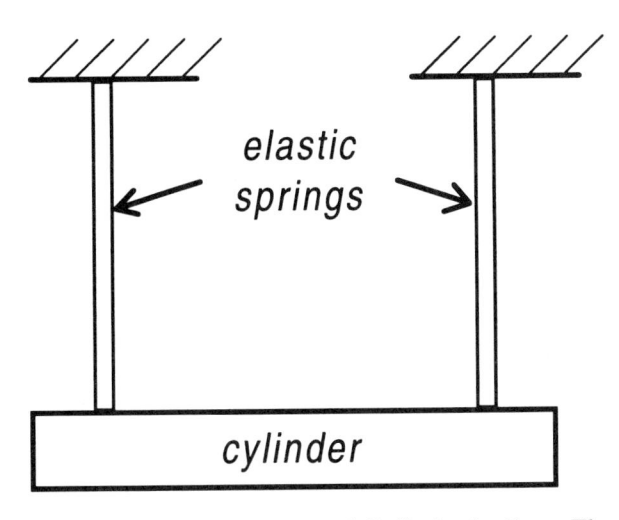

Fig. 2.7: Elastically Mounted Cylinder in Cross Flow

It has been found that if the value of the reduced velocity is less than 10, then there is a strong interaction between the cylinder and the near wake from the flow over the cylinder.

Another important parameter that determines the vortex shedding frequency, f_s is called the Strouhal number, defined as

$$St = \frac{U}{f_s D} \qquad (2.19)$$

The frequency of vortex shedding has been shown to be a linear function of St over a wide range of Re. Thus, it is interesting to seek the relationship between the Strouhal number and Reynolds number. This relationship has been established by model experiments for a circular cylinder. The general behavior of St versus Re is shown in Fig. 2.8. It is generally accepted that $St \approx 0.2$ in the range $2.5 \times 10^2 < Re < 2.5 \times 10^5$. Beyond this range, St increases up to about 0.3 and then, with further increase in Re, the regular periodic behavior of U in the wake behind the cylinder disappears. Around the transition Re value of $10^5 - 5 \times 10^6$ the shedding frequency is not well defined and the Strouhal number has a broad band of values.

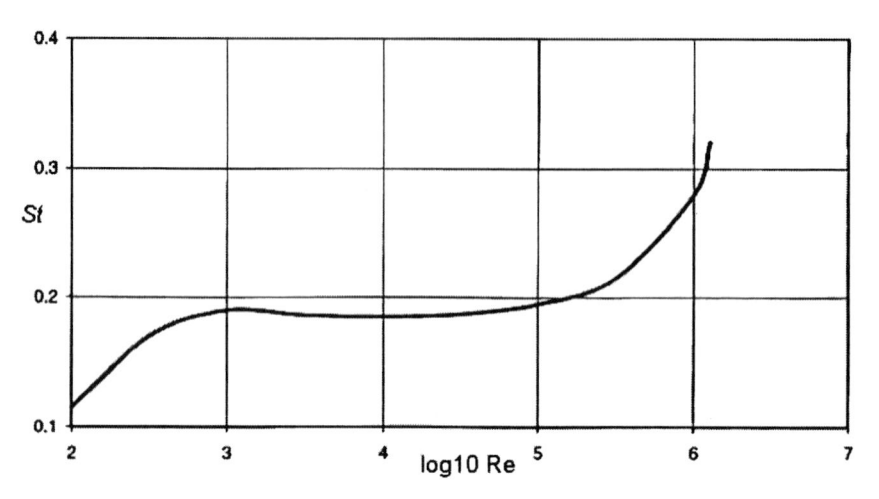

Fig. 2.8: Reynolds vs. Strouhal Number for a Circular Cylinder

2.6 Moving Structures in Fluid

Sometimes, certain physical phenomenon may force a structure to oscillate harmonically in an otherwise calm fluid. Alternatively, the fluid may oscillate harmonically past a stationary structure. It can be demonstrated that the case of an oscillating structure in calm fluid is equivalent and kinematically identical to oscillating fluid flowing past the stationary structure. The inline force (per unit length) experienced by an oscillating cylinder (of diameter D) in fluid is similar to the drag force on a steadily moving cylinder except for the alternating nature of the force.

$$f_D = \frac{1}{2}\rho C_D D |\dot{x}|\dot{x} \qquad (2.20)$$

where x is the harmonic displacement of the cylinder and dot represents derivatives of x. The changing direction of the force is accommodated in the expression of Eq. 2.20 with an absolute sign for one velocity component. Thus the force changes sign with the sign of the oscillating structure velocity.

Chapter 2 Basic Theory of Hydrodynamics 27

For an oscillating cylinder in a calm fluid the inertia force (per unit length) on the cylinder imparted from the fluid is written as

$$f_I = \rho C_A \frac{\pi}{4} D^2 \ddot{x} \qquad (2.21)$$

where C_A is the added (mass) inertia coefficient, and $\ddot{x}$ is the cylinder acceleration. Note that the inertia of the cylinder due to its own motion is not included in Eq. 2.21.

Combining the two force components, the total force per unit length experienced by the cylinder of diameter D due to its motion through the water is described by the following equation:

$$f = m\ddot{x} + C_A \rho \frac{\pi}{4} D^2 \ddot{x} + \frac{1}{2} C_D \rho D |\dot{x}| \dot{x} \qquad (2.22)$$

in which the first term is the inertia of the cylinder itself and m is the mass of the cylinder per unit length.

Since Eq. 2.22 is empirical, the values of the coefficients C_A and C_D are determined experimentally. It has been shown that in oscillatory flow these coefficients are function of Reynolds number (Re) and Keulegan Carpenter number (KC). For forced oscillation, these quantities are defined as

$$Re = \frac{\dot{x}_0 D}{\nu}; \qquad KC = \frac{\dot{x}_0 T}{D} \qquad (2.23)$$

in which the subscript zero represents the amplitude. Sometimes the ratio of Re and KC, known as Stokes viscous parameter is used instead of Re

$$\beta = \frac{Re}{KC} \qquad (2.24)$$

Note from Eq. 2.22 that when inertia force is maximum, drag force is zero and *vice versa*. Hence the coefficients can be determined independently from these times. However, the coefficient values are assumed invariant over a cycle for a given frequency of oscillation and one set of C_A and C_D is computed per cycle. Two general testing methods are followed in the laboratory to determine the hydrodynamic coefficients for the circular cylinder. In the first method the cylinder is oscillated harmonically in otherwise still water. Known frequency and amplitude of oscillation are imposed on the cylinder through a variety of mechanical systems. Secondly, the cylinder is held fixed while the fluid is oscillated harmonically past it. In both cases the flow is one-dimension, except in the vicinity of the cylinder. The kinematic fields in the mechanical oscillation of a structure are controlled and the accuracy with which they are described can be excellent.

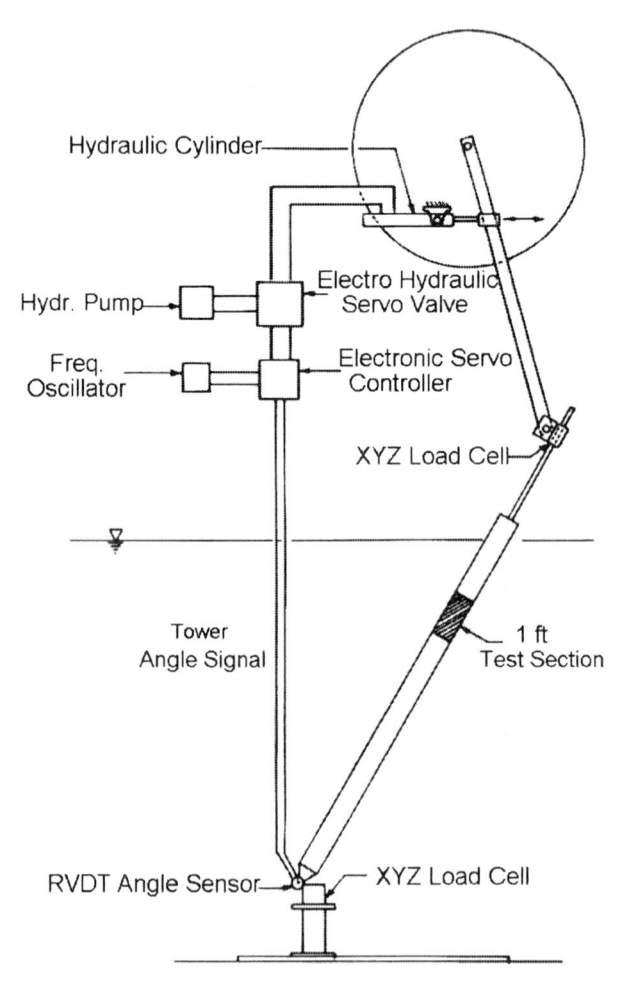

Fig. 2.9: Setup of Mechanical Oscillation Test for a Vertical Cylinder

Chapter 2 Basic Theory of Hydrodynamics 29

A mechanical oscillation test of an articulated cylinder was performed in which the cylinder was rotated about a pin in still water. The vertical cylinder was mounted to a mechanical oscillator on a support structure (Fig. 2.9). In the experimental setup, the motion of the cylinder was measured with a displacement transducer. Since motion is sinusoidal, calculation of its derivatives is straightforward. For example, the cylinder velocity is simply obtained by multiplying the displacement time history by the frequency and shifting it by 90 deg. The driving force of the mechanical oscillator was a hydraulic cylinder, which received fluid from a hydraulic power unit. An electro-mechanical servo valve was used to control the fluid flow to the hydraulic cylinder and thus control the characteristics of the imposed motion. The cylinder pierced the free surface of the water. The force transducer section was held at a constant depth from the free surface. The *KC* and *Re* numbers were varied during the test.

Since there are two unknowns in Eq. 2.22, the common method for determining them is either a Fourier technique or a least square method. One set of C_A and C_D values was computed for each cycle of oscillation, and an average value was determined over 10 cycles of measured data. The values of these coefficients are shown in Figs. 2.10 and 2.11 as functions of *KC* number and *Re* values. The experimental *Re* values ranged from 0 to 40,000. Different symbols are used to distinguish the data points for different *Re* values. Note that the added mass coefficients failed to show clear dependence on the *Re* values and experienced a scatter instead.

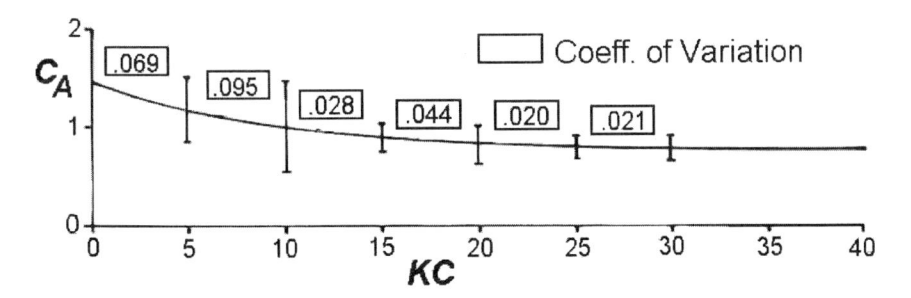

Fig. 2.10: Inertia Coefficients for an Oscillating Vertical Cylinder

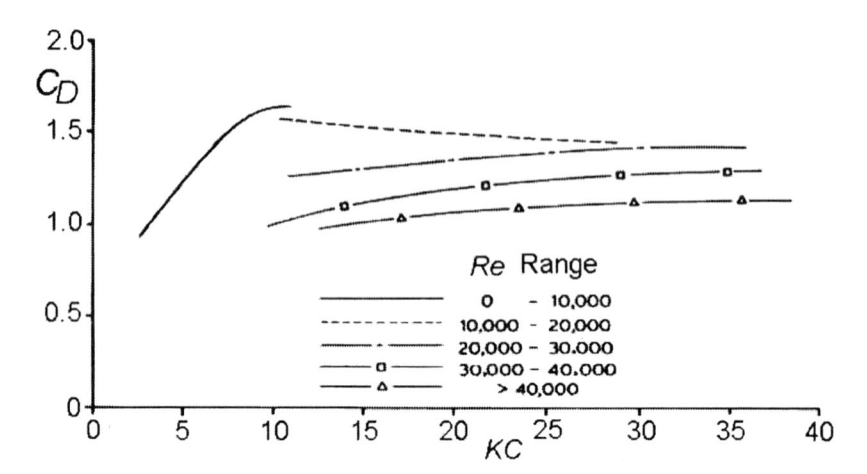

Fig. 2.11: Drag Coefficients for an Oscillating Vertical Cylinder

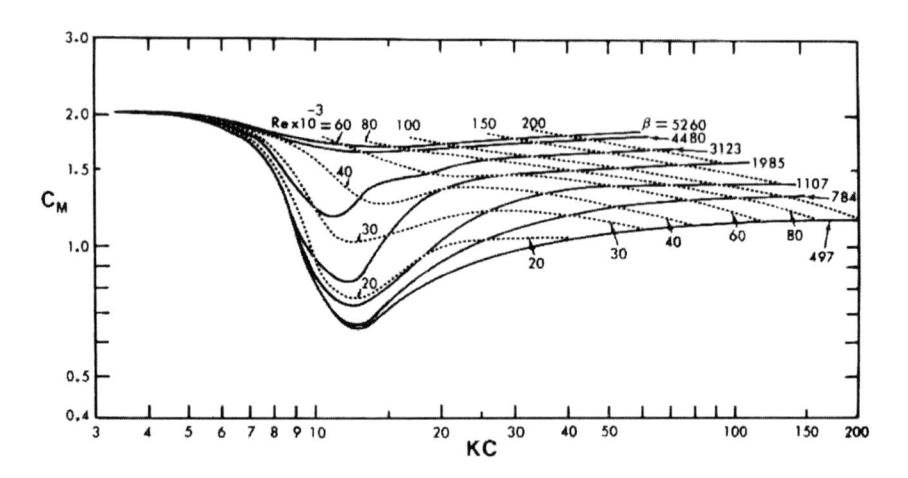

Fig. 2.12: Inertia Coefficients from a Fluid Oscillation Test [Sarpkaya (1976)]

Chapter 2 Basic Theory of Hydrodynamics 31

Similar experiments were conducted by Sarpkaya (1976) in a large water-filled U-tube in which water was oscillated back and forth past a fixed horizontal cylinder placed in the horizontal arm of the U tube. The forces on the cylinder were measured from which the inertia and drag coefficients were determined.

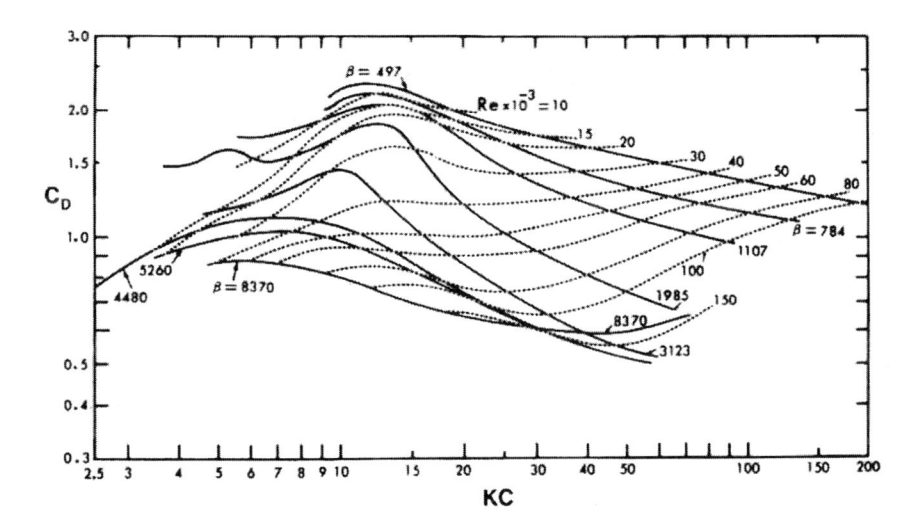

Fig. 2.13: Drag Coefficients from a Fluid Oscillation Test [Sarpkaya (1976)]

The results from one such experiment are shown in Figs. 2.12 and 2.13 respectively for the inertia and drag coefficients. The hydrodynamic coefficients are presented as function of *KC* for different values of *Re*. Note that the variable β is related to the other two dimensionless quantities by $\beta = Re/KC$. The dependence of these coefficients on the quantities *KC* and β, or equivalently, *KC* and *Re* is clear in these figures. Generally, the values of the inertia coefficients increased, when those of the drag coefficients decreased, and *vice versa*.

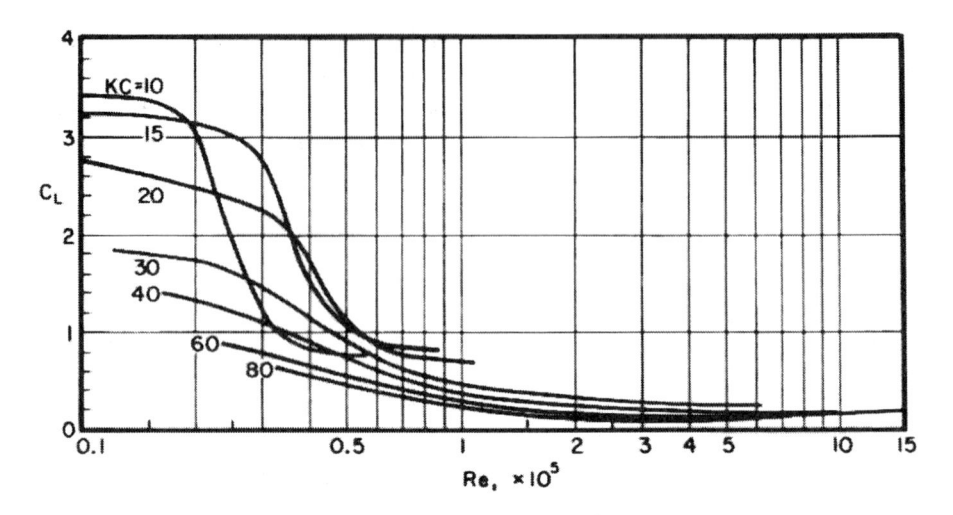

Fig. 2.14: Lift Coefficients from a Fluid Oscillation Test [Sarpkaya (1976)]

The transverse (or vertical in the present case) force on the cylinder due to asymmetric vortex shedding was additionally measured in these tests. As discussed earlier, the lift forces were irregular, having multiple frequencies. Therefore, unlike the drag and inertia coefficients, a single lift coefficient over one cycle may not be determined. The lift coefficient is, generally, presented as an rms or a maximum value over one oscillation cycle. The rms lift coefficients from the U-tube tests are shown in Fig. 2.14 as functions of KC and Re. The frequencies present in the lift force were multiples of one another and related to the number of vortices (eddies) shed from the cylinder. The ratio of the predominant lift force frequency to the imposed oscillatory frequency is plotted in Fig. 2.15 as function of the KC and Re numbers. The larger the number of vortices shed, the higher is the predominant frequency in the measured force. The frequency ratio f_r is found to increase in value from 2 to 15 in the KC number range of 10 and 85. The dependence of the lift force frequency is principally on KC with a weak one on Reynolds number except in the high range of Re. It is clear from this discussion that if the cylinder is free or mounted on springs and allowed to move, then the transverse force will cause the cylinder to

Chapter 2 Basic Theory of Hydrodynamics 33

oscillate in the flow. Moreover, the oscillation will, in general, be irregular and if one of the frequencies fall close to the natural frequency of the spring-mounted cylinder, then a dynamic amplification will take place. This is the source of the vortex-induced vibration and will give rise to lock-in mechanism when the vibration frequency coincides with the natural frequency. This area will be discussed further in Chapter 8.

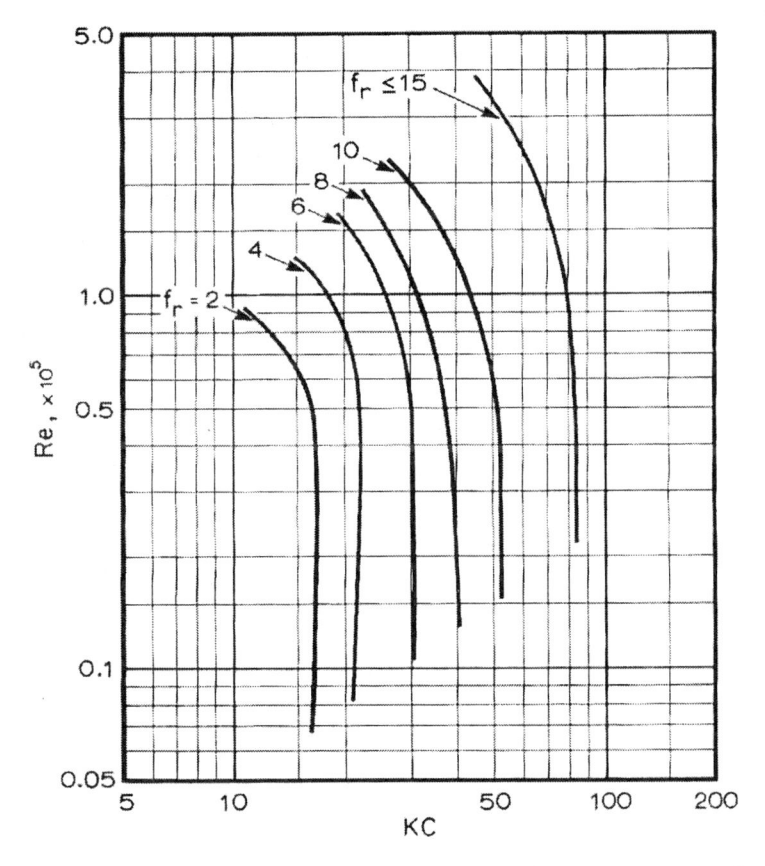

Fig. 2.15: Lift Force Frequency as a Function of *KC* and *Re* [Sarpkaya (1976)]

2.7 Free Surface Flows

A wave creates a free surface motion at the mean water level acted upon by gravity. The elevation of the free surface varies with space x and time t. If the wave is sinusiodal in form then the free surface profile is given by the following simple form:

$$\eta = a\sin(kx - \omega t) \tag{2.25}$$

in which the quantities a, ω and k are constants. The coefficient a is called the amplitude of the wave, which is a measure of the maximum departure of the actual free surface from the mean water level. The quantity ω is the frequency of oscillation of the wave, and k is called the wave number.

Therefore, this equation describes a simple harmonic motion of the free surface, which varies sinusoidally between the limits of $+a$ and $-a$ as shown in Fig. 2.16. The point where the value of the profile is $+a$ is called a crest while the point with the value $-a$ is called the trough. This form is a description of a progressive wave, which states that the form of the wave profile η not only varies with time, but also varies with the spatial function x in a sine form. At time $t = 0$, the spatial form is a sine curve in the horizontal direction.

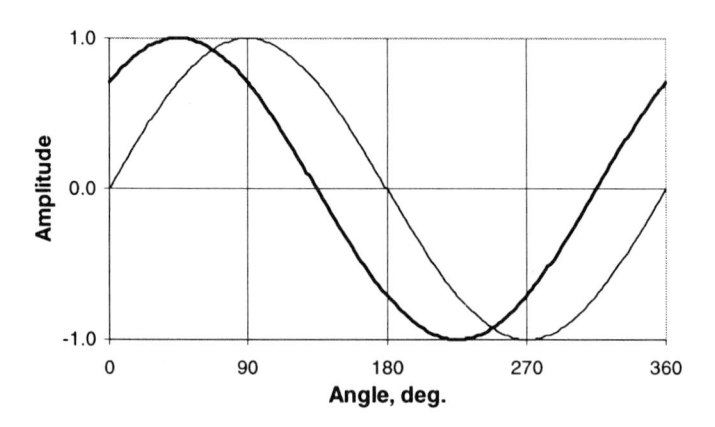

Fig. 2.16: Free Surface Profile in Wave Motion at Two Times t_1 and t_2

Chapter 2 Basic Theory of Hydrodynamics

$$\eta = a \sin kx \tag{2.26}$$

This form is the same at any time t as the wave is frozen at that time. Rewriting Eq. 2.25,

$$\eta = a \sin k(x - \frac{\omega}{k}t) \tag{2.27}$$

This form suggests that the frozen wave profile moves in the direction of the horizontal axis x with a velocity

$$c = \frac{\omega}{k} \tag{2.28}$$

The velocity c is called the celerity or speed of propagation of the progressive wave. One can write the frequency ω as

$$\omega = \frac{2\pi}{T} \tag{2.29}$$

where T is the period of the wave. It is clear that the wave form repeats at each cycle with the period T of the wave. Similarly one can write k as

$$k = \frac{2\pi}{L} \tag{2.30}$$

in which L is the length of the wave. Referring to Fig. 2.16, L is then the distance between two crests or two troughs in the wave form. The wave propagation speed may also be written as

$$c = \frac{L}{T} \tag{2.31}$$

This description of the free surface flow is in two dimensions – horizontal (x) and vertical (y). It is bounded by the free surface on top and the sea floor at a water depth d. In order to describe the fluid particle motion within the wave field, the kinematic and dynamic properties of the particle should be established first. For this purpose it is assumed that the floor is flat so that the water depth is constant at all points along the *x* direction. This is shown in Fig. 2.17.

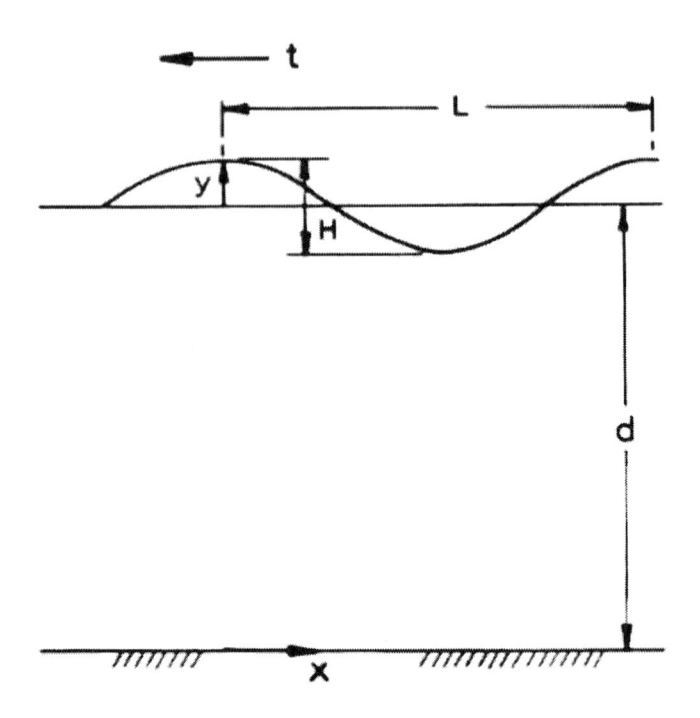

Fig. 2.17: Two-dimensional Wave Motion over Flat Bottom

2.7.1 Kinematical Condition at the Floor

The kinematic condition at the flat bottom states that the vertical component of velocity at any point on the bottom surface should be zero. This is required in order that there is no flow through the bottom or no voids created at the bottom resulting from the fluid flow. Mathematically, it is written simply as

$$v = 0, \quad at \ y = 0 \tag{2.32}$$

Chapter 2 Basic Theory of Hydrodynamics

Thus, the particle velocity at the bottom follows a simple harmonic back and forth motion in the x direction.

2.7.2 Kinematical Condition at the Free Surface

Considering the profile η in Eq. 2.25 and the water depth d, the free surface is given by the identity

$$y - \eta - d = 0 \tag{2.33}$$

Since this surface moves with the fluid, it must be time-invariant and its total time derivative should be zero:

$$\frac{d(y - \eta - d)}{dt} = 0 \tag{2.34}$$

Writing the total derivatives in terms of the partial derivatives with respect to time t and space x, and noting that d = constant, $\frac{\partial x}{\partial t} = u$ and $\frac{\partial y}{\partial t} = v$, this equation becomes

$$\frac{\partial \eta}{\partial t} + u \frac{\partial \eta}{\partial x} = v \tag{2.35}$$

If the flow is assumed irrotational and the expressions for u and v are used in terms of the potential function Φ (Eq. 2.6), then this equation may be written as

$$\frac{\partial \eta}{\partial t} + \frac{\partial \Phi}{\partial x} \frac{\partial \eta}{\partial x} = \frac{\partial \Phi}{\partial y} \tag{2.36}$$

This is the complete form of the free surface kinematic boundary condition in two dimensions. It can be easily extended to three dimensions by introducing a term in the coordinate z to obtain the following form:

$$\frac{\partial \eta}{\partial t} + \frac{\partial \Phi}{\partial x} \frac{\partial \eta}{\partial x} + \frac{\partial \Phi}{\partial z} \frac{\partial \eta}{\partial z} = \frac{\partial \Phi}{\partial y} \tag{2.37}$$

This form of boundary condition (Eqs. 2.36 and 2.37) is nonlinear in the velocity potential (having product terms). For simplicity, many practical fluid-structure interaction problems are solved by linearizing this equation. If it is assumed that the wave amplitude 'a' is small in describing the wave profile, then one can neglect the product terms in the profile and kinematics as second order quantities. In this case, the kinematic boundary condition becomes

$$\frac{\partial \eta}{\partial t} = \frac{\partial \phi}{\partial y} \qquad (2.38)$$

which is satisfied at $y = d$ to first order.

2.7.3 Dynamic Condition at the Free Surface

The dynamic free surface boundary condition describes the pressure field at the free surface at the elevation η. It is obtained by applying the Bernoulli's theorem. Assuming irrotational flow, the pressure p at a point in the fluid is obtained from Eq. 2.13:

$$p = -\rho \frac{\partial \Phi}{\partial t} - \rho g \eta - \frac{1}{2} \rho \left[\left(\frac{\partial \Phi}{\partial x} \right)^2 + \left(\frac{\partial \Phi}{\partial y} \right)^2 \right] + Q \qquad (2.39)$$

Again, if linear theory is applied, the square terms in the velocity potential may be neglected giving the dynamic pressure at the free surface

$$p - p_0 = -\rho \frac{\partial \Phi}{\partial t} - \rho g \eta \qquad (2.40)$$

where p_0 which is the Bernoulli's constant corresponds to the constant atmospheric pressure. At the free surface $y = d + \eta$, the left-hand side will become zero (neglecting surface tension). Then

$$\eta = -\frac{1}{g} \frac{\partial \Phi}{\partial t} \qquad (2.41)$$

which is satisfied at the still water level $y = d$ to first order. This then is the linearized dynamic free surface boundary condition.

2.7.4 Linear Wave Theory

If the equations for the kinematic and dynamic free surface conditions are examined, it is clear that they may be combined at $y = d$ giving the combined linearized free surface boundary condition

$$\frac{\partial^2 \Phi}{\partial t^2} + g \frac{\partial \Phi}{\partial y} = 0 \qquad (2.42)$$

If one uses the boundary value problem with the differential equation for the velocity potential and the boundary conditions at the free surface and the bottom including the crest condition ($a = H/2$) at time $t = 0$, one can derive the expression for the potential:

$$\Phi = \frac{ga}{\omega} \frac{\cosh ky}{\cosh kd} \sin[k(x-ct)] \qquad (2.43)$$

If this expression is substituted in Eq. 2.41, one obtains the surface profile as

$$\eta = a \cos[k(x-ct)] \qquad (2.44)$$

Thus, the wave crest occurs at the location $x = 0$ at time $t = 0$. If Eq. 2.43 is substituted in Eq. 2.42, then the relationship for the wave speed (or equivalently wave length) is obtained. This is commonly known as the dispersion relationship:

$$c^2 = \frac{g}{k} \tanh kd \qquad (2.45)$$

The horizontal and vertical particle velocities in the fluid region are derived from Eq. 2.43 and Eq. 2.6:

$$u = \frac{gka}{\omega} \frac{\cosh ky}{\cosh kd} \cos[k(x-ct)] \qquad (2.46)$$

$$v = \frac{gka}{\omega} \frac{\sinh ky}{\cosh kd} \sin[k(x-ct)] \qquad (2.47)$$

An example of the velocity profiles in an intermediate water depth by linear theory is plotted in Fig. 2.18.

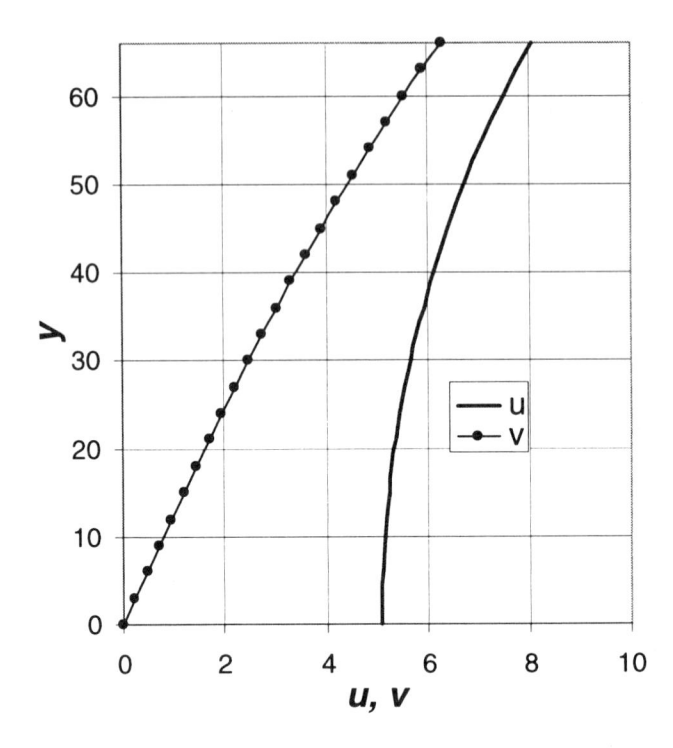

Fig. 2.18: Horizontal and Vertical Particle Velocity Profile with Depth

Note that according to linear theory the free surface boundary condition is met at $y = 0$ and the velocities are defined up to the still water level. The vertical component of velocity is smaller in magnitude than the horizontal component of velocity. Their values approach each other as the water depth increases. Moreover, the vertical component of velocity is zero at the bottom in order to meet the bottom boundary condition. If one examines the expressions in Eqs. 2.46-47, one can see that the two velocities are orthogonal to each other so that when the horizontal velocity is maximum, the vertical velocity component is zero and *vice versa*. Since the path of the particles may be defined by integrating these expressions, it is clear that the path will be defined by an ellipse with the major axis given by the horizontal particle motion double amplitude and the vertical axis will be the vertical

Chapter 2 Basic Theory of Hydrodynamics

motion double amplitude. Since the vertical axis is smaller than the horizontal axis and the two motions are 90 degrees out of phase, the elliptical path results. The particle path for the above example is shown in Fig. 2.19 at a few elevations in the fluid region. The ellipse collapses to an oscillatory motion at the bottom where $v = 0$.

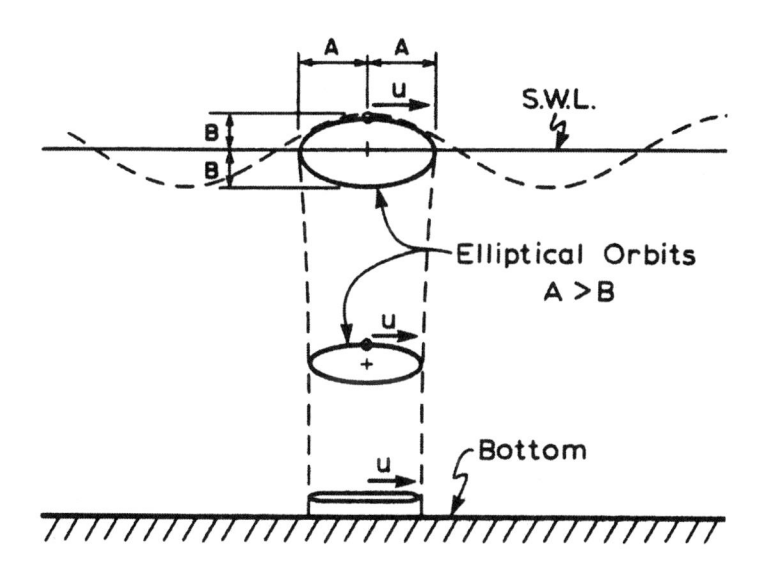

Fig. 2.19: Contour of a Fluid Particle under Linear Theory

The dynamic pressure (excluding the hydrostatic component) on a fluid particle due to linear theory may be obtained from Eq. 2.14 and the velocity potential (Eq. 2.43)

$$p = \rho g a \frac{\cosh ky}{\cosh kd} \cos[k(x-ct)] \qquad (2.48)$$

Thus, the dynamic pressure is in phase with the horizontal particle velocity. Its amplitude will have a similar vertical profile as shown for the horizontal velocity in Fig. 2.18.

2.7.5 Oscillatory Drag and Inertia Forces

Since the wave flow is oscillatory and, in particular, since the linear wave flow follows a simple harmonic motion, the flow around the cylinder will be more complex than the steady uniform flow. In a simplified description one can say that the oscillatory flow over one cycle will change the wake region 'behind' the cylinder every half cycle. As the flow changes direction, the low-pressure region will move from the downstream to the upstream side. Thus the force on the cylinder will change direction every half a wave cycle. Using a form similar to Eq. 2.20, the drag force per unit length of the cylinder in an oscillatory wave flow will have the form:

$$f_D = \frac{1}{2}\rho C_D D |u| u \qquad (2.49)$$

in which the absolute value of the water particle velocity has been used to maintain the direction of the force. Thus the direction of force changes with the direction of velocity. In the above description of force the additional effect of inertia due to change in the acceleration of the flow has not been considered. A fluid particle moving in a simple harmonic motion possesses a momentum. As the fluid particle under wave passes around the circular cylinder it first accelerates to reach the midpoint and then decelerates down the surface. This translates to work done on the cylinder, which introduces a force on the cylinder. The force is proportional to the ambient fluid particle acceleration and the volume of the cylinder. The force per unit cylinder length is written in terms of an inertia coefficient C_M as

$$f_I = \rho C_M \frac{\pi}{4} D^2 \dot{u} \qquad (2.50)$$

where $\dot{u}$ represents the acceleration of the simple harmonic fluid motion.

Combining the effects of water particle velocity and acceleration on the structure, the force per unit length of the cylinder due to a regular wave is computed from the empirical formula commonly known as the Morison formula [Morison, et al. (1950)]:

$$f = \rho C_M \frac{\pi D^2}{4} \dot{u} + \frac{1}{2} \rho C_D |u|u \qquad (2.51)$$

In order to determine the force on a submerged member of a structure in waves the values of the hydrodynamic coefficients must be known. The hydrodynamic coefficients of the member in an oscillatory wave flow are determined experimentally by testing an instrumented, scaled model of the member.

The member, such as a cylinder, is held fixed in the model basin. Progressive waves are generated in the basin and allowed to flow past the cylinder. In this case the flow is two-dimensional such that the free-stream velocity field varies both in the horizontal and vertical directions. The waves, although generally regular, are not necessarily sinusoidal, and may deviate from it through refraction and shoaling effect from the bottom floor. The frequencies of regular waves are invariant from input while the amplitude and shape of the waves are obtained by measuring the free surface profile at the cylinder axis using a probe mounted near the cylinder. The water particle kinematics are derived from this measurement. Water particle velocity is often measured simultaneously at this point. The particle acceleration is seldom measured, but is computed by numerical differentiation of the velocity profile.

Knowing the kinematics of the fluid flow at the instrumented section of the member and the corresponding measured inline loads, the mass and drag coefficients are computed. In general, the coefficients are functions of Reynolds number Re and Keulegan-Carpenter number KC. These quantities in waves are defined by

$$Re = \frac{u_0 D}{v} \qquad (2.52)$$

$$KC = \frac{u_0 T}{D} \qquad (2.53)$$

where u_0 is the maximum horizontal particle velocity due to wave and T is the wave period.

The oscillatory wave also produces a transverse or lift force normal to the wave direction. The form of the lift force is given similar to the drag force

$$f_L = \frac{1}{2}\rho C_L D u^2 \tag{2.54}$$

in which C_L is the lift coefficient.

In oscillatory flow the lift coefficient is a variant with time over a cycle. This is because the transverse force, unlike inline force, is irregular and has multiple frequencies. This is illustrated by the following example in which the forces on a vertical cylinder were measured in a progressive wave in a wave tank. The measured inline and transverse forces on the cylinder are shown in Fig. 2.20 as a function of time. The wave was generated at a single frequency. The inline force, being inertia dominated in this case, follows the same frequency. However, the transverse force, although small, displays clear evidence of multiple frequencies.

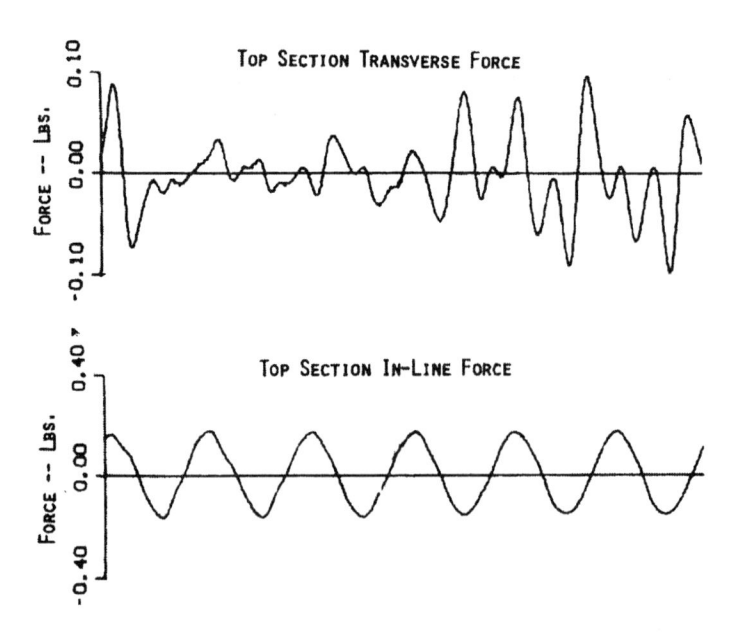

Fig. 2.20: Measured Inline and Transverse Force Time History

Chapter 2 Basic Theory of Hydrodynamics 45

2.7.6 Summary of Hydrodynamic Forces

The common nondimensional quantities that are encountered in hydrodynamics are summarized in Table 2.1.

Table 2.1: Common Dimensionless Quantities in Hydrodynamics

Symbol	Dimensionless Number	Force Ratio	Definition
Fr	Froude Number	Inertia/Gravity	$\dfrac{u_0^2}{gD}$
Re	Reynolds Number	Inertia/Viscous	$\dfrac{u_0 D}{\upsilon}$
Eu	Euler Number	Inertia/Pressure	$\dfrac{p}{\rho u_0^2}$
Ch	Cauchy Number	Inertia/Elastic	$\dfrac{\rho u_0^2}{F_e}$
KC	Keulegan-Carpenter Number	Drag/Inertia	$\dfrac{u_0 T}{D}$
St	Strouhal Number		$\dfrac{f_e D}{u_0}$

As shown, many of these nondimensional quantities are defined in terms of the ratio of forces. The Cauchy number is encountered in the hydroelastic problem in which the structure experiences deformation in shape due to fluid loading. The subscript zero in u represnts amplitude of horizontal velocy. Strouhal number is the nondimensional vortex shedding frequency, while Euler number is encountered in structural components, which are subjected to pressure force.

2.7.7 Combined Steady and Oscillating Flows

Modifications are needed in computing the wave kinematics and associated loading from waves propagating on a superimposed steady current. The wave period is modified to an apparent period by the free-stream current velocity. A current in the wave direction stretches the wavelength and opposing current shortens it. In a reference frame moving in the same direction as the steady current speed, U, waves encounter a structure at a frequency lower than the wave frequency alone, i.e., $\omega_A = 2\pi/T_A$, where T_A is the apparent period seen by an observer moving with the current. The two frequencies are related by the Doppler shift as

$$\omega = \omega_A + kU \qquad (2.55)$$

where k is the wave number. The Doppler effect is computed based on ambient current in the direction of waves. The current profile with the water depth need not be necessarily uniform. The apparent wave period for a prescribed current profile is computed from three simultaneous equations [API RP2A (1979)] assuming that the variation of current with the water depth is known:

$$\frac{L}{T} = \frac{L}{T_A} + U_A \qquad (2.56)$$

$$T_A^2 = \frac{2\pi L}{g \tanh kd} \qquad (2.57)$$

and

$$U_A = \frac{2k}{\sinh 2kd} \int_0^d U_c(y) \cosh(2ky) \, dy \qquad (2.58)$$

Velocity U_A is a weighted mean velocity and is obtained by the integration of the local current over the entire water depth. Thus, it is called an effective in-line velocity. It is used to derive the apparent period, T_A. The quantity $U_c(y)$ is assumed to be the horizontal steady current at an elevation y. If the current is at an angle to the wave, the component in line with the wave should first be computed and this is what should subsequently be used. For a uniform current profile U, the apparent period is given by the curves in

Chapter 2 Basic Theory of Hydrodynamics 47

Fig. 2.21 [API RP2A (1979)] for different values of d/gT^2. For waves propagating in the same direction as the current, T_A is on the order of 10% greater than the wave period T. The apparent wave kinematics may be computed using a two-dimensional wave theory since there is no current in a moving reference frame.

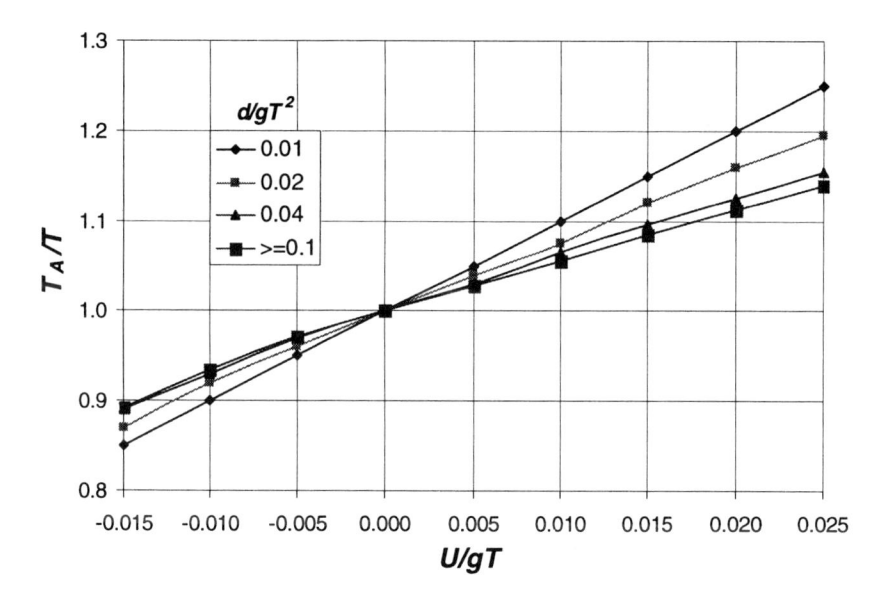

Fig. 2.21: Apparent Wave Period due to Doppler Shift in Steady Current

Once wave-current velocities have been determined in the fixed reference frame, wave loading per unit length on a cylindrical structure is based on the modified Morison equation:

$$f = \rho C_M \frac{\pi D^2}{4} \dot{u}_A + \frac{1}{2} \rho C_D |u_A + U|(u_A + U) \qquad (2.59)$$

where the drag coefficient corresponds to the combined wave-current flows. For all circular cylindrical members, the hydrodynamic coefficients are selected based on wave and current velocities and surface roughness.

The current velocity strength is defined by the ratio r:

$$r = \frac{U_I}{u_0} \tag{2.60}$$

in which U_I is the inline current velocity and u_0 is the maximum wave particle velocity. The quantity r is a ratio of current magnitude and maximum wave amplitude irrespective of direction. Thus r is positive.

When current is present with waves, the value of the hydrodynamic coefficients is altered from its value in wave alone. In order to compensate for the current in the chosen C_D value, the KC value is modified by the following correction factor [API RP 2A (1979)] based on the strength r:

$$C_r = (1+r)2\theta_* / \pi \tag{2.61}$$

where

$$\theta_* = ATAN2(-r, \sqrt{(1-r^2)}) \tag{2.62}$$

Table 2.2: First Order Correction Factor

r	θ	Degrees	Correction Factor
0	1.5708	90.00	1.00
0.05	1.6208	92.87	1.08
0.1	1.6710	95.74	1.17
0.15	1.7214	98.63	1.26
0.2	1.7722	101.54	1.35
0.25	1.8235	104.48	1.45
0.3	1.8755	107.46	1.55
0.35	1.9284	110.49	1.66
0.4	1.9823	113.58	1.77

Chapter 2 Basic Theory of Hydrodynamics

The correction factor is listed in Table 2.2 for different values of r and θ_*. The corrected value of KC should be used to choose the coefficients for the force calculation.

Consider the case of no current, i.e., $r = 0$. The Keulegan-Carpenter number is computed based on the maximum wave velocity only. From Eq. 2.62, $\theta_* = \pi/2$ and the correction factor is equal to one. In this case, the value of C_D will be based on the value in waves alone. For the case $r < 0.4$, the value of KC is modified by the correction factor, C_r as shown in Table 2.2 below.

Note that the value of the correction factor increases with the value of r, since the correction for KC becomes greater with increased strength of current velocity. The correction factors are shown for values of r up to 0.4. When $r > 0.4$, current is strong, so that the drag coefficient for practical purposes is C_{Ds}, the steady-current value.

2.8 Exercises

Exercise 1

Compute the steady drag force and oscillating lift force on a horizontal cylinder of diameter 0.5 m placed in a steady flow at a Reynolds number of 2×10^4. Plot the force vector on the cylinder based on the drag and lift forces.

Exercise 2

Derive the expressions for the particle motion from the linear theory. Show that the contour of this path is an ellipse at any elevation in the fluid. Plot the profile for the case of water depth of 100 ft, a wave period of 10s and height of 20 feet.

Exercise 3

Compute the wave force on a 1m section of a cylinder of diameter 0.1 m located 1m below the still water level in a water depth of 10m. Use linear wave theory for the wave of period 2s and height 1m.

Exercise 4

Compute the wave force for the above case assuming an inline current of 1.0 m/s superimposed on the wave. Show the difference in the force with and without the correction due to the apparent period.

2.9 References

1. American Petroleum Institute, "Recommended Practice for Planning, Designing and Constructing Fixed Offshore Platforms", API-RP2A, Washington, DC, Mar. 1979.

2. Morison, J.R., O'Brien, M.P., Johnson, J.W., and Schaaf, A.S., "The Force Exerted by Surface Waves on Piles", Petroleum Transactions, American Institute of Mining and Metal Engineering, Vol. 4, 1950, pp. 11-22

3. Sarpkaya, T., "In-Line and Transverse Forces on Cylinder in Oscillating Flow at High Reynolds Number", Proceedings on Offshore Technology Conference, Houston, Texas, OTC 2533, 1976, pp. 95-108.

Chapter 3

Basic Theory of Vibration

3.1 Introduction

The vibration of structures deals with their response under the influence of oscillatory forces. Structural vibration is a dynamic phenomenon, which includes the inertia of the structure. In hydrodynamics, the excitation forces arise from the fluid flow around the structure. The oscillatory forces produce the structural oscillation not only in the inline direction but may also induce vibration in the normal or transverse direction. Steady forces may produce oscillatory response to the structure as well. In steady flow the oscillation is primarily in the transverse direction relative to the direction of the steady force.

Structures, having a mass and an elasticity (which may be internal or external to the body of the structure), are capable of experiencing vibration. In most physical structures, vibration is undesirable. One possible exception is a system where the energy of vibration is extracted for useful purposes. Therefore, the determination of vibration in structures and the techniques in avoiding or reducing its underlying causes in a structure is a design challenge. Sometimes, vibrations have a galloping effect producing increasing amplitudes of motions or stresses in a structure, causing instant failure. However, many vibrations may be high frequency low amplitude phenomenon, which allows the structure to fail in high cycle fatigue damage. The structure responds to the frequency of excitation as well as the response frequencies (normal modes or natural frequencies) of the structure. If vibration is undesirable, the structure should be designed such that the natural frequencies of the structure in various modes are kept outside the expected range of frequencies of the environmental excitation to which the structure is exposed.

Vibrations fall into two general classes: free and forced. In the case of free vibration of a structure in a stable condition, no external forces act on the structure. If the structure is displaced from its equilibrium position and released, it will eventually return to its original position of equilibrium. In the process it will go through an oscillation due to its inherent forces from the mass and elasticity about this equilibrium position. The oscillation frequency in this case coincides with the natural frequency of the system. The natural frequency is determined from the mass and the elastic characteristics of the structure and its external mooring system. The system has inherent damping. If the damping in the system is small, then it will take a long time before the structure comes to rest again at its equilibrium position. On the other hand, when damping is large, the structure may not oscillate at all before coming back to rest. In this latter case, no appreciable vibration is expected to occur, even when the structure is excited externally at its natural frequency.

When an external oscillatory force acts on the structure, the structure responds to this force and vibrates with the forcing function. This is known as forced vibration. Unlike free vibration, the forced vibration continues as long as the excitation is present. The frequency of vibration generally coincides with the frequency of the external force. When this frequency also coincides with the natural frequency of the system, the structure is said to be in resonance. This condition generally is detrimental to the structure, especially if the damping is small. The amplitude of structure motion at resonance is large even when the exciting force is small and is limited only by the damping present in the system.

3.2 Periodic Vibration

The periodic vibration is an oscillating motion of a structure about an equilibrium position such that the motion repeats itself exactly at the specific interval of time. This repeat interval is called the period of vibration. The simplest form of periodic motion is a simple harmonic motion in the form of a sinusoidal oscillation. In the mathematical form, this simple harmonic motion may be expressed as:

$$x(t) = X \sin \frac{2\pi t}{T} \tag{3.1}$$

where x describes the position of the structure at a particular time t, T is the period of oscillation and X is the maximum displacement (or the amplitude of oscillation as it is generally called) of the structure from its equilibrium position. It is clear that the structure displacement under simple harmonic motion remains between the limits of $\pm X$ from its equilibrium position as graphically represented in Fig. 3.1. In the figure the amplitude X is shown as 1.

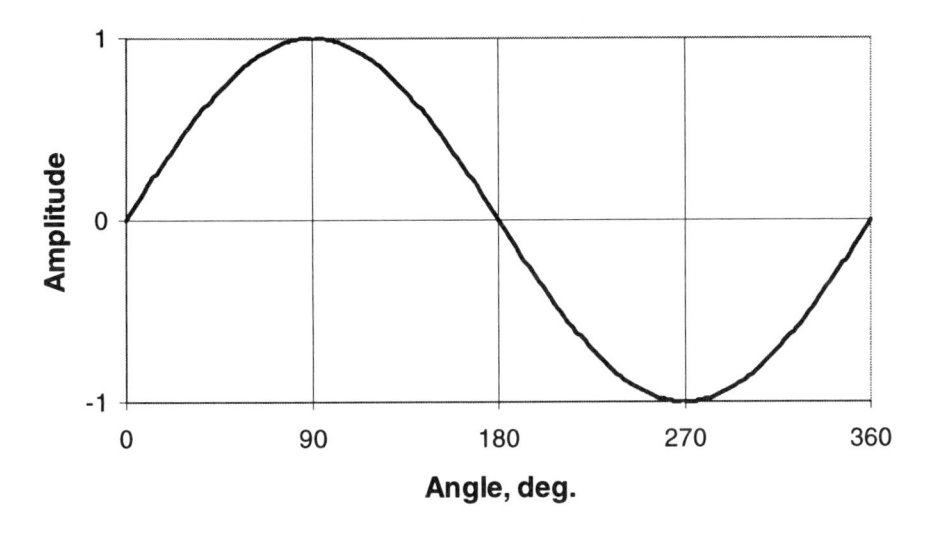

Fig. 3.1: Simple Harmonic Motion

The mathematical equation for the system in Fig. 3.1 is generally written in terms of the frequency rather than period of oscillation:

$$x = X \sin 2\pi f t = X \sin \omega t \tag{3.2}$$

where $f = 1/T$ is the cyclic frequency and $\omega = 2\pi f$ is the angular (or circular) frequency.

It follows from the expression of displacement (Eq. 3.2) that the velocity and acceleration of the structure will have the form

$$\dot{x} = X\omega \cos \omega t = X\omega \sin(\omega t + \pi/2) \tag{3.3}$$

$$\ddot{x} = -X\omega^2 \sin \omega t = X\omega^2 \sin(\omega t + \pi) \tag{3.4}$$

where $\dot{x}$ and $\ddot{x}$ represent the first and second derivatives, respectively. Thus the structure velocity leads the structure displacement by 90 deg (= $\pi/2$) and the structure acceleration is 180 deg (=π) out of phase with the structure displacement. The peaks of the velocity and acceleration are given by $X\omega$ and $X\omega^2$.

Another important parameter in a vibration analysis is the root mean square (rms) value. The variance is given by

$$VAR(x) = \left[\frac{1}{T}\int_0^T x^2(t)dt\right] \tag{3.5}$$

The rms value is then defined as

$$x_{rms} = [VAR(x)]^{1/2} \tag{3.6}$$

In other words, the time history of vibration is first squared, and the mean value of the squared function is taken over one cycle of oscillation through the integration, and the square root of the result gives the rms value of the function $x(t)$. For a harmonic oscillation, where $x(t)$ is given by Eq. 3.1, the expression for x_{rms} becomes:

$$x_{rms} = \frac{1}{\sqrt{2}} X \tag{3.7}$$

Thus the rms value of a harmonic oscillation is related to its peak value by a factor of 0.707 (= $1/\sqrt{2}$).

3.3 Linear Single Degree of Freedom System

The simplest of the dynamic systems is a single degree of freedom system in which a mass attached to a spring describes the system. Consider a system having a mass m, and a spring having a spring constant k and no internal or external damping. The system is schematically shown in Fig. 3.2. The right hand side gives the free body diagram. An undamped system experiences an inertia force from its mass and a restoring force from the spring, the sum of which will resist the excitation force.

Chapter 3 Basic Theory of Vibration

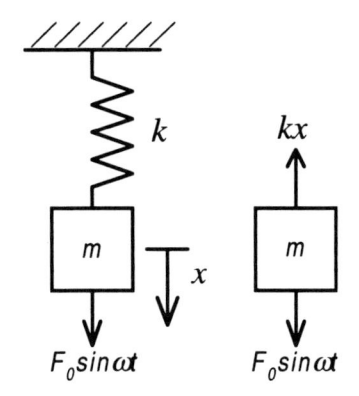

Fig. 3.2: Forced Spring-Mass System

The equation of motion of the mass shown in Fig. 3.2 may be described by its inertia and restoring force, which are equated to the excitation force whose magnitude is F_0. The forcing function in this example is assumed to be a harmonic function having a frequency ω.

$$m\ddot{x} + kx = F_0 \cos \omega t \quad (3.8)$$

Consider first the case of the spring mass system (Eq. 3.8) in which there is no external force:

$$m\ddot{x} + kx = 0 \quad (3.9)$$

The general form of solution for this equation will be harmonic of the form:

$$x = A_1 \cos \omega t + A_2 \sin \omega t \quad (3.10)$$

in which A_1 and A_2 are arbitrary constant values determined by boundary conditions. On substitution of this expression in Eq. 3.9 one obtains,

$$-\omega^2 m + k = 0 \quad (3.11)$$

which gives

$$\omega_n = \sqrt{\frac{k}{m}} \quad (3.12)$$

This frequency ω_n is called the undamped natural frequency of the system. This implies that if the spring-mass system is displaced from its equilibrium position and released, it will undergo a harmonic oscillation whose period will coincide with the natural period of the system, and ideally in the absence of damping, the oscillation would take on a perpetual motion.

Now return to the case of the forced oscillation given in Eq. 3.8. In this case, the displacement will be in phase with the force itself so that it may be written as

$$x = X \cos \omega t \qquad (3.13)$$

When this expression is introduced in Eq. 3.8, the amplitude of displacement may be obtained from

$$X = \frac{F_0}{k - m\omega^2} \qquad (3.14)$$

It is clear that the amplitude will "blow up", approaching infinity as the denominator becomes zero. Equating the denominator to zero, one gets Eq. 3.12, the natural frequency ω_n. It is customary to design a physical system in which the resonant frequency is removed as far as practical from the possible excitation frequency range that may be encountered from the environment.

A physical system, however, is never damping free. The damping may be extremely small in which case the amplitude of oscillation will be quite large even for a small excitation. This is extremely important to note since many systems are excited from nonlinear effect at frequencies near the natural frequency of the system, which may be further from the excitation frequencies. Some of these systems will be described later.

3.3.1 Transient Response

Assume that the structure is in contact with a viscous fluid. The motion of the structure or fluid around it will introduce a viscous force arising from the friction between the structure and the fluid in relative motion. Such forces in their simple form may be represented as a damping force proportional to the velocity $\dot{x}$ of the relative motion:

$$F_d = c\dot{x} \qquad (3.15)$$

Chapter 3 Basic Theory of Vibration

where c is called the linear damping coefficient. This damping is very important in an oscillating structure since it helps limit the excursion of the structure in a resonance situation. This is why it is desirable to physically introduce damping whenever possible with the help of a dashpot in the system.

Now if a linear damper is present in the system, then a damping force is additionally introduced in the system as shown in Fig. 3.3. Then the equation of motion for the displacement of the mass is described similar to Eq. 3.8 by

$$m\ddot{x} + c\dot{x} + kx = F_0 \cos \omega t \tag{3.16}$$

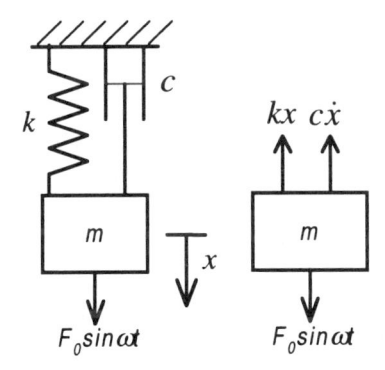

Fig. 3.3: Damped Single Degree of Freedom System

Examine this equation first in the absence of the external force in which the mass m is displaced from its equilibrium position and released. It will give rise to a free vibration of the mass given by

$$m\ddot{x} + c\dot{x} + kx = 0 \tag{3.17}$$

The solution of this equation is termed transient solution or complimentary function. Due to the presence of damping, the free oscillation will eventually dissipate. The solution is assumed of the form:

$$x = X \exp(qt) \tag{3.18}$$

in which the unknowns are X and q. Once this expression is substituted in Eq. 3.17, one gets:

$$X(mq^2 + cq + k)\exp(qt) = 0 \qquad (3.19)$$

Since this equality must be satisfied for all times t, one must have

$$mq^2 + cq + k = 0 \qquad (3.20)$$

This quadratic equation yields two solutions

$$q_{1,2} = -\frac{c}{2m} \pm \sqrt{\left(\frac{c}{2m}\right)^2 - \frac{k}{m}} \qquad (3.21)$$

In terms of q_1 and q_2, the general solution for the free oscillation is

$$x = A_1 \exp(q_1 t) + A_2 \exp(q_2 t) \qquad (3.22)$$

In order to solve for the arbitrary constants A_1 and A_2, one invokes the initial conditions of the mass in its equilibrium position.

The above equations (i.e., Eqs. 3.21-22) are often written in a more practical form as follows. Note from Eq. 3.21 that, in general, there are two distinct solutions related to q_1 and q_2. However, when the quantity under the radical sign becomes zero, there is only one solution for q. By setting this quantity to zero, one can solve for the damping coefficient:

$$c_c = 2\sqrt{km} \qquad (3.23)$$

This damping has a special significance and is termed critical damping coefficient, which will be further explained shortly. The ratio of damping in a linear system and this critical damping is generally known as the damping factor:

$$\zeta = \frac{c}{c_c} \qquad (3.24)$$

The amount of damping in a system is often described by the value of the damping factor. Higher the value of the damping factor, the higher is the damping in the system. The natural frequency of the linear system is given by the quantity ω_n shown in Eq. 3.12 so that

Chapter 3 Basic Theory of Vibration

$$\frac{c}{m} = 2\omega_n \zeta \tag{3.25}$$

Substituting the relationships from Eqs. 3.12 and 3.25 in Eq. 3.21, one obtains:

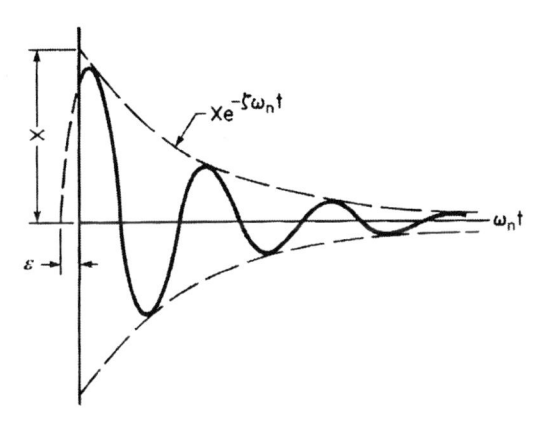

$$q_{1,2} = -\zeta\omega_n \pm i\omega_n\sqrt{1-\zeta^2} \tag{3.26}$$

Fig. 3.4: Decaying Oscillation of a Linearly Damped SDOF System

Note that this expression is valid only for $\zeta \leq 1.0$. Using this expression, the solution for x becomes

$$x = X \exp(-\zeta\omega_n t)\sin[\sqrt{1-\zeta^2}\omega_n t + \varepsilon] \tag{3.27}$$

in which the unknowns now are the amplitude X and phase angle ε. The solution has two parts – an exponential time decay function and an oscillating term, which has a frequency of $\sqrt{1-\zeta^2}\omega_n$. This frequency is different from the undamped frequency and is termed the damped frequency. It is expected that the amplitude of oscillation will diminish in value as time passes, the rate of this decrease being a direct function of the damping factor. This relationship is plotted in Fig. 3.4. It is valid for $\zeta < 1.0$. When $\zeta = 1.0$, the oscillation disappears and the solution behaves as an exponential function (Fig. 3.5). If the damping value is greater than 1.0, no oscillation of the solution is expected.

This is why the unit value of damping ($\zeta = 1.0$) is referred to as critical damping. Systems with $\zeta > 1.0$ are called overdamped systems, while those with $\zeta < 1.0$ are termed underdamped. A critically damped or an overdamped system has an aperiodic or non-oscillatory motion (Fig. 3.5).

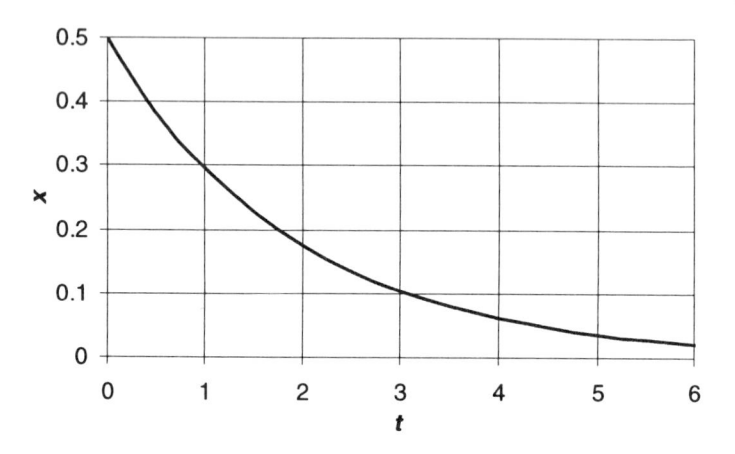

Fig. 3.5: Free Motion of a Critically Damped SDOF System

Most of the practical systems that one encounters have a damping factor less than one. If a weight is hung from a coil spring and displaced, then the oscillation will continue for a long time before it comes to rest. This is an example of an underdamped system, which has a damping factor much less than 1. However, one does encounter systems whose damping factors are equal to or greater than 1.0. For example, the hinged storm doors in a house with the dashpots attached should ideally be adjusted to a damping factor of close to one. This will allow the door to close slowly without oscillating when the door is opened and released. Another example of this type of motion is the heave motion of a floating barge. Since the heave of a flat bottom barge is highly damped, it has a large value of damping factor. In an experiment with a barge model, if the barge is displaced in the vertical direction and released, the measured vertical motion will have little oscillation before coming to rest in its equilibrium position.

Chapter 3 Basic Theory of Vibration

The solution of the equation of motion given by Eq. 3.27 represents harmonic oscillation values in which subsequent amplitudes of oscillation decay exponentially. If two consecutive amplitudes are given by their absolute values $|x_k|$ and $|x_{k-1}|$, then a logarithmic decrement of amplitudes is defined by the difference of their logarithm values.

$$\delta = \ln|x_k| - \ln|x_{k-1}| \quad (3.28)$$

The logarithmic decrement is related to the damping factor by the relation:

$$\zeta = \frac{\delta}{\sqrt{\pi^2 + \delta^2}} \quad (3.29)$$

which may be determined quite accurately by the simple relation:

$$\zeta = \frac{\delta}{\pi} \quad (3.30)$$

For small values of ζ the error in the above simplification is quite small (for example, for $\zeta = 0.1$, the error is about 0.5 percent). Thus, for each pair of consecutive amplitudes, a damping factor may be derived. In practice, however, a practical range of the complete decayed oscillation is utilized to derive one average value of ζ.

The term $x_k \exp(-\zeta\omega_n t)$ represents the curve that can be drawn through the succeeding peaks of the damped oscillation. Strictly speaking, the curve does not pass exactly through the peaks, but the small deviation is usually neglected. If the natural logarithms of these peaks are determined, the quantity $\zeta\omega_n$ represents the slope s of the line that can be drawn through the converted values. The frequency of the damped motion, ω_d, is also obtained from Eq. 3.27, and thus one obtains two equations and two unknowns:

$$s = -\zeta\omega_n \quad (3.31)$$

$$\omega_d = \omega_n \sqrt{1 - \zeta^2} \quad (3.32)$$

The terms on the left-hand side of Eqs. 3.31 and 3.32 are obtained from the free decay curve. Once the values of ω_n and ζ are known from the above equations, the added mass and damping coefficients are computed as follows:

$$m_a = m - m_0 = \frac{k}{\omega_n^2} - m_0 \qquad (3.33)$$

and

$$c = 2m\zeta\omega_n \qquad (3.34)$$

in which m_0 is the mass of the structure.

Therefore, knowing the extinction curve for a moored floating structure, such as from a model test, the damping of the system may be determined by a simple analysis. This is illustrated by an example based on Fig. 3.6. The extinction curve represents the free oscillation of a moored tanker. The displacement of the tanker is m_0 = 38.1 kg/m (25.5 slugs) and the spring constant k = 10.4 kg/m (7.0 lbs/ft). The least square analysis described above of the peaks of the free decay curve gives an added mass coefficient and a damping factor of 0.049 and 0.008 respectively. The natural period between the positive second and third peaks in Fig. 3.6 is measured as 12.3 sec. Then,

$$m_a = 1.17 \text{ slug} \qquad (3.35)$$

and

$$C_A = m_a/m_0 = 0.046 \qquad (3.36)$$

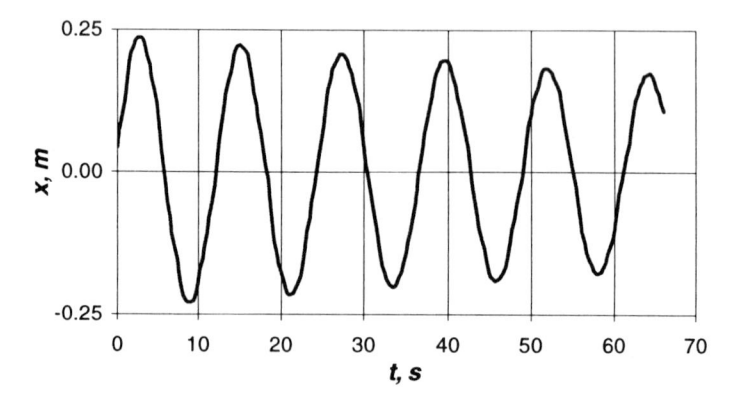

Fig. 3.6: Extinction Curve of the Example Problem

Chapter 3 Basic Theory of Vibration

Also, the amplitudes of peaks 2 and 3 are 0.235 and 0.22 m (0.78 and 0.72 ft) respectively. Therefore,

$$\delta = \frac{1}{2}(\ln 0.235 - \ln 0.22) = 0.03 \qquad (3.37)$$

which reduces to

$$\zeta = \frac{0.03}{\sqrt{\pi^2 + 0.03^2}} = 0.01 \qquad (3.38)$$

This example is based on only two peak values in the extinction curve to determine the two unknowns. In practice, for better estimates, a least square fit of the individual well-defined measured peak values (both absolute positive and negative peaks) are used to compute these quantities.

3.3.2 Forced Linearly Damped System

Now if the structure experiences a harmonic exciting force, the equation of motion for the displacement of the mass becomes:

$$m\ddot{x} + c\dot{x} + kx = F_0 \cos \omega t \qquad (3.39)$$

The solution in this case should still be harmonic as in the undamped case. But the displacement will be leading or lagging the excitation based on the frequency of the forcing function relative to the natural frequency of the system, so that a solution of the form:

$$x = X \cos(\omega t - \varepsilon) \qquad (3.40)$$

should be assumed. In the above equation, X is the amplitude of the harmonic oscillation and ε is its phase angle relative to the excitation. The form of the solution shows that for positive values of ε, the motion will be lagging the excitation force. Substituting the value of x and its first and second derivatives in the equation of motion, Eq. 3.39, one has:

$$-m\omega^2 X \cos(\omega t - \varepsilon) + c\omega X \sin(\omega t - \varepsilon) + kX \cos(\omega t - \varepsilon) = F_0 \cos \omega t \quad (3.41)$$

This equation is valid for all times t. Therefore, the coefficients of sin ωt and cos ωt should be set to zero individually (since they are orthogonal to each other) in order to satisfy this relationship. Hence, the following equalities hold:

$$-m\omega^2 X \cos\varepsilon + c\omega X \sin\varepsilon + kX \cos\varepsilon = F_0 \quad (3.42)$$

and

$$m\omega^2 X \sin\varepsilon + c\omega X \cos\varepsilon - kX \sin\varepsilon = 0 \quad (3.43)$$

Solving for the two unknowns from above two equations, one obtains:

$$X = \frac{F_0}{[(k-m\omega^2)^2 + (c\omega)^2]^{1/2}} \quad (3.44)$$

and

$$\tan\varepsilon = \frac{c\omega}{k-m\omega^2} \quad (3.45)$$

These equations are often normalized in terms of the equivalent static deflection of the system, which is expressed as:

$$X_s = \frac{F_0}{k} \quad (3.46)$$

Using this expression and the natural frequency ω_n and damping factor ζ, Eqs. 3.44-45 can be reduced to the following normalized forms:

$$\frac{X}{X_s} = \frac{1}{[(1-\frac{\omega^2}{\omega_n^2})^2 + (2\zeta\frac{\omega}{\omega_n})^2]^{1/2}} \quad (3.47)$$

and

$$\tan\varepsilon = \frac{2\zeta\frac{\omega}{\omega_n}}{1-\left(\frac{\omega}{\omega_n}\right)^2} \quad (3.48)$$

It is customary to plot these normalized forms of displacement and phase angle instead of Eqs. 3.44 and 3.45. The displacement of the linearly damped system is shown in Fig. 3.7. This is known as the frequency response curve. The normalized displacement is called the magnification factor, and reduces to its static value of X_s at zero frequency. The maximum displacement is found to occur at the damped natural frequency, the amplitude of which depends on the amount of damping in the system. This is also the point where the phase angle (Fig. 3.8) goes through a rapid change from zero to 180 degrees. For zero damping (i.e., undamped system, $\zeta = 0.0$), the maximum amplitude occurs at the undamped natural frequency ($\omega/\omega_n = 1$) and approaches infinity. This, of course, does not exist in nature since some amount of damping is always present, so that the amplitude in practice is finite. When the damping is large, the amplification disappears. For example, at $\zeta > 0.5$, there is practically no "hump" in the magnitude of displacement at the natural frequency.

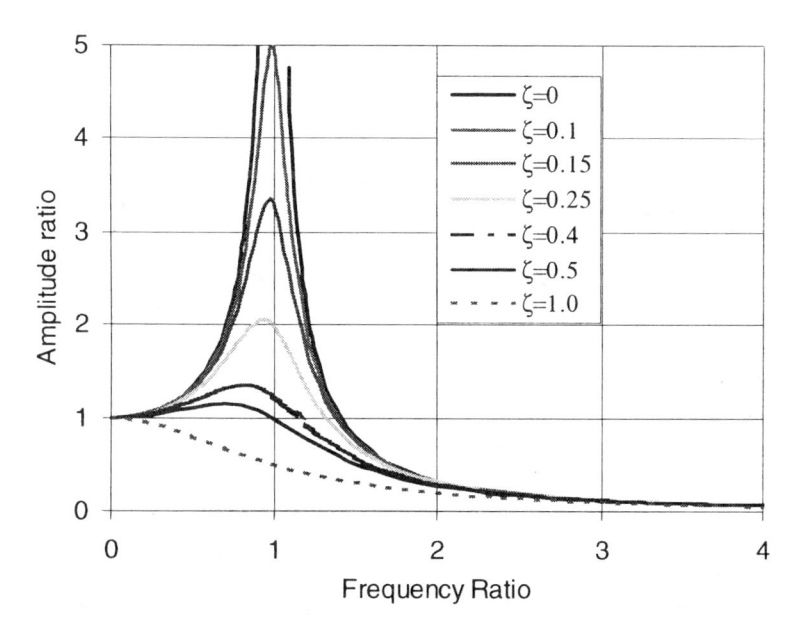

Fig. 3.7: Normalized Displacement Amplitude of a SDOF System

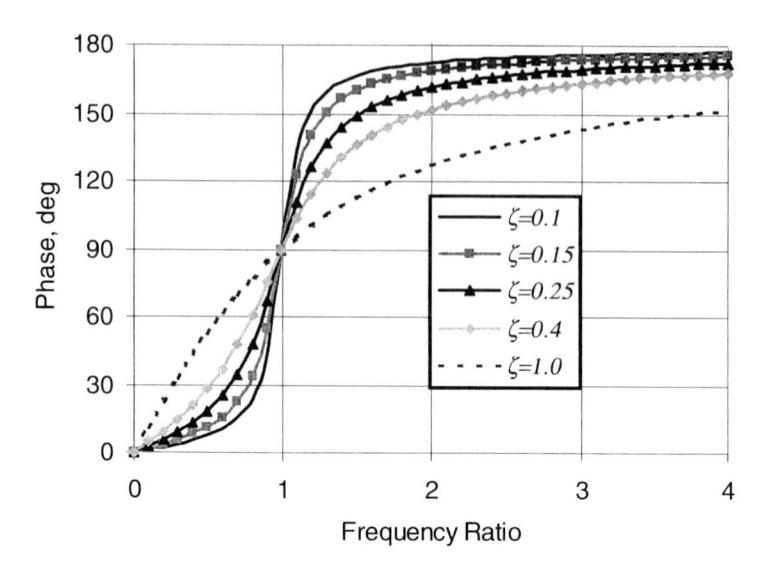

Fig. 3.8: Displacement Phase of a SDOF System

The solution discussed above is the steady state solution. However, when a spring mass system is subjected to an external harmonic excitation at time 0, its initial motion includes the transient oscillation (corresponding to free oscillation discussed in section 3.3.1) due to the initial impact from the excitation. Therefore, the initial time history of motion will show the sum of the transient and forced oscillation. The transient dies out with time (based on the amplitude of system damping) and the steady state harmonic oscillation takes over. An example of a floating moored ship-shaped structure subjected to a regular wave is shown in Fig. 3.9. The excitation frequency corresponding to 10s is far removed from the natural frequency of the system. The surge motion of the structure shows the large transient oscillation superimposed on the sinusoidal oscillation at the imposed wave frequency. The transient oscillation occurs at the damped natural period of the system. In this case, the natural period of the structure in surge is 100s. This oscillation dies slowly since the damping in surge for the floating structure is small. Toward the end of the record, the oscillation is predominantly at the wave period.

Chapter 3 Basic Theory of Vibration

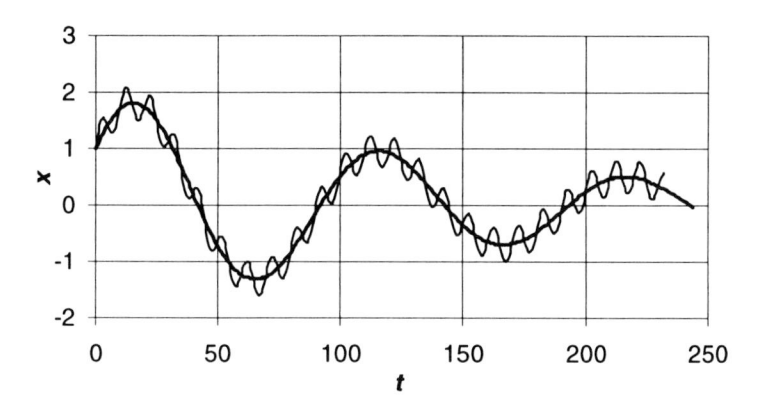

Fig. 3.9: Initial Surge Motion of a Floating Soft-Moored Structure in Waves

However, the natural period in heave for the same structure is much smaller. If the excitation period is close to the natural period, but not equal to it, then the initial response from the periodic excitation will take a form similar to Fig. 3.10. Here the natural period of motion is taken as 12s and the wave period is 10s. The response is shown for a small damping and it takes the form of beating. The beating amplitude builds up slowly to a maximum before reducing again. The beating period is based on the difference frequency between the excitation frequency and the natural frequency. In the example in Fig. 3.10, the beating period is 60s.

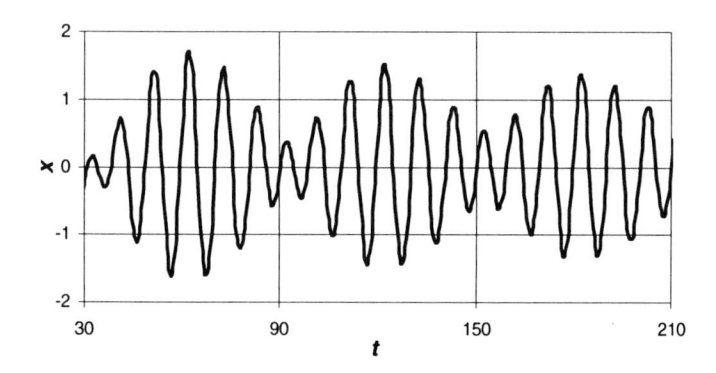

Fig. 3.10: Beating Effect from Two Close Periods

3.3.3 Energy Dissipation

For the linearly damped single degree of freedom system, the steady-state oscillation is given by Eq. 3.40. The net force exerted on the structure by the spring and damping device is written from Eq. 3.39 as:

$$f_e = c\dot{x} + kx = -c\omega X \sin(\omega t - \varepsilon) + kX \cos(\omega t - \varepsilon) \qquad (3.49)$$

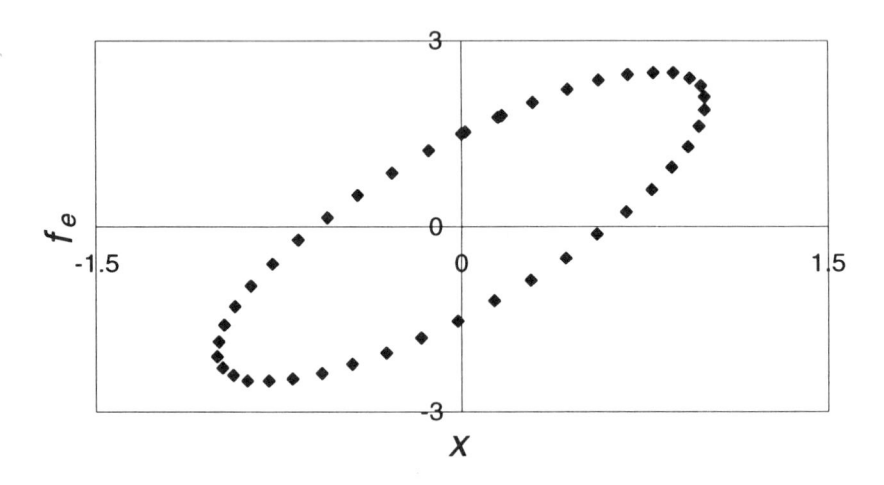

Fig. 3.11: Hysteresis Curve for a Spring and Linear Damper in a System

The relationship between f_e and x may be established from the two equations, 3.40 and 3.49. If f_e is plotted with respect to x, then a figure similar to Fig. 3.11 is obtained. It is elliptical in shape and the area within this ellipse is the energy dissipated (work done) by the system due to the spring and the damper. This is known as hysteresis. Over one cycle, the total energy represented by the ellipse is obtained from the integral:

$$E = \int_0^T f_e \frac{dx}{dt} dt = \pi c \omega X^2 \qquad (3.50)$$

where T is the period of oscillation ($T = 2\pi/\omega$).

3.3.4 Mechanical Oscillation

Forced oscillation is often a common practice to simulate oscillatory forces on a submerged structure in a fluid. Consider the case of a forced oscillation of a structure in equilibrium in a fluid medium. The structure is forced to oscillate sinusoidally in a prescribed direction of a given amplitude and frequency, and the resulting reaction forces from the fluid are recorded. Assuming linear damping, the equation of motion due to sinusoidal oscillation has the form similar to Eq. 3.39. On the assumption that oscillation is described by Eq. 3.40, the solution of Eq. 3.39 is written in Eq. 3.41. After eliminating time t, Eqs. 3.42 and 3.43 are obtained, which reduce to

$$(k - m\omega^2)^2 + (c\omega)^2 = \left(\frac{F_0}{X}\right)^2 \tag{3.51}$$

and

$$c\omega = (k - m\omega^2)\tan\varepsilon \tag{3.52}$$

Equation 3.51 may be equivalently written as:

$$\left(\frac{\omega_n^2}{\omega^2} - 1\right)^2 + \left(\frac{c}{m\omega}\right)^2 = \left(\frac{F_0}{m\omega^2 X}\right)^2 \tag{3.53}$$

Note that in a forced oscillation test, X is the amplitude of oscillation and ε is its phase angle relative to the phase of the force amplitude F_0 (i.e., difference between oscillation phase and force phase). These three quantities are known from the measurement from the mechanical oscillation test. Equations 3.51 and 3.52 can be reduced to the following:

$$k - m\omega^2 = \left(\frac{F_0}{X}\right)\cos\varepsilon \tag{3.54}$$

and

$$c\omega = \frac{F_0}{X}\sin\varepsilon \tag{3.55}$$

The added mass coefficient and damping factor may be derived from the above expressions as

$$C_A = \frac{m - m_0}{m_0} \tag{3.56}$$

where

$$m = \frac{k}{\omega^2} - \left(\frac{F_0 \cos \varepsilon}{\omega^2 X}\right) \tag{3.57}$$

and

$$\zeta = \frac{c}{c_c} = \left(\frac{F_0 \sin \varepsilon}{2\omega X \sqrt{mk}}\right) \tag{3.58}$$

in which the critical damping coefficient $c_c = 2m\omega_n$.

The quantities F_0, X and ε are determined from the simulation or test data by a Fourier series analysis.

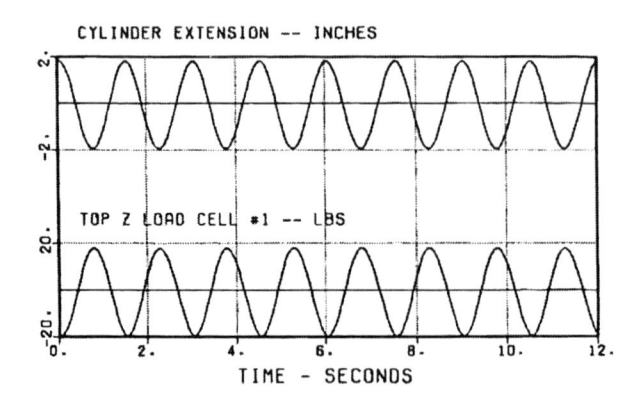

Fig. 3.12: Measured Displacement and Load in a Forced Oscillation Test

A vertical cylinder was tested mechanically in water in the heave mode. Some mechanical noise was present in the data from the vibration of the driving hydraulic cylinder. Both the oscillation and force traces were

Chapter 3 Basic Theory of Vibration 71

digitally filtered through the use of the Fourier series analysis, and only the first harmonic data were retained assuming the response to be linear. A sample trace of measured displacement and load after filtering is shown in Fig. 3.12. The normalized force amplitudes and phase angles are plotted with respect to the oscillation frequency in Figs. 3.13 and 3.14. The force was normalized by $m_0\omega^2 X$ (see Eq. 3.53). Once the quantities on the right-hand side of Eqs. 3.57 and 3.58 are computed, the added mass coefficient and damping factor are known. These are plotted against the oscillation frequency and shown in Figs. 3.15 and 3.16. The range of frequency (0.4 to 0.5 Hz) where the phase angle is significantly different from 0 or 180 degrees is expected to produce the most reliable results in this method. Note that near 0 and 180 degrees, the stiffness or inertia force predominates, and the accuracy in the damping coefficient diminishes.

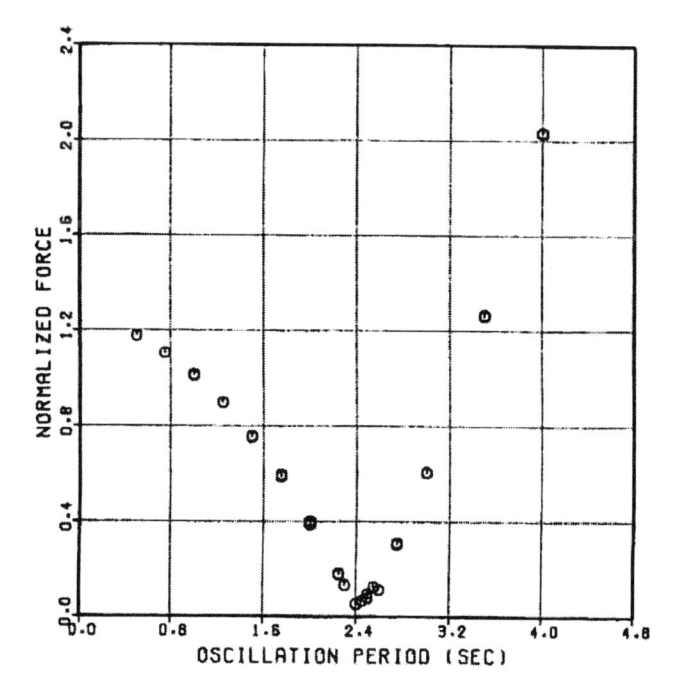

Fig. 3.13: Force Transfer Function

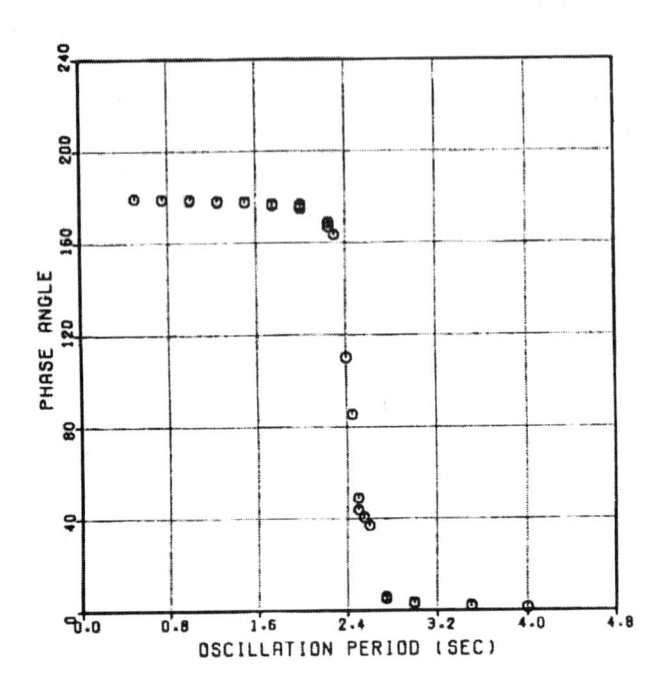

Fig. 3.14: Computed Phase Difference Between Force and Oscillation

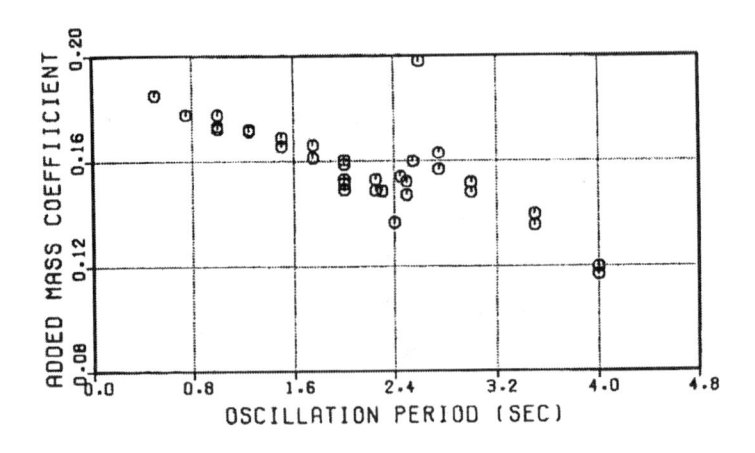

Fig. 3.15: Derived Added Mass Coefficients for Cylinder

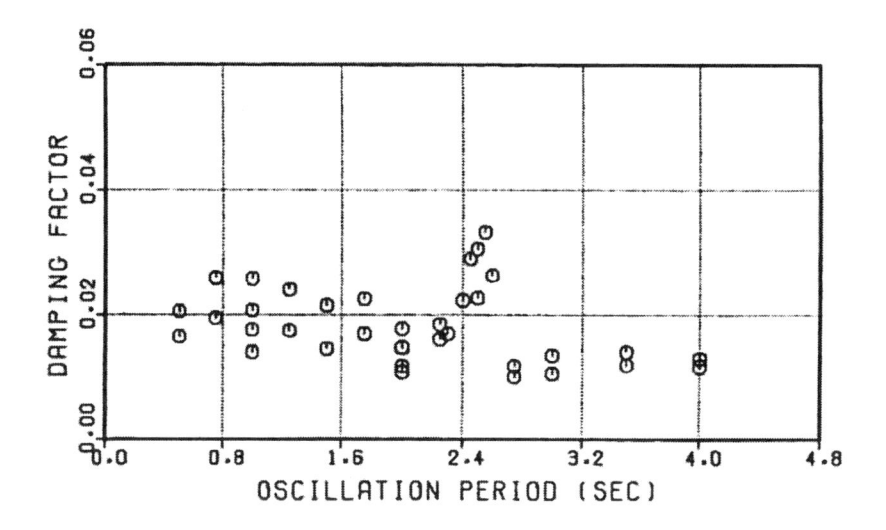

Fig. 3.16: Derived Damping Coefficients for Cylinder

3.4 Nonlinear Single Degree of Freedom System

Many physical systems have components that cannot be described by a simple linear system. In these cases, the equation of motion for the system is nonlinear. Some of these nonlinearities may be linearized for simplicity and still yield a reasonable representation of the physical system. However, in other cases, the true nonlinearity needs to be maintained and included in the description of the system. The nonlinearity appears either in the damping term arising from the structural material and the surrounding medium or in the restoring force term in the resilient support system.

It is possible to obtain an exact solution for a few nonlinear single degree of freedom systems, which may be described by a differential equation. By exact is meant that the solution is obtained either in a closed form or in a mathematical expression that can be evaluated to any degree of accuracy by a numerical means, e.g., a series expression.

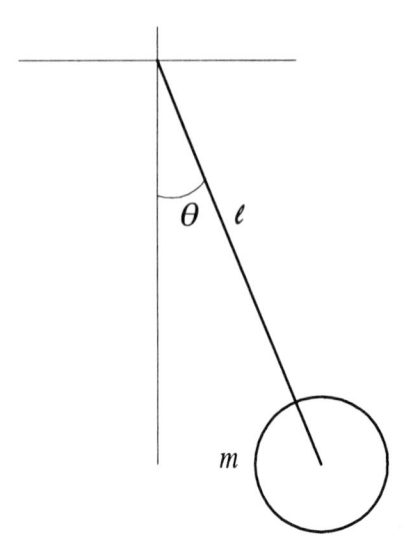

Fig. 3.17: Pendulum with Massless Arm

One of the simplest nonlinear systems is a simple pendulum (Fig. 3.17). In this case the mass of the pendulum is concentrated at the end of a linear spring represented by a rigid massless rod. This system is commonly described by a single degree of freedom system whose undamped equation of free oscillation is commonly approximated for nearly linear cases by

$$ml\ddot{\theta} + mgl\theta = 0 \qquad (3.59)$$

where the angle of oscillation is small. The natural period of oscillation of this system is given by the equation

$$T_n = 2\pi\sqrt{\frac{g}{l}} \qquad (3.60)$$

If large angles are accommodated, then a nonlinear version of this equation must be applied. The exact form of the equation of motion then becomes:

$$ml\ddot{\theta} + mgl\sin\theta = 0 \qquad (3.61)$$

This equation can be solved in terms of elliptic integrals as long as the initial conditions are assumed to be a known displacement angle of the pendulum and zero initial velocity. The values of elliptical integrals are tabulated in mathematical handbooks.

The first equation, however, provides a reasonable representation for most such physical systems. In order to check the accuracy of the linear approximation, the natural periods of the two equations may be compared. The natural period of the nonlinear equation (Eq. 3.61) depends on the initial angle while that of the linear system is independent of the initial angle (Eq. 3.59). The ratio of the two is tabulated in Table 3.1 for a few initial angles. It is clear that even at an angle of 30 degrees, the small angle approximation for the natural period is quite accurate.

Table 3.1: Ratio of Natural Periods of Nonlinear and Linearized Pendulum Problem

Initial angle, deg	$T_n(NL)/T_n(L)$
0	1.00007
5	1.00051
10	1.00191
15	1.00433
20	1.00764
25	1.01206
30	1.01738
35	1.02388
40	1.03132

3.4.1 Phase-Plane Diagram

For many nonlinear vibration problems, it is convenient and beneficial to present the solution in a graphical format. The solution provides a graphical representation of the general behavior of the system in motion. This is particularly useful for a single degree of freedom system.

If the system has one degree of freedom, then it needs two parameters to completely describe its behavior when in motion. These two parameters are used in an orthogonal coordinate system to plot their variations during the dynamic motion of the system. The graphical representation of the system is commonly known as the Phase-Plane representation. For an ordinary dynamic system, these two parameters are normally the displacement and velocity of the system with respect to time. This is illustrated with a simple system. Consider the undamped free single degree of freedom system given by Eq. 3.9 with a solution having the form shown in Eq. 3.10. The velocity of the system may then be obtained from Eq. 3.9 as

$$\dot{x} = -A_1 \omega \sin \omega t + A_2 \omega \cos \omega t \tag{3.62}$$

Write Eq. 3.62 in terms of $y = \dot{x}/\omega$. Then squaring both sides of Eqs. 3.10 and 3.62 and adding the two, the time t is eliminated giving

$$x^2 + y^2 = A_1^2 + A_2^2 = C^2 \tag{3.63}$$

in which C is a constant. This is the equation of a series of concentric circles, which have their common center at the origin (0,0). Each of these curves is called a trajectory. The motion is considered periodic if the phase-plane trajectories are closed and the path is defined within a finite time. For an aperiodic solution the contour about the center will be open. The difficulty of this technique is that the time variable may not be eliminated as simply as was done in this example. However, the power of the technique is clearly evident from this simple example. It provides an insight regarding the stability of a complex physical system.

3.4.2 Nonlinearly Damped Free Vibration

Examine a nonlinear system where the nonlinearity appears in the damping term. This type of nonlinear system is quite prevalent in hydrodynamic problems, particularly arising from the viscous flows in hydrodynamics. Consider a simple spring mass system with a dashpot attached to it. In addition to a linearly proportional dashpot, this dashpot also gives a force that is proportional to the square of the velocity and always opposes the motion. When such nonlinear damping is present, the equation of motion for the damped free oscillation is given by

$$m\ddot{x}(t) + c_1 \dot{x}(t) + c_2 |\dot{x}(t)| \dot{x}(t) + kx(t) = 0 \tag{3.64}$$

Chapter 3 Basic Theory of Vibration

where m = total mass of the structure, c_1 = linear damping coefficient, c_2 = nonlinear damping coefficient, and k = linear spring constant. The quantity x is the displacement of the structure and the dots as before represent derivatives with respect to time.

Similar to the linear equation, the solution for this equation will be a decayed oscillation of the type shown in Fig. 3.4. Since this equation is nonlinear, it is difficult to solve in a closed form. However, useful information may be obtained from this equation if a simplification is made. On the assumption that each half cycle of the decayed oscillation is reasonably sinusoidal, the nonlinear term is linearized by the first term of its Fourier series expansion as

$$|\dot{x}(t)|\dot{x}(t) = \frac{8}{3\pi}\omega_n x_k \dot{x}(t) \tag{3.65}$$

where ω_n is the frequency of oscillation corresponding to the natural frequency of the system and x_k corresponds to the amplitude of the k-th cycle of the decayed oscillation of the structure. In other words, the product of the two represents the amplitude of velocity. By this approximation, the nonlinear dependence of time on the left-hand side of the equation is reduced to a single velocity term. Upon substitution of Eq. 3.65 in Eq. 3.64, a linearized (with respect to time) equation is obtained:

$$m\ddot{x}(t) + c_1\dot{x}(t) + \frac{8c_2}{3\pi}\omega_n x_k \dot{x}(t) + kx(t) = 0 \tag{3.66}$$

Then substituting a single coefficient for the damping term gives:

$$c' = c_1 + \frac{8c_2}{3\pi}\omega_n x_k \tag{3.67}$$

Equation 3.66 becomes the familiar form as in Eq. 3.17, whose solution may be written in the form similar to Eq. 3.27, with ζ replaced by ζ' where ζ' is the damping factor including the linearized nonlinear term,

$$\zeta' = \frac{c'}{2m\omega_n} \tag{3.68}$$

Then, using Eqs. 3.28, 3.67 and 3.68, the logarithmic decrement has the form:

$$\ln\frac{x_{k-1}}{x_{k+1}} = \frac{2\pi}{2m\omega_n}[c_1 + \frac{8c_2}{3\pi}\omega_n x_k] = \frac{T_n}{2}\left[\frac{c_1}{m} + \frac{c_2}{m}\frac{16}{3T_n}x_k\right] \qquad (3.69)$$

where T_n is the natural period of oscillation. In terms of the traditional damping factor, ζ, a more convenient nondimensional form may be written as

$$\frac{1}{2\pi}\ln\frac{x_{k-1}}{x_{k+1}} = \zeta + \frac{4c_2}{3\pi m}x_k \qquad (3.70)$$

which is the equation of a straight line with the left-hand side being the Y-axis and x_k being the X-axis. Thus, knowing the peak values of the oscillation, the points (X,Y) from the measured data may be fitted to a straight line by the least square method. Then, the nonlinear damping coefficient c_2 and damping factor ζ may be obtained from the above equation by fitting a straight line through the least square method. The quantities c_2 and ζ become the least square slope and the intercept of the fitted line. It should be noted that a reasonable number of peaks are required for sufficient accuracy in these estimates. However, for a highly damped system, the decaying amplitudes reduce to a small value rather quickly, and the estimates in these cases are difficult to make.

In the hydrodynamics field, such nonlinear damping is frequently present arising from the vortex shedding phenomenon. An example of this is a moored floating vessel, e.g., a semisubmersible structure undergoing surge oscillation. The semisubmersible generally consists of vertical columns as well as horizontal pontoons, which induce vortex shedding during the (back and forth) oscillating motions. Thus the free surge decay may introduce nonlinear drag type damping.

Assuming that the nonlinear damping term may be represented by the Morison equation drag term:

$$c_2 = \frac{1}{2}\rho A C_D \qquad (3.71)$$

where A = projected cross sectional area of the floating vessel in the direction of flow, Eq. 3.70 reduces to

$$\ln\frac{x_{k-1}}{x_{k+1}} = 2\pi\zeta + \frac{4}{3}\left(\frac{\rho A C_D}{m}\right)x_k \quad (3.72)$$

An example of a semisubmersible and a tanker tested in still water [Chakrabarti and Cotter (1990)] is given in Fig. 3.18. The floating structures were moored with linear springs such that the natural periods varied with the selected spring sets. The vessels were displaced from equilibrium and allowed to oscillate freely. The decay curves were analyzed to obtain the linear damping and the drag coefficient by a least square fitting method outlined above (see Eq. 3.72). The drag coefficients for the tanker and the semisubmersible versus the decay (natural) period are presented in Fig. 3.18. It is clear that the tanker had very small nonlinear damping while the semisubmersible experienced a large nonlinear damping. This is due to the presence of a large number of submerged cylindrical members in the semisubmersible.

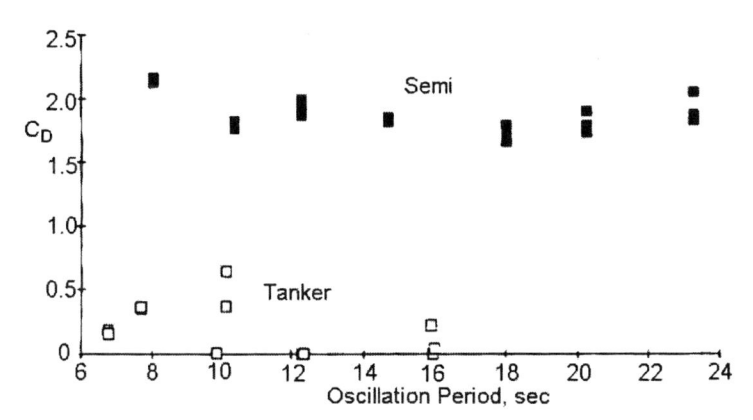

Fig. 3.18: Drag Coefficients for Tanker and Semisubmersible from Decay Curves

The added mass coefficient is computed from the measured natural period, the spring constant k, and the displaced mass m_0, using the formula:

$$C_A = \frac{kT_n^2}{4\pi^2 m_0} - 1 \quad (3.73)$$

3.4.3 Nonlinearly Damped Forced Vibration

Consider now the same nonlinearly damped system, which is forced by a harmonic forcing function.

$$m\ddot{x}(t) + c_1\dot{x}(t) + c_2|\dot{x}(t)|\dot{x}(t) + kx(t) = F_0 \cos\omega t \qquad (3.74)$$

This equation may be handled in a closed form as long as one linearizes the nonlinear damping term by the first term of the Fourier series expansion (Eq. 3.65). Then, the linearized damping term is combined with the linear damping term as shown in Eq. 3.67. The equation of motion becomes

$$m\ddot{x}(t) + c'\dot{x}(t) + kx(t) = F_0 \cos\omega t \qquad (3.75)$$

The solution becomes harmonic (but nonlinear) and obtained from the closed form expression:

$$X = \frac{F_0}{[(k - m\omega^2)^2 + [(c_1 + \frac{8c_2}{3\pi}\omega_n X)\omega]^2]^{1/2}} \qquad (3.76)$$

and the phase angle:

$$\tan\varepsilon = \frac{(c_1 + \frac{8c_2}{3\pi}\omega_n X)\omega}{k - m\omega^2} \qquad (3.77)$$

Since the amplitude of motion appears on both sides of the equation, the solution can only be obtained in an iterative way. The amplitude on the right hand side of Eq. 3.76 is taken as zero in the initial estimate of the motion amplitude. This estimated amplitude is substituted on the right hand side in the next iteration. This process is continued until a convergence is reached. The phase angle then may be obtained from Eq. 3.77. There are many physical systems where such nonlinear damping is present. The above solution is in closed form and provides a reasonable estimate of the response of the nonlinearly damped system.

3.4.4 General Nonlinear Damping

A general form of the free decayed motion may be described by the equation

$$m\ddot{x} + f(x,\dot{x}) + kx = 0 \tag{3.78}$$

In general, the damping term is a function of structure displacement and structure velocity. A general damping function in terms of only the structure velocity is often written as

$$f(x,\dot{x}) = c\dot{x}|\dot{x}|^{\alpha-1} \tag{3.79}$$

where c is the damping coefficient and α is an exponent that determines the linearity of the damping term. If $\alpha = 1$, the damping is linear and the equation of the motion is given by Eq. 3.17. For $\alpha = 0$, the term becomes Coulomb damping

$$f(x,\dot{x}) = c\frac{\dot{x}}{|\dot{x}|} \tag{3.80}$$

For $\alpha = 2$, the damping is quadratic drag type:

$$f(x,\dot{x}) = c|\dot{x}|\dot{x} \tag{3.81}$$

The linear damping arises from the wave radiation and current effect. The Coulomb type damping is caused by the internal friction in the component material of a structure. The quadratic damping appears from the flow separation from the oscillatory motion of the structure. The total damping in a system may be a combination of these various types of damping.

For a linear damping, the damping factor has been shown to be a function of the logarithmic decrement in Eq. 3.30 ($\zeta = \delta/\pi$). For a lightly damped Coulomb model, the damping factor is approximately given by

$$\zeta = \frac{c}{\pi \omega_n^2 X} \tag{3.82}$$

Thus, the damping factor for Coulomb type damping is inversely proportional to the motion amplitude. In structures with Coulomb type damping, when the amplitude of motion is small, the damping factor becomes large and the movement of the structure ceases due to friction.

The damping factor for the quadratic damping is also amplitude dependent. In this case, the damping factor becomes

$$\zeta = \frac{4}{3\pi} cX \qquad (3.83)$$

3.4.5 Nonlinear Restoring Force

Consider a system, which has a nonlinear restoring force function. If x is considered the change in length of the spring in the system from its unstretched length, then the restoring force may be written in a general form as

$$F_r = \chi(x) \qquad (3.84)$$

In practice, the function $\chi(x)$ may be represented by its Taylor series expansion:

$$\chi(x) = k_0 + k_1 x + k_2 x^2 + k_3 x^3 + k_4 x^4 + k_5 x^5 + \ldots \qquad (3.85)$$

Since the restoring force is zero for $x = 0$, one gets $k_0 = 0$. Also, the odd powers of x provide tension and compression of the spring as the sign of x changes, while the even powers always give tension. Since an extension of a spring is considered a tension and a shortening a compression, the general form of nonlinear restoring force becomes

$$\chi(x) = k_1 x + k_3 x^3 + k_5 x^5 + \ldots \qquad (3.86)$$

The undamped free vibration of the system with nonlinear restoring force is written as

$$m\ddot{x} + k\chi(x) = 0 \qquad (3.87)$$

This equation is normalized by writing it in the form

$$\ddot{x} + \kappa^2 \chi(x) = 0 \qquad (3.88)$$

where $\kappa^2 = k/m$. The function $\chi(x)$ is an odd function of x:

$$\chi(-x) = -\chi(x) \qquad (3.89)$$

Chapter 3 Basic Theory of Vibration

Equation 3.88 may be equivalently written in terms of the velocity $\dot{x}$ [Abramson (1976)]:

$$\frac{d(\dot{x})^2}{dx} + 2\kappa^2 \chi(x) = 0 \tag{3.90}$$

If we integrate this equation with respect to x, then the expression for the velocity of the system is obtained as

$$|\dot{x}| = \sqrt{2}\kappa \left[\int_x^X \chi(\alpha) d\alpha \right]^{1/2} \tag{3.91}$$

in which the integration is carried out from x to X where X is the displacement when $\dot{x} = 0$. This equation may be integrated with respect to time t in order to obtained the relationship between x and t:

$$t - t_0 = \frac{1}{\sqrt{2}\kappa} \int_0^x \frac{d\xi}{\left[\int_\xi^X \chi(\alpha) d\alpha \right]^{1/2}} \tag{3.92}$$

in which t_0 corresponds to the time when $x = 0$. The expression for x may be obtained in terms of time t by inverting this relation. The period of oscillation (i.e., free vibration period for the system) may be derived as the time from zero displacement to maximum displacement:

$$T_n = \frac{4}{\sqrt{2}\kappa} \int_0^X \frac{d\xi}{\left[\int_\xi^X \chi(\alpha) d\alpha \right]^{1/2}} \tag{3.93}$$

A few special cases [Abramson (1976)] are taken here for specific description of the function $\chi(x)$. If the restoring force is a pure power of the displacement:

$$\chi(x) = x^n \tag{3.94}$$

then the expression for the period of vibration becomes:

$$T_n = \frac{4}{\kappa} \sqrt{\frac{n+1}{2}} \int_0^X \frac{d\xi}{\left[X^{n+1} - \xi^{n+1} \right]^{1/2}} \tag{3.95}$$

Normalizing the independent variable by X, we have

$$T_n = \frac{4}{\kappa\sqrt{X^{n-1}}}\left(\sqrt{\frac{n+1}{2}}\int_0^1 \frac{d\hat{\xi}}{[1-\hat{\xi}^{n+1}]^{1/2}}\right) \tag{3.96}$$

Note that the expression within the parenthesis is only dependent on n. By writing this function as $\psi(n)$, Eq. 3.96 becomes:

$$T_n = \frac{4}{\kappa\sqrt{X^{n-1}}}\psi(n) \tag{3.97}$$

The values of the function $\psi(n)$ may be evaluated numerically for different values of n and are shown in Table 3.2. The function increases in value with the exponent n.

Table 3.2: Numerical Values of the Function $\psi(n)$

n	$\psi(n)$
0	1.4142
1	1.5708
2	1.7157
3	1.8541
4	1.9818
5	2.1035
6	2.2186
7	2.3282

The above method may be expanded for a restoring force function, which is a polynomial in the power of the displacement x. For example [Abramson (1976)], if a binomial function is chosen:

$$\chi(x) = x^n + \beta x^m \tag{3.98}$$

where the exponents n and m are positive, then the expression for the period becomes:

Chapter 3 Basic Theory of Vibration

$$T_n = \frac{4}{\kappa\sqrt{X^{n+1}}} \left(\sqrt{\frac{n+1}{2}} \int_0^1 \frac{d\hat{\xi}}{\left[(1+\overline{\beta})-(\xi^{n+1}+\overline{\beta}\xi^{m+1})\right]^{1/2}} \right) \qquad (3.99)$$

where $\overline{\beta}$ is given by

$$\overline{\beta} = \beta X^{m-n}\left(\frac{n+1}{m+1}\right) \qquad (3.100)$$

Eq. 3.99 may be evaluated to any desired accuracy for given values of n, m and β.

Now for another example, return to Eq. 3.88 where

$$\chi(x) = \sin x \qquad (3.101)$$

Introducing this function into Eq. 3.92, one obtains the relationship between time t and displacement x:

$$t - t_0 = \frac{1}{2\kappa} \int_0^x \frac{d\xi}{\left[\sin^2\frac{X}{2} - \sin^2\frac{\xi}{2}\right]^{1/2}} \qquad (3.102)$$

If one considers $x = X$ and $t_0 = 0$, then the integral in Eq. 3.102 becomes the complete elliptic integral of the first kind.

$$K(\alpha) = \int_0^{\pi/2} \frac{d\xi}{\left[1 - \sin^2\alpha \sin^2\xi\right]^{1/2}} \qquad (3.103)$$

The period of vibration for this case is given by

$$T_n = \frac{1}{\kappa} K\left(\frac{X}{2}\right) \qquad (3.104)$$

The values of elliptic integrals are tabulated in the handbook of mathematical functions [Abramowitz and Stegun (1964)].

The above example of a specific nonlinear restoring force may be important in the practical hydrodynamic application. Often a structure placed in a fluid flow is held in place by mooring lines that have nonlinear spring characteristics. These "springs" may be represented by one of the above mathematical expressions. In these cases, the free vibration period of the system may be determined exactly by the equations presented here. Note that since the restoring forces are nonlinear, unlike the linear springs, this period is dependent on the amplitude of oscillation.

3.4.6 Duffing's Equation

The nonlinear restoring force in a spring mass system in a particular degree of freedom sometimes fits a term having an nth power of the displacement. The equation of motion in this case for a harmonically excited structure may take the following form:

$$m\ddot{x} + c\dot{x} + k(x + \kappa^2 x^n) = F_0 \cos(\omega t - \alpha) \qquad (3.105)$$

in which the restoring force is composed of a polynomial with a linear and a nonlinear term of nth power.

The usual method of solving this equation is to apply the Raleigh-Ritz averaging technique. The Raleigh-Ritz method is based on approximately solving the equation of motion on the average. If the differential equation in Eq. 3.105 is represented as

$$E(t) = 0 \qquad (3.106)$$

an average solution is sought over one cycle of the motion by choosing a weighting function $w(t)$. Then the integral of the product of the differential equation and the weighting function $w(t)$ over one cycle is equated to zero:

$$\int_0^{2\pi} E(t)w(t)dt = 0 \qquad (3.107)$$

The solution obtained from this integration is the cycle-averaged approximation. The weighting function is often represented by a trigonometric cosine and sine function in succession. This provides two algebraic equations relating the amplitude and phase of motion for a single degree of freedom system. In this case the solution is considered to be harmonic, being of the form:

$$x = X \cos(\omega t - \varepsilon) \qquad (3.108)$$

In order to apply this technique, Eq. 3.106 is written as:

Chapter 3 Basic Theory of Vibration 87

$$E = m\ddot{x} + c\dot{x} + k(x + \kappa^2 x^n) - F_0 \cos(\omega t - \alpha) = 0 \tag{3.109}$$

whose weighted integrals over one cycle are equated to zero, expressed as:

$$\int_0^{2\pi} E \cos\varepsilon \, d\varepsilon = 0 \tag{3.110}$$

and

$$\int_0^{2\pi} E \sin\varepsilon \, d\varepsilon = 0 \tag{3.111}$$

where ε is substituted for ωt. Once the expressions for E and x are substituted in Eqs. 3.110 and 3.111, one obtains:

$$-\omega^2 mX \cos\varepsilon + c\omega X \sin\varepsilon + k(X + \kappa^2 \Delta(n) X^n) \cos\varepsilon = F_0 \cos\alpha \tag{3.112}$$

$$-\omega^2 mX \sin\varepsilon + c\omega X \cos\varepsilon + k(X + \kappa^2 \Delta(n) X^n) \sin\varepsilon = F_0 \sin\alpha \tag{3.113}$$

These are two equations in two unknowns X and ε. The solutions for X and ε are obtained from Eqs. 3.112-113 as follows:

$$\left[k(1 + \kappa^2 \Delta(n) X^{n-1}) - \omega^2 m\right]^2 + [c\omega]^2 = \left[\frac{F_0}{X}\right]^2 \tag{3.114}$$

and

$$\tan\varepsilon = \frac{[c\omega]}{\left[k(1 + \kappa^2 \Delta(n) X^{n-1}) - \omega^2 m\right]} \tag{3.115}$$

The function $\Delta(n)$ is a numerical function of n only. For $n = 1$, $\Delta(n) = 1$. The value of the function $\Delta(n)$ is obtained from the integral:

$$\Delta(n) = \frac{4}{\pi} \int_0^{\pi/2} \cos^{n+1}\alpha \, d\alpha \tag{3.116}$$

The integration may be carried out for different values of n giving:

$$\Delta(n) = \begin{cases} \dfrac{4}{\pi} & \text{for } n=0 \\ \dfrac{1.3.5...(2j-1)}{2.4.6...(2j)} 2 & \text{for } j=1,2,...\text{and } 2j=n+1 \\ \dfrac{2.4.6...(2j)}{1.3.5...(2j-1)} \dfrac{4}{\pi} & \text{for } j=1,2,...\text{and } 2j=n \end{cases} \qquad (3.117)$$

The value of $\Delta(n)$ is plotted versus n in Fig. 3.19.

Equation 3.105 is the general form for the Duffing's equation. Quite often the restoring force curve fits the cubic form well. For the special case of $n=3$, the value of Δ is ¾ and Eq. 3.114 becomes:

$$\left[k(1+\kappa^2 \frac{3}{4} X^2) - \omega^2 m\right]^2 + [c\omega]^2 = \left[\frac{F_0}{X}\right]^2 \qquad (3.118)$$

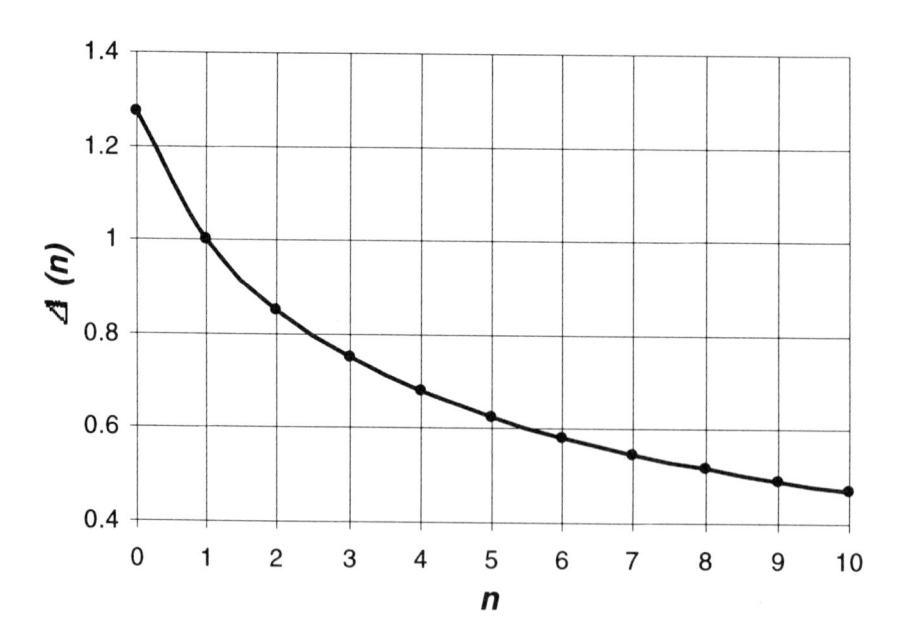

Fig. 3.19: Numerical Values of $\Delta(n)$ for Different Values of n

Chapter 3 Basic Theory of Vibration

Dividing both sides of this equation by k^2 and noting Eq. 3.34, one finds:

$$\left[(1+\frac{3}{4}\kappa^2 X^2)-\frac{\omega^2}{\omega_n^2}\right]^2 + \left[2\zeta\frac{\omega}{\omega_n}\right]^2 = \left[\frac{F_0}{kX}\right]^2 \qquad (3.119)$$

where ω_n corresponds to the natural frequency of the linear system. This relationship is plotted in Fig. 3.20 for the nondimensional response $\left[\sqrt{3/4}(\kappa X)\right]$ as a function of the nondimensional frequency, ω/ω_n for a damping factor of $\zeta = 0.1$. The response is plotted for different values of the nondimensional force given by $S = \left[\sqrt{3/4}(\kappa F_0/k)\right]$.

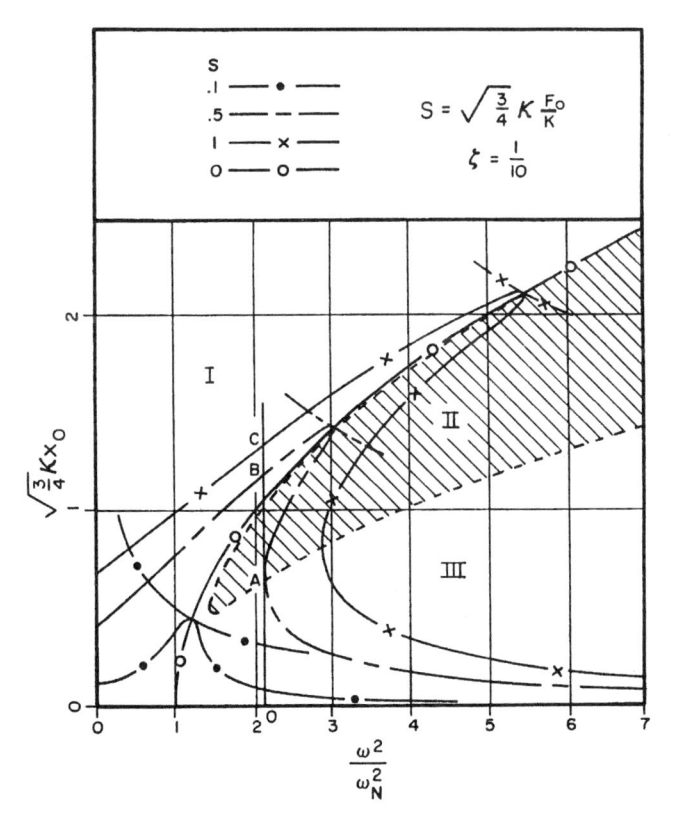

Fig. 3.20: Response of a SDOF System with Nonlinear Restoring Force

The curve for the case S = 0 gives the zero-excitation case as the locus of the peaks of the response. The response peaks increase with the value of the nondimensional force. For this case three separate regions may be defined. These are marked as regions I, II, III in the figure. Regions I and III are considered stable regions, which merge together as the excitation becomes small. Region II is an unstable region for the excitation, which is bounded by the above locus. At S = 0.1, the solution is stable. However, as the value of S increases, instability appears in the response. Consider the frequency along the vertical line denoted by OC. The response increases in magnitude as S increases to a value given by A. At this point, if the excitation increases even slightly above S = 0.5, the response jumps to a new value at B. Beyond this value, the response again steadily increases to C. The reverse phenomenon will occur with the decrease in S, except that the path from C to B will differ from the path from B to C, forming a closed loop. The solution will come back to A with a decrease in S. A physical example of the effect of this solution will be discussed in Chapter 7.

3.4.7 Nonlinear Damping and Restoring Force

Many physical systems have a nonlinear restoring force as well as a nonlinear damping force present simultaneously. A general form for this case will be written in terms of an angular equation of motion, such as the roll motion of a ship. For a single degree of freedom system, the equation may be written as

$$I\ddot{\phi} + B(\dot{\phi}) + C(\phi) = M(t) \qquad (3.120)$$

in which ϕ is the angular roll motion, dots represent time derivatives, I is the total moment of inertia in roll, M is the amplitude of the exciting moment, ω is the excitation frequency and t is time. The damping and restoring moments are left as open functions of the angular velocity and angular displacement respectively. On the assumption that the functions B and C are odd (see Eq. 3.89), this equation may be solved in a general way by applying the Raleigh-Ritz method described earlier.

Consider a practical example of the roll motion of a long floating structure such as a ship in waves, where such systems are encountered. The

Chapter 3 Basic Theory of Vibration

ship is assumed to be equipped with bilge keels (thin flat plates) at its beams. While the other degrees of motion of a ship may be described reasonably well by a linear equation, this is not true for the roll motion. The equation of motion in roll may be written in general as Eq. 3.120 in which the wave exciting moment is given by the linear term $M \cos \omega t$.

The restoring term in Eq. 3.120 is often given in a polynomial form in ϕ (see Eq. 3.86). The damping term B may be represented by the odd function

$$B(\dot{\phi}) = B_1 \dot{\phi} + B_2 |\dot{\phi}|\dot{\phi} + B_3 \dot{\phi}^3 \tag{3.121}$$

in which the first term is linear, the second term represents quadratic drag and the third term is cubic. Quite often, the total damping is approximated by

$$B(\dot{\phi}) = B_{eq} \dot{\phi} \tag{3.122}$$

The coefficient B_{eq} denotes an equivalent damping coefficient, which is a linear function of angular velocity. In terms of the above coefficients, it is expressed as

$$B_{eq} = B_1 + \frac{8}{3\pi} B_2 (\omega \phi_0) + \frac{3}{4} B_3 (\omega \phi_0)^2 \tag{3.123}$$

where ϕ_0 is the amplitude of roll. The total damping coefficient is composed of the individual components

$$B_{eq} = B_f + B_e + B_w + B_L + B_{BK} \tag{3.124}$$

in which the component damping coefficients are as follows: B_f = hull skin friction damping, B_e = hull eddy shedding damping, B_w = free surface wave damping, B_L = lift force damping, and B_{BK} = bilge keel damping. Explicit empirical expressions for these components will be given in Chapter 7.

Here, consider another practical nonlinear problem and examine the stability question for this equation. In mooring an offloading tanker in the open water, an upright buoyant tower is often used. The tower is attached to the tanker by a single hawser and when subjected to waves, articulates about a universal joint near the bottom of the ocean. This system is commonly known as a Single Point Mooring (SPM) system. The equation of motion of the articulated tower may be described by the following equation:

$$I\ddot{\psi} + B_1\dot{\psi} + B_2\dot{\psi}^3 + C_1\psi + C_2\psi^3 = M(t) \qquad (3.125)$$

where M is the moment about the tower U-joint arising from the excitation. If a symmetric stiffness function is assumed (which is unrealistic for a hawser, which may become slack) and the excitation is considered harmonic, then Eq. 3.125 becomes Duffing's equation and the solution has earlier been obtained analytically.

A special case of this form is often encountered in practice where the damping consists of a linear term, and the restoring force is composed of a linear and a cubic term:

$$I\ddot{\psi} + B_1\dot{\psi} + C_1\psi + C_2\psi^3 = M(t) \qquad (3.126)$$

This equation is normalized in the following way. A critical angle, ψ^* is defined such that at this angle, the restoring moment is zero.

$$\psi^* = (C_1/C_2)^{1/2} \qquad (3.127)$$

Then normalizing ψ by ψ^*, i.e., $\hat{\psi} = \psi/\psi^*$, the equation of motion, Eq. 3.126 may be reduced to

$$\ddot{\hat{\psi}} + a\dot{\hat{\psi}} + \hat{\psi} - \hat{\psi}^3 = \hat{M}(\varepsilon) \qquad (3.128)$$

in which $\varepsilon = \omega_n t$, $a = B_1/(I\omega_n)$, $\psi^* = \omega_n \psi^*$, $\hat{M}(t) = M(t)/(I\omega_n^2 \psi^*)$ and $\omega_n = (C_1/I)^{1/2}$. At this point the stability of the motion caused by Eq. 3.128 will be examined. For this purpose, the damping and the exciting force terms are dropped from Eq. 3.128 without loss of generality. Then, one has

$$\ddot{\hat{\psi}} + \hat{\psi} - \hat{\psi}^3 = 0 \qquad (3.129)$$

The solution of this equation is obtained in terms of the Jacobian elliptical integral, sn.

$$\hat{\psi} = A\,sn(\omega_1 \varepsilon + K(m)|m) \qquad (3.130)$$

where A is the amplitude of oscillation and ω_1 and m are given by

$$\omega_1 = (1 - A^2/2)^{1/2} \qquad (3.131)$$

and

$$m = \frac{A^2}{2-A^2} \tag{3.132}$$

Also, the complete elliptical integral is defined as

$$K(m) = \int_0^{\pi/2} \frac{d\alpha}{(1-m\sin^2\alpha)^{1/2}} \tag{3.133}$$

The natural period of oscillation is obtained from the relation:

$$T_N = \frac{4K(m)}{(1-A^2/2)^{1/2}} \tag{3.134}$$

Note that as A approaches zero, $K(m)$ approaches $\pi/2$ so that T_N takes on the value of 2π. Defining $\omega_N = 2\pi/T_N$, the normalized natural frequency is given by

$$\frac{\omega_N}{\omega_n} = \frac{\pi(1-A^2/2)^{1/2}}{2K(m)} \tag{3.135}$$

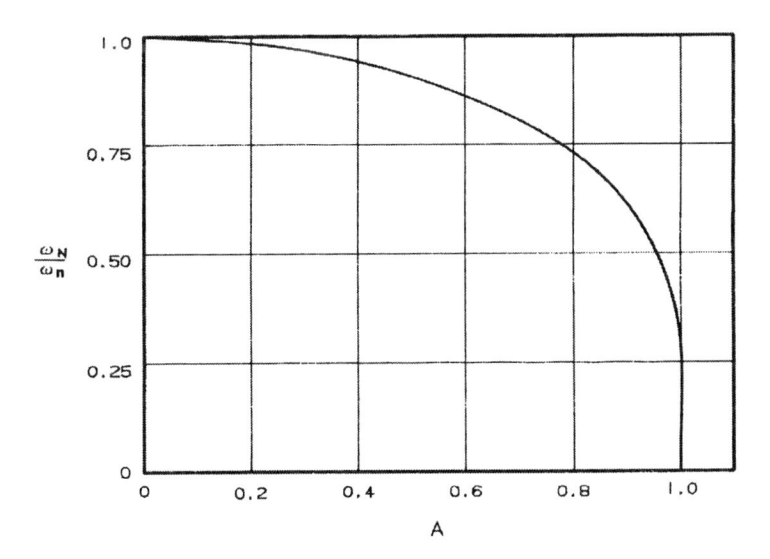

Fig. 3.21: Amplitude vs. Frequency of Oscillation [Roberts (1982)]

In the above expression as A approaches zero, ω_N approaches ω_n. Equation 3.135 is plotted in Fig. 3.21. The value of ω_N is relatively unchanged from ω_n at low values of A, but becomes quite different at larger values of A, becoming tangent to the line $A = 1.0$ at its limit. Beyond $A > 1.0$, the solution is unstable. The phase-plane diagram for Eq. 3.130 is given in Fig. 3.22. The shaded area bounded by the value of $A = \pm 1.0$ is the stable region. The arrows indicate the orbital path in the ($\hat{\psi}$, $\dot{\hat{\psi}}$) plane.

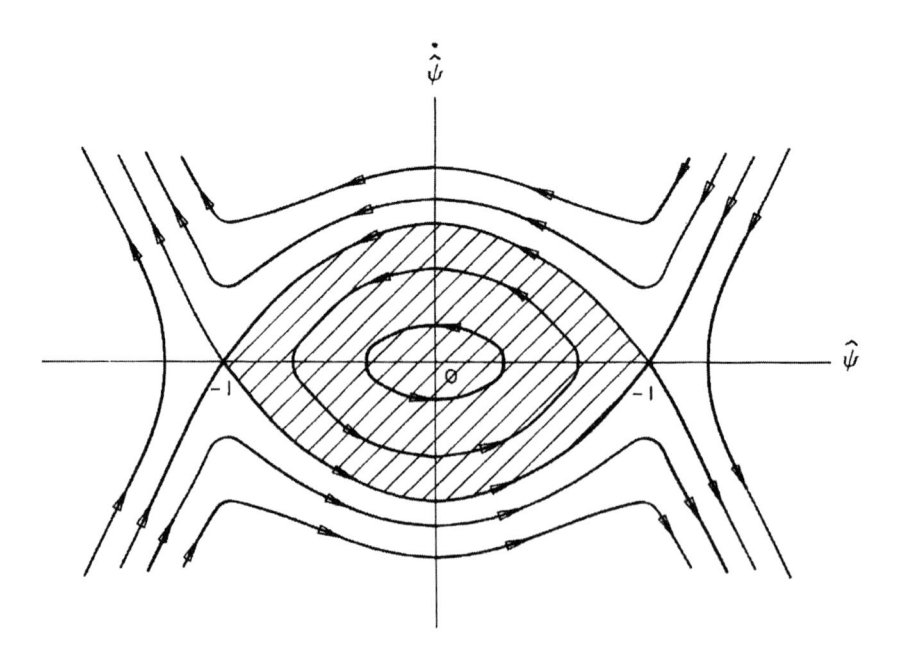

Fig. 3.22: Phase-Plane Diagram Showing Stability of Solution [Roberts (1982)]

3.5 Vibration of a Rigid Body

In the earlier sections, the concept of the single degree of freedom of a system was described. Within this concept, the system was assumed to carry out oscillations in only one direction, while being restrained in the other directions by the design of the system. Several practical systems have such type of oscillations. One example is a vertical pendulum having a hinge type joint restricting its freedom in the normal direction. Structures mounted on foundations having springs allowing motions in only the vertical directions is another example. Even when a structure is free to move in other directions, quite often its motion in a principal direction is most important for its design and survival. Such systems may be analyzed by one of the methods described in the earlier sections.

However, most structures, especially those subjected to fluid flow, have multiple degrees of freedom. Their analysis will involve multiple motions in different planes. In order to study the motions of a rigid body, normally two coordinate systems are chosen. A fixed coordinate frame of reference is written in terms of a rectangular Cartesian coordinate system, X, Y, and Z. The second coordinate system is fixed with its origin at a convenient point on the rigid body, normally its center of mass, and it is allowed to move with the body. This set of coordinates is given by x, y, and z. The two coordinate systems are assumed to coincide at the equilibrium position of the body.

The rigid body is, in general, free to experience six degrees of freedom (DOF) motion. Three of these motions are translational in the three directions of the coordinate system X, Y, and Z. For these motions, the body coordinate system moves parallel to the fixed coordinate system. The three other motions are angular in which case the body coordinate system rotates with the body about the fixed coordinate system. Sometimes, one or more of these 6 DOF motions are restrained by the boundary condition of the body. The translational motions along X, Y and Z are shown here by a, b, c (Fig. 3.23) and the rotational motions about these axes by α, β, γ respectively.

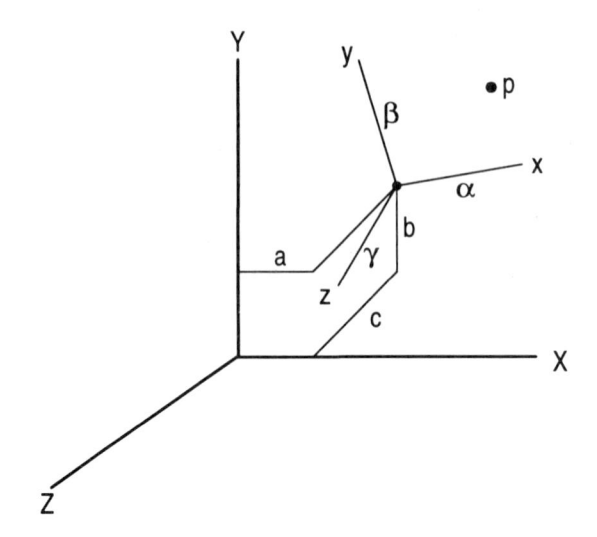

Fig. 3.23: Fixed and Moving Coordinates for a Rigid Body Motion

Since the two sets of coordinate systems are related to each other by the angular motions, the transformation matrix between the two may be expressed by the trigonometric functions of the rotational angles.

$$\begin{bmatrix} X \\ Y \\ Z \end{bmatrix} = \begin{bmatrix} \cos\beta\cos\gamma & -\cos\beta\sin\gamma & \sin\beta \\ \sin\alpha\sin\beta\cos\gamma + \cos\alpha\sin\gamma & -\sin\alpha\sin\beta\sin\gamma + \cos\alpha\cos\gamma & -\sin\alpha\cos\beta \\ -\cos\alpha\sin\beta\cos\gamma + \sin\alpha\sin\gamma & \cos\alpha\sin\beta\sin\gamma + \sin\alpha\cos\gamma & \cos\alpha\cos\beta \end{bmatrix} \begin{bmatrix} x \\ y \\ z \end{bmatrix}$$

(3.136)

In order to solve the problem in closed forms, the rotation is often assumed to be small. This assumption is quite good for many practical problems as will be demonstrated throughout this book. Thus the rotations are commutative. In other words, the order of the component rotation in achieving the resulting position of the body is immaterial. In this case the transformation matrix becomes:

$$\begin{bmatrix} X \\ Y \\ Z \end{bmatrix} = \begin{bmatrix} 1 & -\gamma & \beta \\ \gamma & 1 & -\alpha \\ -\beta & \alpha & 1 \end{bmatrix} \begin{bmatrix} x \\ y \\ z \end{bmatrix} \qquad (3.137)$$

Therefore, the displacements (Xp,Yp,Zp) in the fixed coordinate system of a point p (x_p,y_p,z_p) on the body are obtained from

$$X_p = a - y_p\gamma + z_p\beta \tag{3.138}$$

$$Y_p = b + x_p\gamma - z_p\alpha \tag{3.139}$$

$$Z_p = c - x_p\beta + y_p\alpha \tag{3.140}$$

For floating structures, these six motions have special names. The translational motions in the X, Y, Z directions are known as surge, sway and heave motions, respectively. The angular motions about the X, Y, Z directions are called roll, pitch and yaw, respectively. These are defined for a floating tanker in Fig. 3.24.

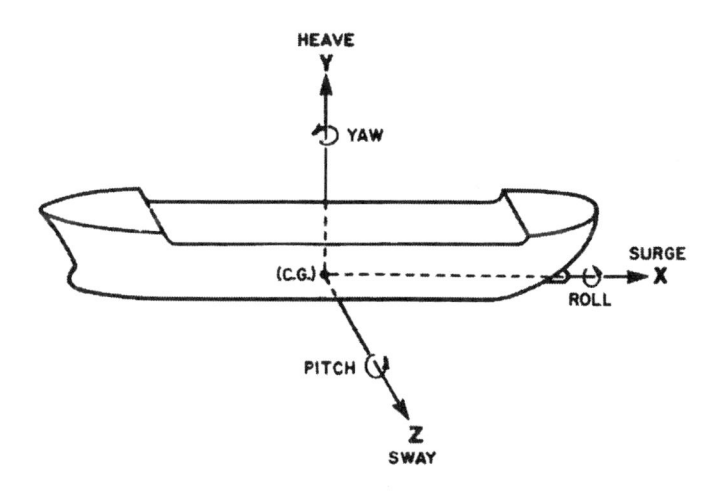

Fig. 3.24: Six DOF Motions of a Floating Structure

In order to write the equation of motion for the structure, the mass and the moments of inertia of the structure should first be known. Assuming that the structure has a single mass density of ρ_s, the total mass of the structure may be obtained by the integration of the elemental volume dV of the body of the structure over the structure volume:

$$m = \rho_s \int_V dV \tag{3.141}$$

If the body is composed of multiple elements, each having different mass densities, then the right hand side of the above equation may be summed over multiple segments of the same integral form having different densities. For a floating structure this may include contained mass, such as stored oil, drilling fluid or ballast water. Note that for a freely floating structure, the displacement of the structure in its equilibrium position is equal to its mass. Therefore, if the structure is of a rectangular form of length L, breadth B, and draft D, then the mass of the structure is given by

$$m = \rho L B D \tag{3.142}$$

The center of mass or the center of gravity of the structure is computed from the integral representing the moment of the mass of the form:

$$r_0 = \frac{1}{m} \rho_s \int_V r \, dV \tag{3.143}$$

The quantity r in Eq. 3.143 is the distance to the elemental volumes dV and may be substituted by x, y or z for the Cartesian coordinate system. The moments of inertia of the structure about its axes are computed from the second moment of the elemental mass as

$$I_{xx} = \int_m (y^2 + z^2) \, dm \tag{3.144}$$

$$I_{yy} = \int_m (z^2 + x^2) \, dm \tag{3.145}$$

$$I_{xx} = \int_m (x^2 + y^2) \, dm \tag{3.146}$$

The products of inertia are the cross terms of the second moment type:

$$I_{xy} = \int_m xy \, dm \tag{3.147}$$

Chapter 3 Basic Theory of Vibration

$$I_{yz} = \int_m yz\,dm \tag{3.148}$$

$$I_{zx} = \int_m zx\,dm \tag{3.149}$$

For a symmetric structure, the product of inertia about its axes vanishes. Also there is always a unique set of axes for any structure about which the product of inertia vanishes. These axes are called principal axes. For a symmetric structure, the principal axes coincide with the conventional axes through the center of the body.

Upon examination of Eqs. 3.143-146 it may be seen that the moments of inertia may be equivalently written as

$$I_{xx} = k_{xx}^2 m \tag{3.150}$$

$$I_{yy} = k_{yy}^2 m \tag{3.151}$$

$$I_{zz} = k_{zz}^2 m \tag{3.152}$$

in which the quantities k_{xx}, k_{yy} and k_{zz} are termed the radii of gyration. For hydrostatic properties of floating structures, it is customary to specify the radii of gyration rather than the moments of inertia. They are also useful measures to remember when one attempts to ballast the model of a prototype structure in water with ballast weights in order to achieve the scaled moment of inertia.

The equations of motion of the rigid structure are written from Newton's second law of motion. Once the structure mass and mass moments of inertia are computed or measured, the equations of translational motions are obtained for the three orthogonal directions in a matrix form as

$$m \begin{bmatrix} \ddot{x} \\ \ddot{y} \\ \ddot{z} \end{bmatrix} = \begin{bmatrix} F_x \\ F_y \\ F_z \end{bmatrix} \tag{3.153}$$

where the double-dot represents the second derivative of the variable with respect to time. Thus the left-hand side of the matrix is the structure inertia term in the XYZ directions, and the right-hand side is the force matrix (vector). Similarly, the rotational equations of motion are written in a matrix form using the moment and product of inertia and the rotational accelerations:

$$\begin{bmatrix} I_{xx} & -I_{xy} & -I_{xz} \\ -I_{xy} & I_{yy} & -I_{yz} \\ -I_{xz} & -I_{yz} & I_{zz} \end{bmatrix} \begin{bmatrix} \ddot{\alpha} \\ \ddot{\beta} \\ \ddot{\gamma} \end{bmatrix} = \begin{bmatrix} M_x \\ M_y \\ M_z \end{bmatrix} \quad (3.154)$$

in which M_x, M_y and M_z are the moments due to external forces on the structure. If the products of inertia terms are zero, then the equations reduce to a form similar to the translational equations of motion.

3.6 Linear Multi-Degree Freedom System

It is often the case that multiple elements of a structure are interconnected by springs. In this case each component is allowed to move relative to the other components resulting in multiple degrees of freedom that are coupled. This section will consider such systems and will derive methods of analyzing multiple degrees of freedom.

3.6.1 Two Degree-of-Freedom System

The concept of the single degree of freedom system of Fig. 3.2 is extended by introducing a second mass, m_2 with an additional spring, k_2 as shown in Fig. 3.25. In this two-mass system m_1 and m_2, a harmonic force given by $F_0 \cos \omega t$ is applied to m_1. The mass m_1 displaces x_1 from its equilibrium position while m_2 displaces x_2. Therefore, the spring k_1 stretches an amount given by x_1 while k_2 stretches by the amount x_2-x_1. If it is assumed that there is no damper in the system, then the system may be described mathematically by the following two equations:

$$m_1 \ddot{x}_1 + k_1 x_1 - k_2 (x_2 - x_1) = F_0 \cos \omega t \quad (3.155)$$

$$m_2 \ddot{x}_2 + k_2 (x_2 - x_1) = 0 \quad (3.156)$$

Chapter 3 Basic Theory of Vibration

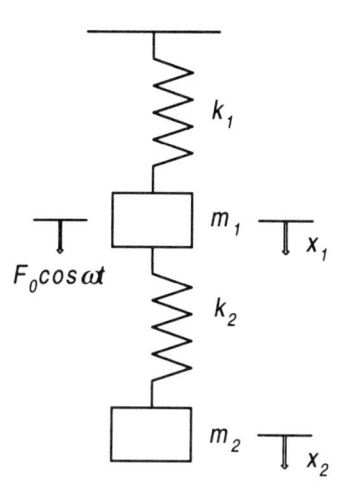

Fig. 3.25: Two Degrees of Freedom System

Note that m_2 does not have any external force acting on it. Moreover, as in the case of the undamped single degree of freedom system, the motions will be in phase with the forcing function. The solutions may then be written as

$$x_1 = X_1 \cos \omega t \tag{3.157}$$

and

$$x_2 = X_2 \cos \omega t \tag{3.158}$$

Substituting these values for the displacements in Eqs. 3.155-156 and eliminating the ωt term, one gets

$$(k_1 + k_2 - \omega^2 m_1)X_1 + k_2 X_2 = F_0 \tag{3.159}$$

$$-k_2 X_1 + (k_2 - \omega^2 m_2)X_2 = 0 \tag{3.160}$$

The above equations may be expressed in a matrix form as follows:

$$\begin{bmatrix} k_1 + k_2 - \omega^2 m_1 & k_2 \\ -k_2 & k_2 - \omega^2 m_2 \end{bmatrix} \begin{bmatrix} X_1 \\ X_2 \end{bmatrix} = \begin{bmatrix} F_0 \\ 0 \end{bmatrix} \tag{3.161}$$

Since this is a coupled system, one would expect two natural frequencies to be associated with the system. It is clear that the motion will "blow up", i.e., experience an unbounded increase, when the determinant of the 2 x 2 matrix on the left-hand side is set to zero:

$$\begin{vmatrix} k_1 + k_2 - \omega^2 m_1 & k_2 \\ -k_2 & k_2 - \omega^2 m_2 \end{vmatrix} = 0 \tag{3.162}$$

Dividing the first row by k_1 and the second row by k_2 and introducing the natural frequencies of the masses 1 and 2 respectively, one has

$$\omega^2_{11} = \frac{k_1}{m_1} \tag{3.163}$$

and

$$\omega^2_{22} = \frac{k_2}{m_2} \tag{3.164}$$

which results in

$$\left[1 + \frac{k_2}{k_1} - \left(\frac{\omega}{\omega_{11}}\right)^2\right]\left[1 - \left(\frac{\omega}{\omega_{22}}\right)^2\right] - \frac{k_2}{k_1} = 0 \tag{3.165}$$

Using Eqs. 3.163-164 to eliminate k_2/k_1 and introducing μ to represent the ratio of the masses:

$$\mu = \frac{m_2}{m_1} \tag{3.166}$$

Eq. 3.165 may be written as

$$\left(\frac{\omega_{22}}{\omega_{11}}\right)^2 \left(\frac{\omega}{\omega_{22}}\right)^4 - \left[2 + \mu\left(\frac{\omega_{22}}{\omega_{11}}\right)^2\right]\left(\frac{\omega}{\omega_{22}}\right)^2 + 1 = 0 \tag{3.167}$$

This is an equation of an elliptical form. It is a function of the mass ratio and the individual natural frequencies of the two masses. Thus, it has

Chapter 3 Basic Theory of Vibration

two solutions corresponding to the two natural periods of the system. The values of the frequency ratio ω/ω_{22} are shown as a function of the mass ratio μ in Fig. 3.26 for $\omega_{22}/\omega_{11} = 1$. Note that for every value of μ there are two values of the resonant frequencies corresponding to the two-mass system.

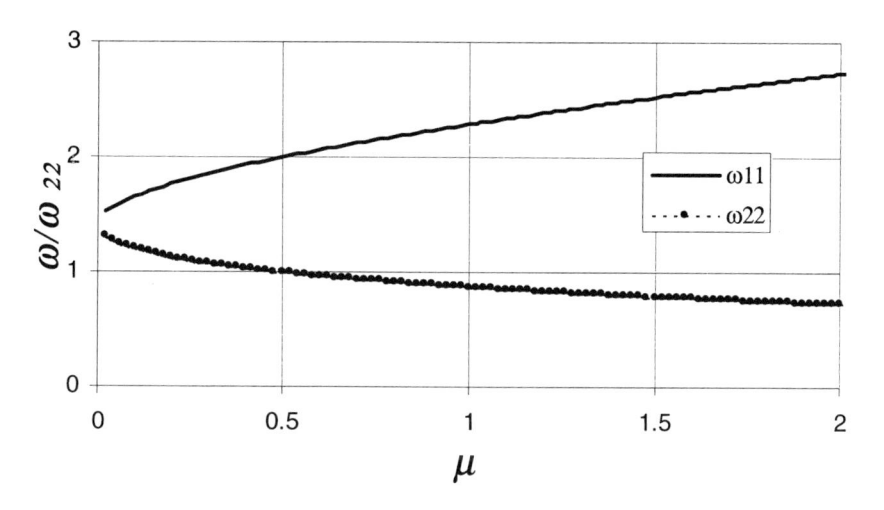

Fig. 3.26: Ratio of Natural Frequency to ω_{22} for a Two DOF System

The solution due to the forced oscillation is obtained from Eq. 3.161. Writing an expression for the static deflection of the mass m_1 due to the force F_0 as

$$X_{st} = \frac{F_0}{k_1} \qquad (3.168)$$

the solution for X_1 and X_2 are

$$X_1 = \frac{\left[1 - \left(\frac{\omega}{\omega_{22}}\right)^2\right] X_{st}}{\left[1 + \frac{k_2}{k_1} - \left(\frac{\omega}{\omega_{11}}\right)^2\right]\left[1 - \left(\frac{\omega}{\omega_{22}}\right)^2\right] - \frac{k_2}{k_1}} \qquad (3.169)$$

and

$$X_2 = \frac{X_{st}}{\left[1 + \frac{k_2}{k_1} - \left(\frac{\omega}{\omega_{11}}\right)^2\right]\left[1 - \left(\frac{\omega}{\omega_{22}}\right)^2\right] - \frac{k_2}{k_1}} \quad (3.170)$$

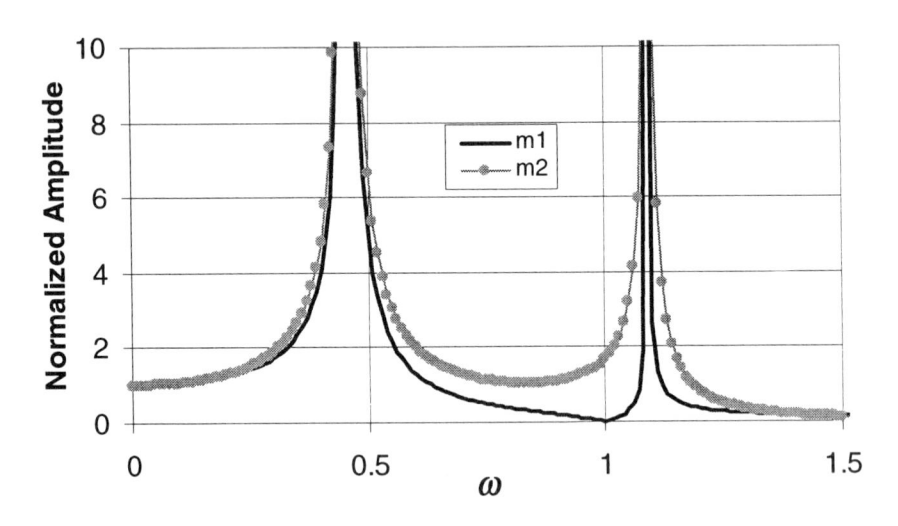

Fig. 3.27: Normalized Displacement of Masses m_1 and m_2

An example for a mass ratio μ value of 0.3 is considered. In this case, the natural frequencies are taken as $\omega_{11} = 0.5$ and $\omega_{22} = 1.0$. The responses of the masses are shown in Fig. 3.27. The displacements are normalized by the static displacement from the force of amplitude F_0. Note that both masses experience two natural frequencies. Since damping is absent in this example, the solutions approach infinity at the natural frequencies. Note, however, that the coupled natural frequencies are different from the single body natural frequencies for the two individual masses.

Returning to the case of the linearly damped two-mass system, the two equations of motion are

$$m_1 \ddot{x}_1 + c_1 \dot{x}_1 + k_1 x_1 - k_2(x_2 - x_1) = F_0 \cos \omega t \quad (3.171)$$

Chapter 3 Basic Theory of Vibration

$$m_2\ddot{x}_2 + c_2\dot{x}_2 + k_2(x_2 - x_1) = 0 \tag{3.172}$$

The solutions for this case are expected to have a phase shift from the forcing function and therefore, are written as

$$x_1 = X_1 \cos(\omega t + \varepsilon_1) \tag{3.173}$$

and

$$x_2 = X_2 \cos(\omega t + \varepsilon_2) \tag{3.174}$$

where ε_1 and ε_2 represent the phase differences between the motion x_1 and the force F_0 and between x_2 and F_0 respectively. Upon substitution of these expressions in Eqs. 3.171-172, one has

$$(k_1 + k_2 - \omega^2 m_1) X_1 \cos(\omega t + \varepsilon_1) + c_1 X_1 \omega \sin(\omega t + \varepsilon_1) \\ + k_2 X_2 \cos(\omega t + \varepsilon_2) = F_0 \cos \omega t \tag{3.175}$$

and

$$-k_2 X_1 \cos(\omega t + \varepsilon_1) + (k_2 - \omega^2 m_2) X_2 \cos(\omega t + \varepsilon_2) \\ + c_2 X_2 \omega \sin(\omega t + \varepsilon_2) = 0 \tag{3.176}$$

Equating the coefficients of $\sin \omega t$ and $\cos \omega t$ in the above two equations, four independent equations in four unknowns are obtained. These equations may be expressed in a matrix form as follows:

$$\begin{bmatrix} k_1 + k_2 - \omega^2 m_1 & c_1 \omega & k_2 & 0 \\ c_1 \omega & -(k_1 + k_2 - \omega^2 m_1) & 0 & -k_2 \\ -k_2 & 0 & k_2 - \omega^2 m_2 & c_2 \omega \\ 0 & k_2 & c_2 \omega & -(k_2 - \omega^2 m_2) \end{bmatrix} \begin{bmatrix} X_1 \cos \varepsilon_1 \\ X_1 \sin \varepsilon_1 \\ X_2 \cos \varepsilon_2 \\ X_2 \sin \varepsilon_2 \end{bmatrix} = \begin{bmatrix} F_0 \\ 0 \\ 0 \\ 0 \end{bmatrix}$$
(3.177)

The solutions for X_1, X_2, ε_1 and ε_2 are obtained from the inversion of the 4 x 4 matrix above and solving for the amplitudes and phases for the displacements of masses m_1 and m_2. The results will look much like Fig. 3.27 with the exception that the solutions will be bounded at the natural frequencies, depending on the amount of damping in the system.

3.6.2 Multi-Mass System

The above concept may be directly extended to a series of spring-connected masses. Consider N masses in the system, each constrained to move only in the same single direction. This direction is considered to be the x-axis. For simplicity in the mathematical development, assume that the spring-mass system is undamped and each mass is subjected to a harmonic force. Then the equation of motion of the system may be written in tensor notation as

$$m_j \ddot{x}_j + \sum_{k=1}^{N} k_{kj} x_k = F_j \cos \omega t \qquad (3.178)$$

in which the subscript j varies from 1 to N, while the summation on k depends on the number of springs in the system. Thus k_{jk} denotes the equivalent spring force per unit amplitude acting on the jth mass for the kth displacement. For example, for a two mass system, for $j=1$, it is k_1+k_2. The force F_j is the force acting on the mass j which is zero for $j=2$ for the example two mass system above. If the free vibration of the system is examined, so that the forcing function on the right hand side of the equation is zero, then the equations of motion become:

$$m_j \ddot{x}_j + \sum_{k=1}^{N} k_{kj} x_k = 0 \qquad (3.179)$$

whose general solution will be of the form:

$$x_j = X_j \cos \omega t \qquad (3.180)$$

When this expression is substituted in Eq. 3.179, the following matrix equation is obtained:

$$\sum_{k} [-\omega^2 m_k \delta_{kj} + k_{kj}] X_k = 0 \qquad (3.181)$$

This set of equations has N unknowns, X_k, which will exist if the determinant is identically zero. Therefore, one gets

$$\begin{vmatrix} \omega^2 m_1 - k_{11} & -k_{12} & . & -k_{1n} \\ -k_{21} & \omega^2 m_2 - k_{22} & . & . \\ . & . & . & . \\ -k_{n1} & . & . & \omega^2 m_N - k_{NN} \end{vmatrix} = 0 \qquad (3.182)$$

Note that the matrix in Eq. 3.181 is symmetric. Therefore, there are N^2 spring stiffness components of which $(N^2+N)/2$ are independent. The determinant is an Nth order algebraic equation, which should have N solutions. The solutions are all real. These N solutions in ω give the natural frequencies ω_n ($n = 1,...N$) of the system at which the system will oscillate when disturbed from its equilibrium position. The particular frequencies at which the system will be excited depend on the initial disturbance. A structure vibrating at one of the natural frequencies, ω_n has a characteristic pattern that is known as its normal mode of vibration at the given frequency. The displacements of the normal mode will be X_j for the frequency ω_{nj}, which satisfy Eq. 3.181. If the displacement along the structure is plotted for this case, it gives the mode shape of the structure for this natural frequency or normal mode.

3.6.3 Multiple Degree-of-Freedom System

This technique may be extended to all six degrees of freedom of multiple modules of a floating structure that are connected to one another by linear springs. The individual modules within the structure are allowed to move in six degrees of freedom and the springs are assumed to be acting both in the translational and rotational directions. Assuming linear damping in the system, the coupled equations of motion of the individual modules in the structure may be written in complex form as

$$m_k^i \ddot{x}_k^i + \sum_{j=1}^{N} \sum_{l=1}^{6} \left(M_{lk}^{ij} \ddot{x}_l^j + c_{lk}^{ij} \dot{x}_l^j + k_{lk}^{ij} x_l^j \right) = F_k^i \exp(-i\omega t); k = 1,2..6 \quad (3.183)$$

where N is the number of modules in the floating structure. The complex algebra introduced here is mathematically efficient to handle this problem. The forcing function is harmonic and the real part of the complex quantity is the excitation. The quantity m_k^i is the mass or moment of inertia, and x_k^i is the displacement of module i in the k-th direction. The quantities M_{lk}^{ij}, c_{lk}^{ij}, and k_{lk}^{ij} are the added mass, damping and restoring force coefficients, respectively for the module i acting in the k-th mode. The second subscript j provides the coupling term and implies that added mass and damping coefficient may arise in other modes from the imposed mode. The restoring force of the modules includes the internal and external stiffness arising from the buoyancy, the mooring lines as well as connectors among the modules. The stiffness contributions from all preceding terms in the above equation have been assumed to be linear in Eq. 3.183.

Assuming harmonic excitation for the six degrees of freedom motion of the form of right-hand side of Eq. 3.183, the equation of motion is reduced to the following matrix equation:

$$\sum_{l=1}^{6}[-\omega^2(m_{kl}^i + M_{kl}^{ij}) - i\omega c_{kl}^{ij} + k_{kl}^{ij}]X_l^j \exp(i\varepsilon_l^j) = F_k^i; k = 1,2..6$$

(3.184)

in which i and j vary from 1 to N for N modules in the system. The solutions are given by the amplitudes X_k^i and phases ε_k^j. This matrix equation may be solved in the frequency domain by the inversion of the $6N \times 6N$ matrix in Eq. 3.184. Application of the frequency domain analysis necessitates that the solution remains linear. An example of such a multi-moduled system will be given in Chapter 6.

3.7 Nonlinear Multiple Degree of Freedom Systems

There are many physical systems where the response of the structure may not be simply described by linear terms. Some of the nonlinear terms that enter into the equation of motion of these systems are the nonlinear damping term, nonlinear restoring force term and nonlinear forcing term. Many of these terms are large enough that linearity similar to what was introduced earlier in section 3.6 will not yield a satisfactory solution. In these cases, the complete equation should be considered. This has already been demonstrated for a single degree of freedom system. For a nonlinear multi-degree of freedom system, the equation of motion may be generalized to include nonlinear damping and restoring force terms.

Chapter 3 Basic Theory of Vibration

The flow separation from the submerged modules provides additional structural damping. The damping term from the flow separation at a module has the form $c_2|\dot{x}_k|\dot{x}_k$ (where $c_2 = \frac{1}{2}\rho C_D A$, which is the term found in drag damping due to the motion of a structure, A being the projected area of the structure in water). The subscript in the velocity term corresponds to the six degrees of motion of the structure.

Since this term makes the equation nonlinear, a simplification may be introduced so that a closed form frequency domain solution may be derived. Considering that the desired solution is still harmonic and the system nonlinearity is small, the approximation in Eq. 3.66 is introduced. Then the equation of motion becomes

$$m_k^i \ddot{x}_k^i + \sum_{l=1}^{6}(M_{lk}^{ij}\ddot{x}_l^j + c_{1lk}^{ij}\dot{x}_l^j + \frac{8}{3\pi}c_{2lk}^{ij}|\omega X_l^j|\dot{x}_l^j + k_{lk}^{ij}x_l^j)$$
$$= F_k^i \exp(-i\omega t); k = 1,2,...6 \qquad (3.185)$$

Note that this set of equations is harmonic, and the solution in the frequency domain can be obtained in the same manner as in the linear equation.

$$\sum_{l=1}^{6}[-\omega^2(m_{kl}^i + M_{kl}^{ij}) - i\omega c_{1kl}^{ij} - i\omega\frac{8}{3\pi}c_{2kl}^{ij}|\omega X_l^j|$$
$$+ k_{kl}^{ij}]X_l^j \exp(i\varepsilon_l^j) = F_{0k}^i; k = 1,2..6 \qquad (3.186)$$

However, the amplitude of the structure velocity remaining on the left-hand side with the drag damping term prevents a one-shot solution. Therefore, a simple iterative technique is needed. In the first step, a linear harmonic solution for $X_k^j = 0$ on the left-hand side is obtained. In the second step, this first iteration value is introduced on the left-hand side and the process repeated. Generally, 2 to 3 iterations produce convergence.

Experiments with a floating vertical cylinder in free surface waves have shown that this nonlinear damping term is necessary for predicting the motions near the natural frequency. This is illustrated in Fig. 3.28, in which the pitch motion of a floating vertical caisson held in place with a pair of horizontal soft linear spring lines was measured in a test with regular waves. The wave periods were chosen such that they excited the pitch natural period of the system. The motion was analyzed using Eq. 3.186 with and without the presence of nonlinear damping term. The drag coefficient for the cylinder for the nonlinear damping term was chosen from the available experimental data for the appropriate Reynolds number and Keulegan Carpenter number (see Chapter 2). It is clear that the nonlinear damping term is needed in this case, at least near the natural period of the cylinder in pitch in order to match the measured data.

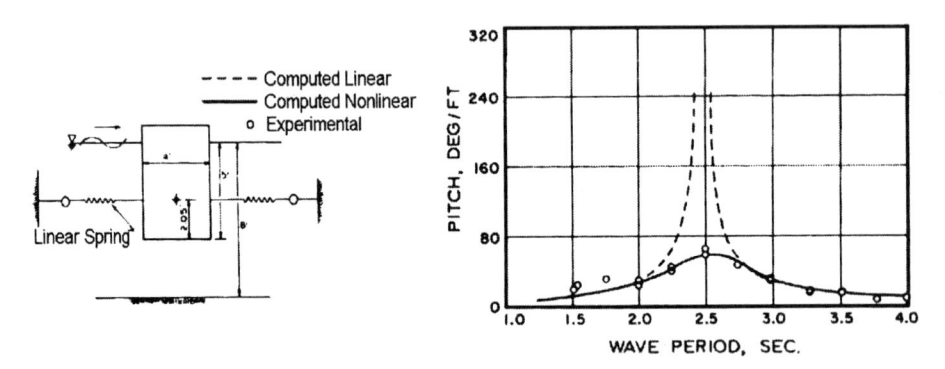

Fig. 3.28: Pitch Motion of a Floating Vertical Caisson in Waves

3.8 Continuous Systems

A system, which has a finite number of degrees of freedom, is called a discrete system. These are the types of system that have been described so far. However, a continuous system has infinite number of degrees of freedom. A continuous structure is defined as a structure that experiences large enough displacements along its length so that the structure undergoes shape changes when subjected to external loading. These structures are generally assumed to be homogeneous and isotropic so that they obey Hooke's law within their elastic limits. This section will address the vibration problem of such structures. The problem is mathematically complicated and normally numerical methods are needed to determine the mode shapes of these structures. The mathematical background that is employed in addressing such problems shall be described. However, their numerical solution will not be discussed in detail. Specific examples will be included in Chapter 9 for practical problems of fluid structure interaction where a more detailed analysis will be taken up.

3.8.1 Modes of Vibration

For a simple spring mass system in which the mass is a single point mass and the spring is external to the mass and acting in one direction, a single degree of freedom system generates. When such a system is subjected to a harmonic force, then the system is said to have only one mode of vibration. The frequency of vibration of the mass in this case coincides with the frequency of excitation.

An unconstrained rigid body will have six degrees of freedom when subjected to a general excitation. For a harmonic excitation, this body will have six modes of motion. For a linear system the frequency of each degree of motion will generally be the same frequency as the harmonic excitation.

Many structures and systems require a mathematical description for vibration analysis that includes continuously distributed masses and elasticity. In order to determine the shape of this elastic structure under loading, one needs an infinite number of coordinates at each particle on the structure. Thus, the structure possesses an infinite number of degrees of freedom. The number of degrees of freedom of a system is the number of elastic movements of masses. A structure that is treated as a beam will have an infinite number of degrees of freedom or resonance. In discreet mass system, all points within one mass will move in phase. However, for a continuous system, this is not a requirement any more.

Most practical structures can not be idealized as simple point masses. In fact, all masses possess elastic elements. In analyzing these structures, it is a common practice to idealize the structure as a series of discreet masses with no elasticity and these masses are connected to each other with idealized elastic elements having no masses. These structures may then be analyzed as N lumped masses having displacements, but no deformation, as the deformation is theoretically addressed by the attached elastic elements. In this case, the structure can assume various mode shapes in a given direction due to harmonic excitation. If the number of lumped masses is N, then it will have $N-1$ natural modes, which are, sometimes simply called modes.

3.8.2 Vibration of an Elastic Structure

Consider a simple example of an elastic structure in the form of a small tube or rod and examine its vibration under an axial load [Thompson (1948)]. The tube or rod is assumed to be uniform along its length. The tube will experience a strain along its length due to the applied axial force at its end. The strain will change as a function of distance x and time t along the length of the tube. As a continuous structure, the tube will have an infinite number of natural modes of vibration. The longitudinal distribution of this strain will be different with a particular mode shape. Consider an element dx of the tube at a distance x from the end. If the strain at x is ε, the strain at the point $x + dx$ will be given by $\varepsilon + \dfrac{\partial \varepsilon}{\partial x} dx$ as shown in Fig. 3.29.

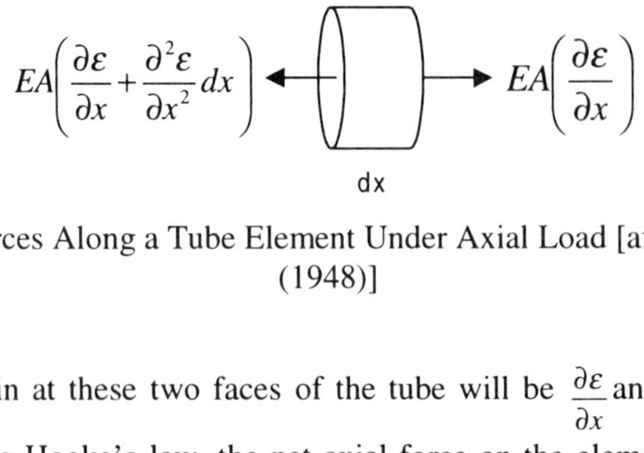

$$EA\left(\frac{\partial \varepsilon}{\partial x} + \frac{\partial^2 \varepsilon}{\partial x^2} dx\right) \longleftarrow \quad \longrightarrow EA\left(\frac{\partial \varepsilon}{\partial x}\right)$$

dx

Fig. 3.29: Forces Along a Tube Element Under Axial Load [after Thompson (1948)]

The unit strain at these two faces of the tube will be $\dfrac{\partial \varepsilon}{\partial x}$ and $\dfrac{\partial \varepsilon}{\partial x} + \dfrac{\partial^2 \varepsilon}{\partial x^2} dx$. Then applying Hooke's law, the net axial force on the element of the tube will be

$$AE\left(\frac{\partial \varepsilon}{\partial x} + \frac{\partial^2 \varepsilon}{\partial x^2} dx - \frac{\partial \varepsilon}{\partial x}\right) = AE \frac{\partial^2 \varepsilon}{\partial x^2} dx \qquad (3.187)$$

in which A is the cross-sectional area of the tube (or rod) and E is the Young's modulus of elasticity of the material. This force may be equated to the inertia of the element of the tube, which gives

Chapter 3 Basic Theory of Vibration   113

$$\rho_s A dx \frac{\partial^2 \varepsilon}{\partial t^2} = AE \frac{\partial^2 \varepsilon}{\partial x^2} dx \qquad (3.188)$$

in which the mass density of the tube material is given by ρ_s. From Eq. 3.188, the equation of strain along the length of the tube with time becomes:

$$\frac{\partial^2 \varepsilon}{\partial t^2} = \frac{E}{\rho_s} \frac{\partial^2 \varepsilon}{\partial x^2} \qquad (3.189)$$

This equation of motion will describe the general shape of the tube as a function of time. For simplicity in this analysis, consider the case of the tube vibrating at one of its principal modes in a simple harmonic motion. The initial condition of the tube under this motion is its principal mode shape so that it passes through the same shape at the beginning of each cycle. The strain in the tube will be a function of x and will have a maximum value given by $S(x)$. Then the solution may be written by the harmonic expression:

$$\varepsilon = S(x) \sin \omega t \qquad (3.190)$$

If this expression is substituted in the differential equation (Eq. 3.187) and the time function is eliminated, then a second order ordinary differential equation for the maximum strain is obtained as a function of the distance x.

$$\frac{d^2 S}{dx^2} + \frac{\omega^2 \rho_s}{E} S = 0 \qquad (3.191)$$

The general solution for this type of differential equation is given by

$$S = A_1 \cos \omega \sqrt{\frac{\rho_s}{E}} x + A_2 \sin \omega \sqrt{\frac{\rho_s}{E}} x \qquad (3.192)$$

Combining Eqs. 3.190 and 3.192, the general expression for the strain on the tube as a function of x and t is obtained in terms of constants A_1 and A_2 as

$$\varepsilon = \left[A_1 \cos \omega \sqrt{\frac{\rho_s}{E}} x + A_2 \sin \omega \sqrt{\frac{\rho_s}{E}} x \right] \sin \omega t \qquad (3.193)$$

The constants are evaluated based on the boundary conditions for the tube, for example, end conditions being free at both ends, fixed at both ends or one end fixed. There are several practical cases that may fall in these categories, such as submerged pipelines near the ocean floor, and drilling risers attached to offshore structures subjected to current and waves.

If the differential equation is written as

$$\frac{\partial^2 \varepsilon}{\partial t^2} = c^2 \frac{\partial^2 \varepsilon}{\partial x^2} \qquad (3.194)$$

then it becomes the well-known wave equation. The quantity c is known as the speed of the wave. Refer to Chapter 2 for similarity of this expression with the linear wave theory.

If the frequency of the excitation force happens to coincide with the particular deflection mode, then the structure is said to be in resonance and the vibration mode takes on a form of a standing wave. Thus the resonance of a structure is associated with a (continuous) mode shape. An example of the first few mode shapes of a beam with different end boundaries is shown in Fig. 3.30. The example considers a simple beam undergoing transverse vibration. Several boundary conditions are applied to the two ends of the beam.

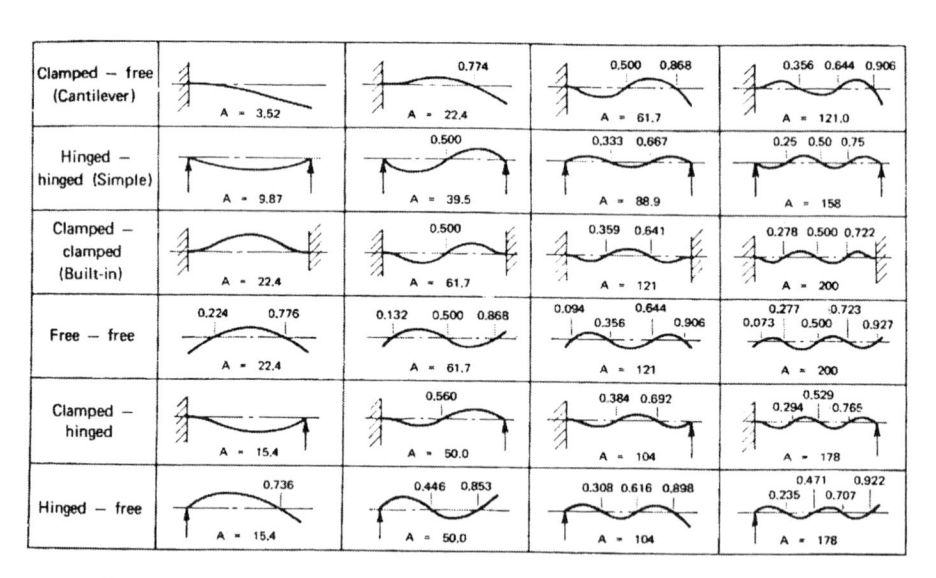

Fig. 3.30: Mode Shapes of a Beam with Various End Constraints

Chapter 3 Basic Theory of Vibration 115

Following Thompson (1948), the end connections included in the table in Fig. 3.30 are clamped-free, hinged-hinged, clamped-clamped, free-free, clamped-hinged and hinged-free. The expression for the natural frequency of vibration is given for the various modes in terms of a coefficient C based on the end connections and mode shape as $\omega_n = C\sqrt{EI/(\rho A l^4)}$ where l is the length of the beam. The node point of zero deflections are indicated in the figure as a fraction of the length from the left-hand support.

3.9 Self-Excited Vibration

Unlike a fixed rigid structure, if the structure is allowed to move in the flow field, such as a long flexible structure or a spring-mounted rigid structure, then the structure will vibrate normal to the free stream at or near the frequency of vortex shedding. If the structure vibrates at a frequency far removed from the harmonics or subharmonics of the vortex shedding frequency, then the vibration of the structure does not influence the wake field appreciably. If the flow velocity is increased, then it is possible that the structure will influence the vortex shedding frequency and actually cause a shift of the stationary vortex shedding frequency toward the vortex shedding frequency coinciding with the natural frequency of the vibration of the structure. This phenomenon is known as the lock-in, and the frequency of vibration is called the locked-in frequency.

The lock-in effect is possible when the vibration frequency of the structure coincides with the harmonic, superharmonic or subharmonic of the shedding frequency. This locked-in resonance vibration transfers energy from the wake region into the structure, which produces significant oscillation of the structure. This is generally known as the self-excited vibration and is one of the main causes of structural failure. Many drilling risers hung from a drilling rig in a field of high ocean currents have experienced such vibrations and have actually failed resulting from such vibrations.

It is, therefore, important to be able to predict such a phenomenon analytically so that the situation may be avoided in a design. In order to solve the complete problem, one should solve the time-dependent Navier-

Stokes equation in the presence of the structure. The solution describes the pressure field around the structure considering the flow separation, wake and vortex field and the coupling effect between the structure and the fluid. However, this problem is very complex for conditions with high Reynolds number, and only limited success has been achieved at low Reynolds numbers where the viscosity effect is neglected. Usually, simplification is used to seek approximate solutions for such fluid-structure interaction problems. This area of numerical analysis will be discussed in further detail in Chapter 8.

3.10 Exercises

Exercise 1

Derive the expressions of the mass and moments of inertia of a hollow right circular cylinder of inside radius R_i and outside radius, R_o and height, h about an axis at the center of its circular bottom. Compute the mass and moments of inertia for a cylinder made of steel for $R_i = 99$ ft, $R_o = 100$ ft and $h = 50$ ft.

Exercise 2

Generate the nondimensional response of a damped single degree of freedom system using linear damping factor of $\zeta = 0.1$ for a linear and linear plus cubic restoring force system. For the latter case use an S value of 1.5. Compare the results at a frequency twice that of the natural frequency. Discuss the stability of the results at this point for the nonlinear restoring force.

Exercise 3

Generate the response of a nonlinearly damped single degree of freedom system described by Eq. 3.74. Use linearization method to arrive at a closed form solution. Assume a natural period of the system to be 12 sec. Consider the linear damping factor to be 0.1 and the nonlinear damping factor 0.05.

Chapter 3 Basic Theory of Vibration   117

Exercise 4

Apply Rayleigh-Ritz method to obtain the general solution of the differential equation for a nonlinear single degree of freedom system described by Eq. 3.125 which is subject to harmonic excitation.

Exercise 5

For a double spring-mass system in series, in which the natural frequencies are equal, derive the expression for the natural frequencies in terms of the mass ratio. Explain the physical implications of this system.

Exercise 6

Consider a system of two simple pendulums connected together by a spring of constant k. The spring is unrestrained when the pendulums are in the vertical position. The setup is shown in Fig. 3.31. Write the equations of motion of the pendulums in terms of the angular displacement of the pendulums. Derive the expressions for the natural frequencies of the system. Show that the amplitude ratio for the displacements is unity in the case of the motion of the two equal masses at the first natural frequency, so that the spring has no effect on the motion of the pendulums and does not undergo extension or compression. On the other hand, the motions are out of phase at the second natural frequency.

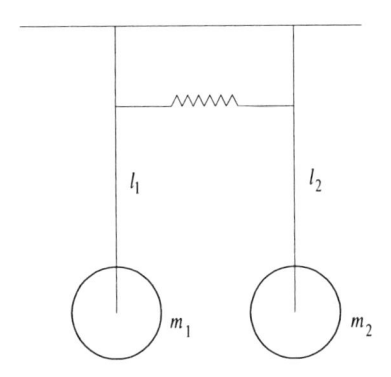

Fig. 3.31: Oscillations of a Coupled Double Pendulum

Exercise 7

Determine the natural frequencies of the double spring-mass system shown in Fig. 3.25 in which the weight of the masses m_1 and m_2 are 3.86 lb. and 1.93 lb., respectively. The springs are k_1 = 20 lb./in and k_2 = 10 lb./in. Describe the two modes of vibration by locating the nodes.

Exercise 8

Find the frequency of vibration and the expression for strain for a uniform rod, which is subjected to an axial harmonic load. Assume, as the boundary condition, that both ends of the rod are unrestrained.

3.11 References

1. Abramowitz, M. and Stegun, I.A., Handbook of Mathematical Functions, Government Printing Office, Washington, DC, 1964.

2. Abramson, N., "Nonlinear Vibration", Shock and Vibration Handbook, Harris, C.M. and Crede, C.E. (Editor), McGraw-Hill Book Co., Inc., 1976.

3. Chakrabarti, S.K. and Cotter, D.C., "Damping Coefficient of a Moored Semisubmersible in Waves and Current", Proceedings on Nineth International Offshore Mechanics and Arctic Engineering Symposium, Houston, TX, Vol. I, Part A, February 1990, pp. 145-152.

4. Roberts, J.B., "Effect of Parametric Excitation on Ship Rolling Motions in Random Waves", Journal of Ship Research, Vol. 26, No. 4, Dec. 1982, pp. 246-253.

5. Thompson, W.T., Mechanical Vibration, Prentice Hall, Inc., 1948.

Chapter 4

Experiments in Hydrodynamics and Vibration

4.1 Introduction

Vibration testing of mechanical systems and machinery generally involves testing of prototypes for the identity of problems related to vibration and the associated component fatigue. However, for larger structures in a flowing fluid environment, vibration testing is generally limited to a small-scale model. This is because of the practical size limitations of the structures to be tested, and the available testing facility limitations. The cost and complexity necessarily limit prototype testing. For certain vibration testing of mechanical systems, vibration analyzers that provide analog signals are sufficient. On the other hand, in hydrodynamic small scale testing, it is important to gather digital data, which may be further analyzed to determine the sources of large motions or vibrations. It helps in the design of the prototype system and in taking possible remedial actions to reduce the motion or vibration.

Therefore, in vibration testing within hydrodynamic systems, there are several aspects of modeling to consider. The first and foremost is the scaling of the measured data to the full-scale system before the data may be interpreted for application and further action. This requires knowledge of the applicable scaling laws. The second critical aspect is the choice of a proper scale in order that the quantities of interest may be measured accurately and the model may represent the essential elements of the system. Of course, the instruments selected to measure the environment and the responses of the model system should be carefully chosen so that the data has the required engineering accuracy. The data collection system should also be carefully designed, including the amplification of signal, frequency response and filtering of analog data for unwanted noise before digital conversion takes place.

This chapter will introduce the need and purpose of model testing, as well as the common scaling laws that are used in the vibration analysis of structures subjected to fluid forces. Different types of model testing facilities including wind and current generating facilities, and wave basins will be described. Short descriptions of a few commercial current and wave testing facilities will be given. The techniques of wind and current generation will also be addressed. The instruments used in the measurement of structure motion and vibration are described. In particular, examples of the design of load measuring instruments are presented. The data collection and the associated data analysis will be discussed with particular emphasis on the quality of data. A few examples of actual vibration testing are presented giving their objectives, testing techniques and level of success.

4.2 Planning a Model Test

For hundreds of years, models have been used as a working plan from which prototype structures have been designed, modified and constructed. Ship models have long enjoyed the usefulness of modeling for various purposes, such as determining the placement of their cargo and ballast. Working mechanical models came into use during the industrial revolution. Systematic hydraulic scale model testing goes back to the nineteenth century.

One of the first and foremost tasks in planning a scale model test is to investigate the modeling laws required for the system in question. This section addresses the scaling parameters that are important in designing a vibration test in fluid flow, and covers a few key areas to consider in replicating a prototype structure for a physical model test.

Many modeling approaches are followed in the study of natural systems. The most important of these are physical models and mathematical models. A physical model may be a scaled hardware model or computer analog model. Physical models are a close representation of reality in which a prototype system is duplicated as closely as possible in a (generally) smaller scale. The purpose of the model is to approximate and anticipate the prototype behavior through certain prescribed modeling laws. This chapter is limited to the case of scaled physical models and the scaling laws associated with the physical models used in hydrodynamics.

4.2.1 Benefits of Model Testing

One of the principal benefits of model testing is that valuable information is provided for the real structure at relatively low cost, which can be used to predict the potential success of the prototype. The physical model provides qualitative insight into a physical phenomenon, which is not fully understood. The use of models is particularly advantageous when the analysis of the prototype structure is very complicated. In other applications, models are often used to verify simplified assumptions which are involved (or inherent) in most analytical solutions, including higher-order effects. An example of this is the discovery during model testing of slow drift oscillation of a moored floating tanker in waves which led to the derivation of the theory of the second-order oscillating drift force and the associated motion. Model test results are also used to determine empirical coefficients that may be directly used in a design of the prototype. The model test of a new concept in an unknown territory is essential for exploring unexpected properties. The ultimate goal of a model test is to obtain reliable results that can be scaled up, and to correlate these results with an analytical tool for design.

4.2.2 Requirements of a Facility

It is important that the model test is conducted in a suitable facility that can perform the tests at a reasonable scale so that accurate measurements can be taken. The model testing facility should include the following capabilities:

- Physical facility,
- Model building facility,
- Instrumentation,
- Hardware to simulate environment,
- Hardware and software to record and analyze data, and
- Knowledgeable personnel familiar with all aspects of model testing.

The hydrodynamic model tests for the purpose of studying the structural vibration require a facility that can generate steady and oscillatory fluid flows. Often times the structure is moved instead of the fluid. For the simulation of the steady flow this can be achieved in a towing tank where the submerged model is towed in an otherwise still water. Similarly, the submerged structure may be oscillated in a fluid with a mechanical system to simulate wave action.

4.3 Description of Test Facilities

This section describes the facilities that may be used in vibration testing of a hydrodynamic structure. The facilities include those that simulate the steady and oscillating flows of fluid past the structure in model scales. The requirements of these types of facilities are discussed. Some of the existing facilities that are suitable for testing the motions and vibrations of structures in fluid are described. This is not intended to be a comprehensive listing of facilities. Only selected representative facilities are mentioned here with the understanding that there are many similar facilities in existence around the world.

4.3.1 Current Generating Facility

In modeling current in a facility, the uniformity and distribution of current should be carefully investigated. The generation of current is simplified if water is circulated in a closed loop in the facility. This is often achieved by pumping water from one end of the tank to the other and through a piping system. If a false bottom or a piping system exists under the floor of the facility, underwater pumps can circulate the water in a loop above and below the floor. Counter-current is generated by reversing the direction of flow. If an installed current generation is not available, local currents are often generated by placing current generators in the basin. These may take the form of a series of hoses with an external water source or a series of portable electric outboard motors. Uniformity of flow is achieved by proper location and flow rates from the individual current generator. Flow straighteners, such as tube bundles, may be installed in the tank to stabilize the flow.

For some purposes, it may be sufficient to model flow past a model using local current generation. There are several methods used to generate local current in the general area of the model in the test basin. One method provides a manifold to produce a flow over the width and depth of the model in the basin. The manifold may consist of small diameter plastic pipes of adequate size and number through which flow can be generated. The manifold is supported on a structure seated on the floor or hung from a bridge above and placed upstream of the model. The flow is created and controlled by a pump. The water is drawn through an intake pipe from the basin. Individual control valves are provided in the pipe manifold at each elevation so that the flow through them may be individually controlled. This method has the following advantages and disadvantages over other more expensive systems:

- It is easy and inexpensive to build and may custom fit the test requirements
- The current in the region of the model is reasonably steady.
- It is possible to generate some vertical shear in the current profile by selectively throttling the flow.
- The current profile is limited in cross section and is only generated over a chosen depth and over the width of the model region.
- The hardware to generate current is not necessarily transparent to flow and may have some influence on the overall flow.
- Undesirable turbulence may be present in the current due to local generation. On the other hand, it may better simulate the prototype situation and minimize the effect of distortion in the model Reynolds number, which typically appear in model scale.

Shear current is a common occurrence in the ocean. In a shear current, the surface current velocity is generally higher than the bottom current. The current profile with depth is represented by a linear or bilinear shear current. Near the bottom, the boundary layer effect provides a parabolic profile reducing to zero current at the bottom. In deep water, in particular, the shear current may have a significant influence on a submerged structure. One example of such a structure is a tensioned riser, or a riser bundle (group of risers).

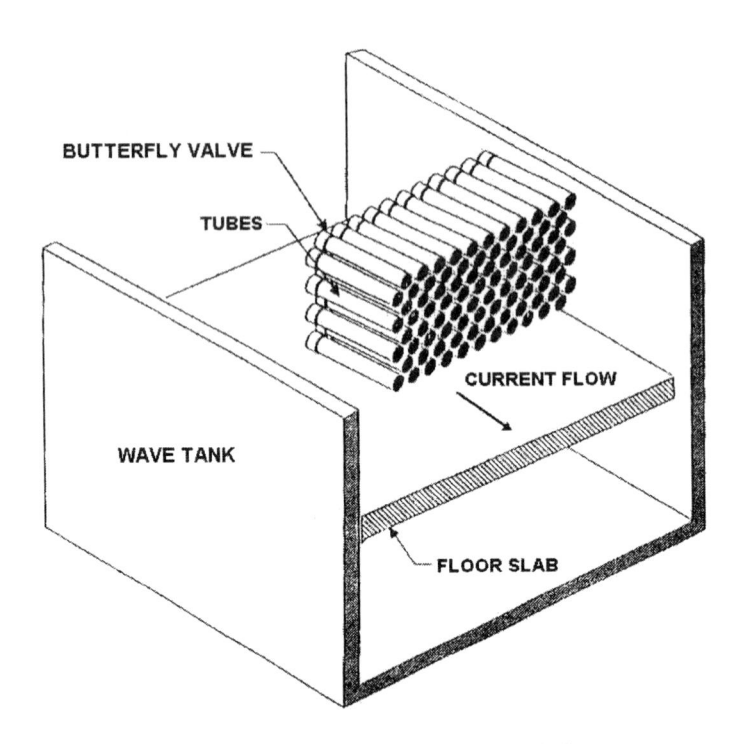

Fig. 4.1: Shear Current Generation Scheme

In order to generate shear current of a given profile, the flow at various elevations must be differentially controlled to provide the desired profile in the test tank. One method of shear current generation may use a flow control mechanism at the inlet end of each flow straightener, such as, butterfly valves (Fig. 4.1). Alternately, materials of varying porosity, e.g., foam or sponge, may be inserted in select tubes to reduce the flow by a desired amount. Thus, by varying the restriction of flow through the various rows of tubes in the flow straightener continuously with depth, a positive or negative (linear) or wedge-shaped shear current may be generated. A bilinear shear current or simultaneous vertical and horizontal shear current profiles can be created using the same technique.

In one such test, flow straighteners consisting of smooth 6 in (102 mm) diameter, 4 ft (1.2 m) long cylinders were placed spanning a 4 ft (1.2 m) deep by 10 ft (3 m) wide channel. Uniform fluid flow was created through the flow straighteners.

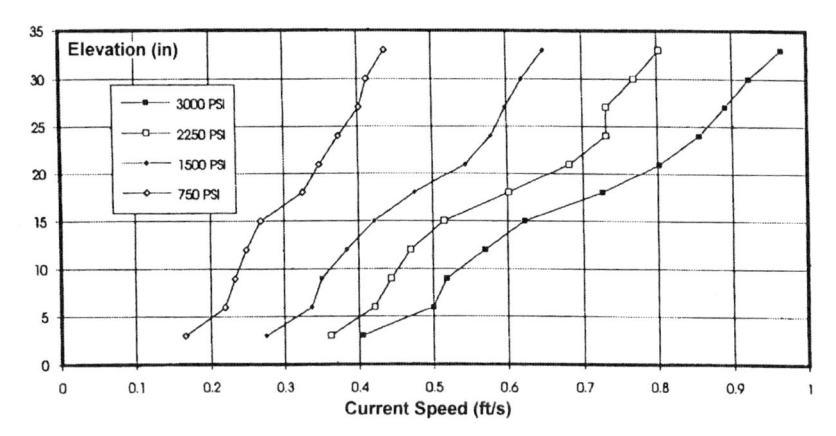

Fig. 4.2: Positive Shear Currents

Shear currents were generated by providing increasing levels of friction to flow at prescribed elevations. This friction was achieved by placing a mesh in front of the flow straighteners. This mesh consisted of a polyethylene cloth with a diamond shaped mesh pattern. The cloth was commercially available in 36 in (914 mm) wide rolls with nominal openings of 1/8 in (3 mm), 3/16 in (5 mm), 1/4 in (6 mm), 1/2 in (13 mm), 3/4 in (19 mm) and 1 in (25 mm).

Table 4.1: Mesh Pattern for Shear Current

ELEVATION (in)	MESH OPENINGS (in)			
30-36	1			
24-30	¾			
18-24	½			
12-18	1/4	¾		
6-12	3/16	1/4		
0-6	1/8	1/8	1/8	1/8
BACK-TO-BACK LAYERS	1	2	3	4

126   *Theory and Practice of Vibration and Hydrodynamics*

The cloths were cut into 6 in (152 mm) widths, and various combinations of mesh openings were tried at each elevation. The added resistance provided by this cloth and the subsequent shear pattern developed proved to be highly nonlinear, and a trial and error approach was applied. The final mesh pattern is shown in Table 4.1. This mesh pattern was then vertically reversed to produce a negative shear. These shear currents are shown in Figs. 4.2 and 4.3 for various hydraulic settings of the current generator.

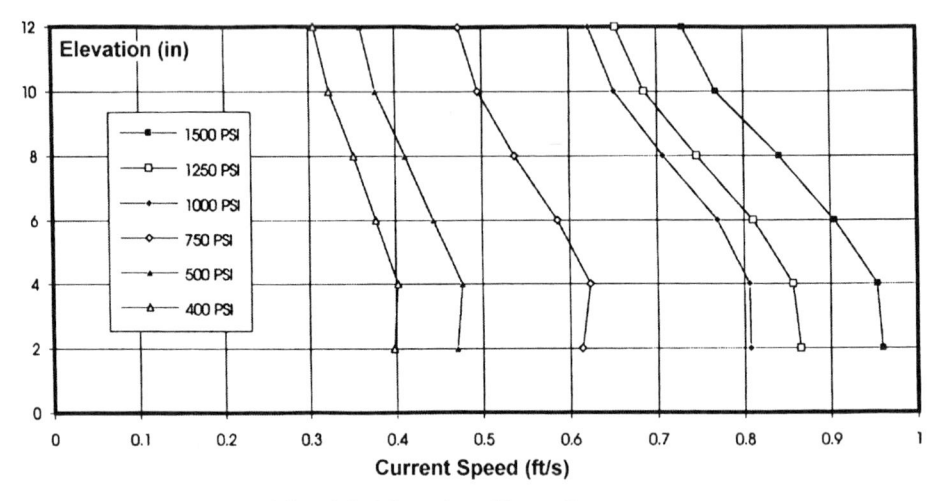

Fig. 4.3: Negative Shear Currents

Examples of facilities that are suitable for current testing are

- St. Anthony's Falls laboratory, Minnesota, USA which has a natural flow of water through a gradient.
- Iowa Hydraulic Institute at the Univ. of Iowa, Iowa City, Iowa, USA where piping systems exist to generate flow in the test section using pumps.
- MARIN Seakeeping Basin, Wageningen, Netherlands where the flow is created in the basin with an external piping system in a closed loop.

- Naval Surface Warfare Center Super Cavitation Tunnel, Memphis, TN, USA where a vertical closed loop tunnel of square cross-section exists to circulate the water at a high speed through the test section.

4.3.2 Towing Tanks

A towing tank should have a towing carriage that is capable of moving at a steady speed and be able to carry a large displacement structure. The availability of a towing carriage allows for the simulated current test. Setup for flow visualization and flow measurement in the vicinity of the structure is also possible from a towing carriage.

The simulation of current by towing the model in still water is a viable alternative to current generation past a stationary structure. Many tests are done today in which towing the structure fixed to the towing carriage simulates the current. For moored structures, an alternative to including the current on the moored system is to tow the entire structure with the mooring system attached to it. In this case, the structure with the mooring line installed is mounted on a framework, which is attached to the carriage. The framework of the structure must be sturdy, but reasonably transparent so that its overall effect on the flow field is minimal. Sometimes, the members of the framework are faired in the direction of flow to minimize flow separation.

In towing tests, the simulation of current is possible for a uniform current. The uniformity of the simulated current velocity, however, is superior to other methods of direct physical simulation of current in a basin. The simulated current is also turbulence-free. If shearing effect on the structure is important, it may be introduced by isolating the lower part of the hull from the flow by introducing an obstruction in front of the structure. One such concept that has been successfully applied, is to design a streamlined box of appropriate height which fully encloses part of the structure and which is mounted to the carriage. This part of the structure is then shielded from the fluid flow as the carriage moves. This provides a velocity gradient between the upper part and the lower part of the structure. By varying the height of the box, the desired location of this gradient is achieved. Perforations of varying degree may be introduced down the face of the box to simulate a gradual decay of the shear flow, much like a real case. These simulation methods will, however, introduce turbulence in the area immediately ahead of the model structure.

The towing tank's carriage is generally fitted with an instrumented staff for performing towing tests. The staff is arranged to accommodate any combination of roll, pitch and heave motion of the model during a test. The instrumented towing staff is mounted on the towing carriage and is used to tow the model. A dynamometer at the end of the staff attached to the model measures the resistance and lateral force. The staff is located at or near the CG of the model that is free to move in the heave direction, but restrained in surge, sway, roll and yaw. A two staff arrangement (Fig. 4.4) is sometimes used in which a second staff is located aft of the CG of the model and is free to heave and pitch. Often both staffs are rigidly connected to the structure model, which is then restrained from motion. Vertical potentiometers are mounted on the heave staff to measure the trim of the structure (and pitch) and its dynamic sinkage (and heave). In a two staff arrangement, it is preferred to instrument both staffs in this way.

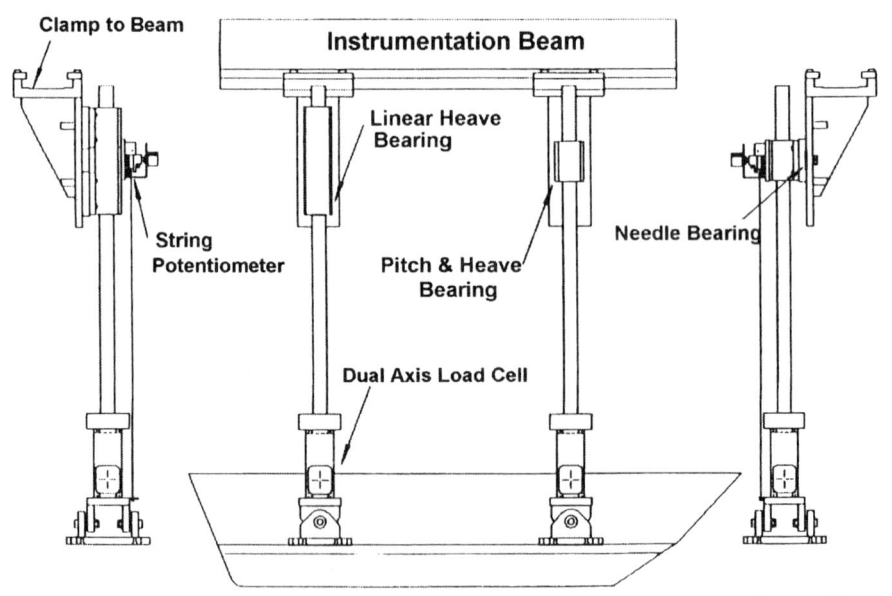

Fig. 4.4: Instrumented Towing Staff (courtesy of Offshore Model Basin, Escondido, CA)

To insure that only the resistance forces are measured by the dynamometer, bearings are used to eliminate undesired restraints, which might cause cross talk in the load cells. A hinge point is generally provided between the staff and the dynamometer in order that the towing force is applied along the thrust line. The hinge point also allows the model to pitch and is designed to restrain the model in sway and yaw. It insures that the bending moments are not measured. Steel shafts and linear bearings are used to allow freedom in the heave direction of the vessel. Each dynamometer is conveniently mounted on the instrumentation beam on the towing carriage.

Examples of facilities that are suitable for towing te sts of models are

- David Taylor High Speed Towing Tank, Bethesda, MD, USA
- Numerous University Laboratory Towing Tanks (e.g., US Naval Academy, Annapolis, MD and Univ. of Mich., Ann Arbor, MI).

4.3.3 Planar Motion Mechanism

Many structures undergo simple harmonic motion or random vibration under fluid flow. It is often desirable to simulate this motion with a mechanical system in which the actual motion of the model may be prescribed through the input function and the reactive forces are measured. A Planar Motion Mechanism (PMM) is used for forced oscillations of moderately heavy models. The PMM is mounted above the fluid surface on a carriage and allows tests of mechanical simulation of models of submerged structures. There are several variations in the design of a PMM. Some are quite complicated.

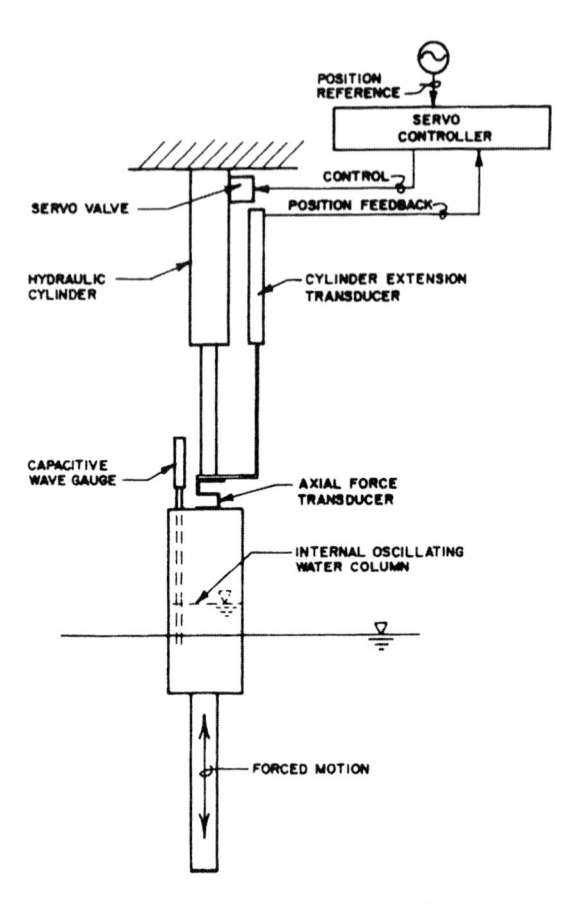

Fig. 4.5: Mechanical Oscillation of a Caisson

A single axis oscillator is used to force models in a single direction. The oscillator is driven by a servo controlled hydraulic cylinder and is capable of generating velocities and the full stroke displacements of the mechanism up to its design value. The control system of the oscillator is identical to that used in the planar motion mechanism. The position reference signal used to control the extension of the cylinder is generated by a microcomputer from a stored digital time series. The system can generate harmonic motions or random motions containing wide-band frequency components.

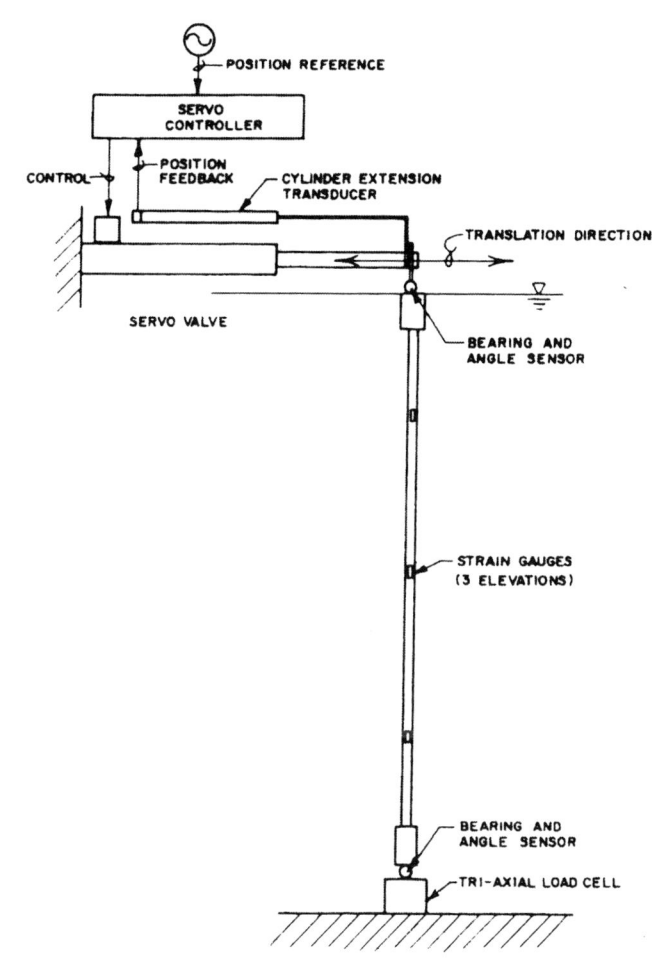

Chapter 4 Experiments in Hydrodynamics and Vibration 131

Fig. 4.6: Mechanical Oscillation of a Riser

Figures 4.5 and 4.6 illustrate two tests that have been performed using a single axis oscillator. Figure 4.5 shows a test setup for a submerged vertical caisson resembling a spar offshore production platform in heave. The caisson includes a resonant chamber, which permits flow of fluid in order to introduce additional damping in heave. The caisson was driven vertically in still water to confirm that it was properly tuned so that the water inside the cylinder oscillated 180 deg. out of phase with the motion of the caisson. The load cell at the top of the caisson was used to measure the vertical load generated by the oscillating water column. A wave staff was used to determine the free surface elevation of the water in the chamber.

The second test shown in Fig. 4.6, on the other hand, used the oscillator to induce a horizontal motion at the top of a flexible vertical riser. The riser was one component of a deepwater structure and was modeled to scale its stiffness along its length. The riser was instrumented with strain gages at three elevations to measure the bending stresses induced by the forced motion. Single period sinusoidal motions and random multi-frequency motions were produced. The stresses of the riser undergoing various bending modes were monitored through measurements.

4.3.4 Wave Generating Facility

Unlike currents, the wind-generated seas produce a changing free surface, which may be modeled by mechanical wave machines. The small-scale model testing of an offshore structure is generally performed at a wave-generating basin. The facility should have the capability of generating the required environment accurately. The generated wave energy in the facility should match the energy distribution of the sea waves at a reasonable scale. In the natural sea state, the period range of the power spectrum of waves having appreciable energy content varies from about 5 seconds to 25 seconds.

There are many types of wave generating devices employed in model basins. They may be classified in two general categories: active and passive. The active generators consist of mechanical devices of various types that displace a volume of water. By controlling the movement of the device, the wave profile is created. There are two main classes of mechanical type wavemakers that are installed in a basin to generate waves. One type moves horizontally in the direction of wave propagation in a simple harmonic motion (to generate sinusoidal waves) and has the shape of a flat plate driven as a paddle or a piston. The other type moves vertically through the water surface in a simple harmonic motion and has the shape of a wedge. The passive wavemakers, on the other hand, have no moving parts in contact with the water. They use a pulsating air column on the water surface generated by a blower to generate controlled oscillations in the water.

The wedge type of wave generators are best for high frequency waves but have limited success in generating low frequency waves. The OTRC facility at Texas A&M University, College Station, Texas uses this technique of wave generation. On the other hand, the pneumatic (passive) wavemaker can generate low frequency waves quite well but is limited to about 1 Hz frequency at the high frequency end due to the quick response time needed. This method is used at the MASK facility at the David Taylor Model Basin (DTMB) in Carderock, Maryland.

A single flapper extending from the top to the bottom of the tank can be hinged near the bottom to generate waves through rotation, e.g., the Offshore Model Basin (OMB) at Escondido, California. A wide frequency range (approximately, 0.5 - 10 seconds) can be best accomplished with a double articulated mechanical flapper. The top flapper addresses waves between the frequencies of 1.0 Hz to 2.0 Hz, while the bottom flapper addresses the lower frequency waves. Note that a random wave of broad-banded spectrum requires a large frequency range. In this case, both flappers are usually operational. An example of this type of facility is the one at the Hydraulic Laboratory at the US Naval Academy in Annapolis, Maryland.

Most of the wave tanks built in the sixties are two-dimensional, i.e., they are capable of generating waves that travel in one direction only. The long period random waves or distant swells in the ocean exhibit such unidirectional behavior. However, local wind-generated ocean waves are generally multi-directional. In order to generate multi-directional waves, the wave basins are required to be equipped with a multi-segmented wavemaker and must have widths comparable to their lengths. Many modern facilities are capable of producing multi-directional waves.

These basins are often capable of generating current as well. It is advantageous to simulate the wave and current generation in a single basin (preferably in different directions) so that they may be combined to investigate their coupled interaction with the structure model. Examples of facilities that have capabilities of generating multi-directional random waves coupled with current are OTRC at Texas A & M, TX, USA, MARINTEK at Trondheim, Norway and MARIN at the Netherlands.

4.3.5 Random Wave Generation

The sinusoidal waves are easy to generate based on a single frequency of oscillation of the wave generator. Ocean waves, however, have a continuous range of frequencies. Therefore, the frequency range at which appreciable energy exists should be duplicated in the model wave generation. There are several available formulas that describe the energy density distribution $S(\omega)$ of random waves as functions of frequency. The following are examples of two of the most commonly used frequency spectral formulations that represent the energy of ocean waves (named after their developers):

Pierson-Moskowitz Wave Spectrum is described as:

$$S(\omega) = \alpha g^2 \omega^{-5} \exp(-1.25[\omega/\omega_p]^{-4}) \tag{4.1}$$

JONSWAP Wave Spectrum is given by:

$$S(\omega) = \alpha g^2 \omega^{-5} \exp(-1.25[\omega/\omega_p]^{-4}) \gamma^{[\exp\{-(\omega-\omega_p)^2/(2\sigma^2\omega_p^2)\}]} \tag{4.2}$$

where ω = circular wave frequency, ω_p = frequency at which the spectrum peaks, g = acceleration due to gravity, α = Phillips constant, γ = peakedness parameter and σ = width parameter, where $\sigma = \sigma_A$ for $\omega \leq \omega_p$, $\sigma = \sigma_B$ for $\omega > \omega_p$.

The constant, α, is normally taken as 0.0081, the peakedness parameter varies from 1 to 7 with a mean value $\gamma = 3.3$ and the width parameters σ_A and σ_B are 0.07 and 0.09, respectively. In general, these are dependent on the significant wave height and peak period. For a fully developed sea, the JONSWAP spectrum reduces to the Pierson-Moskowitz spectrum (with $\gamma = 1.0$).

Waves often have sufficient energy content in the range of 5 to 20 seconds to generate significant motion response in the structure. The laboratory should be capable of generating random wave time history in a model scale. Normally, 200 to 1000 frequency components are chosen over the frequency band of the spectrum in generating the time history.

If the ocean wave is broad-banded, it is efficient to consider a white noise spectrum in a (physical or computational) model to analyze the structural response. A white noise spectrum denotes a wave spectrum with nearly uniform energy over the full range of wave frequency of interest.

The generation of white noise with a significant amount of energy over a wide band of frequencies is a difficult task at any model basin. Therefore, the overall energy level of the generated waves is necessarily low and the response of the structure is expected to be small. The reliable range of frequencies for the structure response may be determined by choosing an area with high value of coherence. The coherence function varies from 0 and 1 and establishes close relationship between the excitation and response. The areas of low coherence value of less than 0.5 are eliminated from the derived response. Generally, this method provides reasonable accuracy and has the advantage of obtaining the structure response versus frequency from one single test run.

4.3.6 Wind Tunnels

Many structures experience wind loads, such as, tall commercial buildings in a downtown area or the exposed deck of an offshore structure. Often, these structures are tested in wind tunnels to determine wind loads in a model scale. The effect of wind on the exposed offshore platform is an important design consideration. When the wind flows over the platform, its flow pattern changes, introducing a pressure difference and a net force on the platform superstructure. Adverse wind conditions also affect several operational efficiencies of the platform, such as drilling operations, transport operations, etc. Since this book is primarily concerned with incompressible fluids and air is compressible, this area is considered to be outside the general scope of this book. However, wind tunnel testing is not limited to just modeling wind loads. Unlike limited current strengths or towing speeds in a test facility, a wind tunnel is capable of generating very high wind speeds. Since Reynolds number distortion is a common occurrence in a Froude model, this mode of testing may be undertaken if Reynolds scaling is desired. The following briefly describes wind tunnel testing.

While the wind loads on a structure may be quite important for a particular design, e.g., the exposed deck of a floating moored structure, the associated scaling problems limit their use. If the wind load alone is important, e.g. on the superstructure of an offshore platform, wind tunnel tests may be performed on these portions of the structure model. In these cases, it is easier to achieve Reynolds number similarity. The conventional aerodynamic wind tunnel generating laminar flow is generally unsuitable for offshore platform tests. A realistic assessment of these problems is made [Littlebury (1981)] by model studies in a boundary layer wind tunnel.

Geometric similarity of airflow in the boundary layer over a solid boundary is obtained in terms of several nondimensional parameters, such as wind speed profile, intensity of turbulence and length scale ratio of turbulence. A typical scale for an offshore platform for a wind tunnel test is between 1:100 and 1:200. To generate a stable boundary layer within this range of scale, the boundary layer is developed upstream of the flow development section so that a reasonably stable boundary layer may be achieved at the test section. A barrier of spires may be used for this purpose, shaped in such a way that the required amount of shear is produced. The tunnel floor is roughened to stabilize the turbulent shear flow. For a model scale of 1:100 to 1:200, it is virtually impossible to strictly maintain the Reynolds similitude. However, for bluff bodies representing the offshore platform, the boundary layer flow effect is independent of the Reynolds number so that this distortion in Reynolds scaling may not be important. The tunnel wind speed is often about 100 m/s. An electronic balance (dynamometer) is normally fitted in the tunnel to measure the forces and moments acting on the model structure.

An example of an excellent wind tunnel facility that is suitable for high speed tests on models is the facility at the Danish Maritime Institute (DMI), Horsolm, Denmark.

Sometimes, it is preferable that the wind effect on a floating structure is simultaneously generated with waves. The wind generation in a wave basin is often accomplished using a bank of fans located just above the water surface near the model. They may be conveniently placed on the face of a traveling bridge used to house the instrumentation and controls. In this case, the model superstructure must be accurately modeled. The wind loads on the structure may be particularly important in the design of such structures as a floating moored structure. However, it should be emphasized that these loads in the model system are limited by the associated scaling problems. Wind loads being a function of Reynolds number which is (an order of magnitude) smaller in the model compared to the prototype Re, a scale distortion exists. Therefore, it is possible that the prototype wind effect falls in the turbulent region while the corresponding model wind effect is closer to a laminar region. In this case, the model test results may be considered conservative. While wind velocity is often taken as a steady value, the wind spectrum may be important in some applications. The frequency range of a wind spectrum is quite broad-banded, often covering a range from 0.005 to 1 Hz. The overhead fans in a wave basin are designed to handle this broad band wind spectrum through the positioning of the fans relative to the model and control of rotational speeds and throttling of the individual fans in the bank.

4.4 Modeling Laws

In order to achieve similitude between the model and the real structure, the following three criteria must be satisfied:

- Geometric similitude
- Hydrodynamic similitude (Froude, Reynolds and Strouhal)
- Structural similitude

Hydrodynamic scaling laws are derived from the ratio of forces commonly encountered in a hydrodynamic model test.

4.4.1 Dimensional Method

One of the simplest methods in arriving at the scaling laws is the dimensional analysis. In the dimensional method, the unknown forces, F on two homologous fluid elements are equated to the sum of all the terms from the flow-governing quantities arranged in power products each having the dimension of F. Then taking the ratio of F forces, the scaling laws are derived. This is illustrated by the following example.

Consider that the force F on a structure in a fluid stream depends solely on density ρ, diameter D, gravity g, velocity U, dynamic viscosity μ and Young's modulus due to elasticity E. Thus, there are 6 independent variables in a mass (M), length (L) and time (T) system. The number of possible independent products having the dimension of F will be $6 - 3 = 3$. To form these, any triad $\rho^x D^y U^z$ of the six independent quantities is chosen and multiplied in succession by the remaining variables g, μ and E. Thus, the 3 nondimensional quantities P_1, P_2 and P_3 are

$$P_1 = \rho^a D^b U^c g; \quad P_2 = \rho^d D^e U^f \mu; \quad P_3 = \rho^g D^h U^i E \tag{4.3}$$

The dimensions of each product is equated to the dimension of $[F] = [ML/T^2]$. For example, the first one yields

$$\left(\frac{M}{L^3}\right)^a L^b \left(\frac{L}{T}\right)^c \frac{L}{T^2} = \frac{ML}{T^2} \tag{4.4}$$

in which the right hand side is the dimension of force. On equating the indices of L, M, and T in Eq. 4.4 successively, one gets three simultaneous equations as follows:

$$-3a + b + c + 1 = 1; \; a = 1; \; -c - 2 = -2 \tag{4.5}$$

Thus, $a = 1$, $b = 3$, $c = 0$ and $P_1 = \rho D^3 g$. Similarly, $P_2 = DU\mu$ and $P_3 = D^2 E$. Then, a general resistance equation for the specified dynamical conditions may be written as

$$F = (\rho D^2 U^2) \varphi\left(\frac{gD}{U^2}, \frac{\upsilon}{DU}, \frac{E}{\rho U^2}\right) \tag{4.6}$$

where φ represents a functional relationship among the nondimensional quantities within the parenthesis, and $\mu = \rho \upsilon$, υ being kinematic viscosity. The first term within parenthesis is the Froude number, the second one is Reynolds number and the last quantity is the Cauchy number.

The dimensional method appears to be a straightforward algebraic operation requiring less knowledge about the physical system. In reality, however, it requires adequate judgment about the physical quantities involved in the dynamical system and their relative importance. Knowledge and understanding of the physical phenomena is necessary in order to consciously neglect those that are of secondary importance and to determine important scale effects.

The typical interaction problem for structures with current or waves involves Froude number, Reynolds number and Keulegan-Carpenter number. For vibrating structures in fluid medium, the Strouhal number is important. The Cauchy number plays an important role for an elastic structure.

4.4.2 Froude Similitude

The Froude law is the most appropriate scaling law for the fluid-structure tests. The physical model is usually scaled following Froude's Law. Define the Froude number Fr as

$$Fr = \frac{u^2}{gD} \quad (4.7)$$

Assuming a model scale factor of λ and geometric similarity, the Froude model must satisfy the relationship:

$$\frac{u_p^2}{gD_p} = \frac{u_m^2}{gD_m} \quad (4.8)$$

where the subscripts p and m stand for prototype and model respectively. According to geometric similarity, the model linear dimensions will be scaled linearly with the scale factor,

$$l_p = \lambda l_m \quad (4.9)$$

Then, from Eq. 4.8 the velocity scales as:

$$u_p = \sqrt{\lambda} u_m \quad (4.10)$$

Table 4.2: Froude Scaling of Structure/Hydrodynamic Parameters

Quantity	Prototype	Model	Scale factor
Mass	m_p	m_m	λ^3
Force	F_p	F_m	λ^3
Acceleration	$\dot{u}_p$	$\dot{u}_m$	1
Time	t_p	t_m	$\sqrt{\lambda}$
Frequency	f_p	f_m	$1/\sqrt{\lambda}$
Pressure	p_p	p_m	λ

The relations of a few other important variables are derived from dimensional analysis and their scale factors from prototype to model scales are included in Table 4.2.

4.4.3 Reynolds Similitude

The Reynolds number is also equally important in many cases. However, Reynolds similarity is quite difficult, if not impossible, to achieve in a small-scale model. Simultaneous satisfaction of Fr and Re is even more difficult. If a Reynolds model is built, it will require that the Reynolds number between the prototype and the model be the same. Assuming that the same fluid is used in the model system, this means that:

$$u_p D_p = u_m D_m \tag{4.11}$$

If a scale factor of λ is used in the model, then this equality is satisfied if

$$u_m = \lambda u_p \tag{4.12}$$

In other words, the model fluid velocity must be λ times the prototype fluid velocity. In general, this is difficult to achieve, especially if a small-scale experiment is planned. In automobile drag tests, since a large-scale model $(\lambda = 1 \text{ to } 2)$ is often used, this scaling is easier to achieve. If a Froude model is used, then the relationship for the Reynolds number is

$$Re_p = \lambda^{3/2} Re_m \tag{4.13}$$

The larger the scale factor, the larger the distortion in the Reynolds scaling. For a scale factor of $\lambda = 44$, the prototype Re is 292 times that of the model Re. The consequence of this difference is that while the prototype flow regime is turbulent, that of the model may necessarily become laminar. Experiments have shown that the flow characteristics in the boundary layer are most likely to be laminar at $Re < 10^5$, whereas the boundary layer is turbulent for $Re > 10^6$. The drag coefficients in a laminar flow are normally higher than those found in a turbulent flow field (Fig. 2.4). The scaling of forces, then, is generally expected to be conservative if no corrective measure is taken in scaling. On the other hand, if nonlinear drag damping is present from the model velocity, the Reynolds distortion will actually produce a smaller model response compared to the expected full-scale response. In spite of this limitation, the resistance of a floating structure in transit or in the presence of current is commonly determined using a towing test of a small-scale model. In this case two different scaling laws apply, namely, the Froude law and the Reynolds law. These laws cannot be satisfied simultaneously. It is most convenient to employ Froude scaling and to account for the Reynolds disparity by other means discussed below. It is obvious that a larger model produces a smaller distortion. However, the magnitude of the Reynolds number distortion is still quite large.

In order to rectify this distortion in Reynolds number modeling it is a common practice to artificially stimulate turbulence in a model test by introducing roughness in the flow approaching the surface of the model. The increased turbulence in effect allows modeling in a more realistic flow environment and reduces the distortion of *Re* in small scale testing. Thus the simulated turbulence is expected to reduce the drag force and produce values close to the expected prototype values. For example, for a ship model this stimulation is introduced by attaching studs, pins or sandpaper near its bow. It works because once the flow regime is turbulent, the drag effect is only weakly dependent on the Reynolds number. The stimulation is also accomplished by using a submerged grid made of a frame of wires (screen) or of wood strips. The frame is located a few meters in front of the model on the same carriage used to tow the model, thus causing turbulence in the flow just ahead of the model. This addition provides the effect of a higher Reynolds number on the resistance, so the simulation of flow is closer to the prototype in spite of the *Re* distortion.

A test series employed a model semisubmersible type floating structure subjected to a uniform current. In the test, current was simulated using constant speed towing. In order to introduce turbulence in the flow to be experienced by the model, a screen was introduced in front of the model. This is shown in Fig. 4.7. The screen was mounted on the towing carriage spanning the beam of the model and towed ahead of the model at the same speed as the model. The test was carried out with and without the presence of this screen.

Fig. 4.7: Screen in Front of the Model as an Artificial Turbulence Stimulator

The test results in Fig. 4.8 provide the average values of the drag forces on the model with and without the turbulence stimulator. At low towing speeds, the stimulator failed to produce appreciable turbulence in the flow and the two cases of with and without the stimulator are similar. This is evident at the speed of 0.5 ft/s (0.15 m/s). The effect of the stimulator is clearly seen in the plot for higher towing velocities. The forces in the presence of the stimulator are considerably lower than those in its absence. This effect is directly attributable to lower values of drag coefficient, as found in a turbulent flow experienced by the prototype Reynolds number.

For equality of both Froude and Reynolds number, a fluid should be chosen whose kinematic viscosity is lower by a factor of about $\lambda^{3/2}$ to that of water. When λ is small, a fluid of lighter viscosity may be used for water. When λ is large, the above equality is almost impossible to achieve.

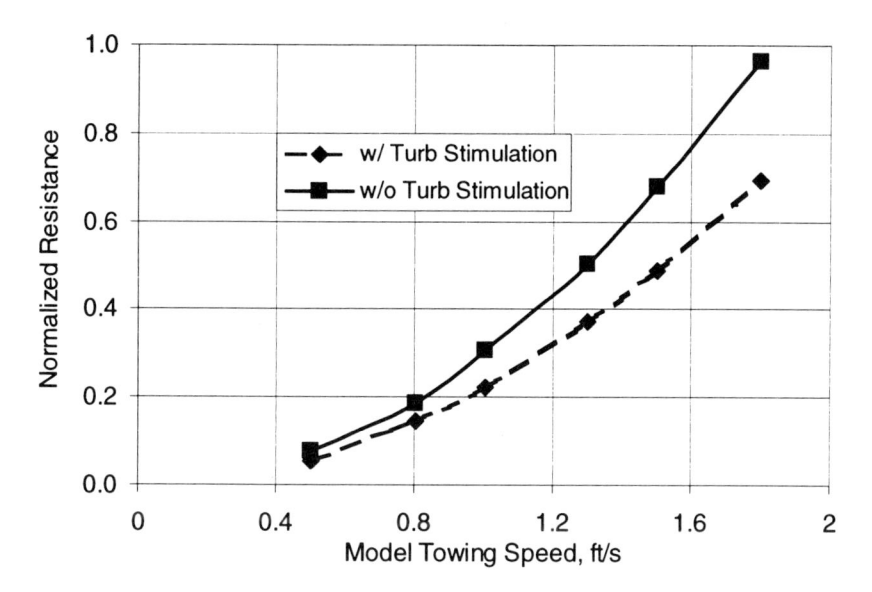

Fig. 4.8: Normalized Drag Force With and Without Stimulator

Another method of correcting for the distortion in the Reynolds number in which Froude scaling is used is in the scaling of the test data. In ship or barge towing resistance tests, model data is separated into potential and viscous and corrections are made in the friction factor based on the respective Reynolds number between the model and the prototype. If this difference is ignored in scaling, the (scaled up) prototype data will often be conservative. This area, of course, needs careful study while scaling up data, particularly for evaluating the effect of hydrodynamic damping.

4.4.4 Strouhal Similitude

Strouhal (as well as Keulegan-Carpenter number) similitude provides the similitude of the unsteady fluid flow. It requires that

$$\frac{U_p}{f_{sp}D_p} = \frac{U_m}{f_{sm}D_m} ; \quad \frac{u_p T_p}{L_p} = \frac{u_m T_m}{L_m} \qquad (4.14)$$

where f_s is the vortex shedding frequency, L is a characteristic length of the structure and T is the period of oscillation. Note that a Froude model satisfies the Strouhal (as well as Keulegan-Carpenter number) similitude simultaneously.

Therefore, in a Froude model, one need not worry about the Strouhal or Keulegan Carpenter number similitude. It has been shown that the oscillatory drag and inertia coefficients depend on both the Reynolds number and Keulegan Carpenter number. In the small scale testing, the Froude model assures the similitude regarding the Keulegan Carpenter number.

4.4.5 Cauchy Similitude

In traditional model testing, the model is considered a rigid structure and the model deformation and associated interaction with waves or current flow is considered insignificant. While this approach is generally acceptable for many structures, for a structure that is long compared to its cross section or for mooring components in conjunction with a floating structure, such simplification is not acceptable. This coupling of external load with structure response is termed hydroelasticity.

Hydroelasticity deals with the problems of fluid flow past a submerged structure in which the fluid dynamic forces depend on both the inertial and elastic forces on the structure. It is well known that for long slender structures, the stiffness of the structure is important in measuring the vibration response of the structure model in steady and oscillating flows.

Model tests of structures are often carried out to determine stresses imposed on their members due to deformation from the external forces, for example, from current or waves. In this case, the elastic properties of the prototype should be maintained in the model. Therefore, in these instances the Cauchy similitude is desired in addition to the Froude similitude [LeMéhauté (1965)].

Consider the longitudinal bending of a long beam. It is desired that this bending should be the same between the model and the prototype. Then,

$$\left(\frac{Mc}{EI}\right)_p = \left(\frac{Mc}{EI}\right)_m \tag{4.15}$$

Chapter 4 Experiments in Hydrodynamics and Vibration

in which the product EI represents the flexural rigidity of the beam where E = Young's modulus, I = section moment of inertia, M = vertical bending moment, and c = distance to the outer fiber of the beam from its neutral axis. Noting that the bending moment in a Froude model scales as λ^4, the Cauchy similitude requires that the bending stiffness of the model must be related to that of the prototype by the relation:

$$(EI)_p = \lambda^5 (EI)_m \tag{4.16}$$

Froude's law requires the deflection of the model to be $1/\lambda$ times the deflection in the prototype; also, stress must be similarly related, such that, $\sigma_p = \lambda \sigma_m$. For example, for a cantilever beam, the maximum deflection is given by $\delta_{max} = Fl^3/(3EI)$, where F is the load at the end of the cantilever of length l. Therefore, Eq. 4.16 satisfies Froude's law for this relationship. Since the section moment of inertia satisfies the relationship:

$$I_p = \lambda^4 I_m \tag{4.17}$$

one must have from Eq. 4.16:

$$E_p = \lambda E_m \tag{4.18}$$

to achieve Cauchy similitude. Thus, the Young's modulus of the model material should be $1/\lambda$ times that of the prototype. For example, assuming that the prototype is made of steel whose Young's modulus is $E_p = 30 \times 10^6$ psi and $\lambda = 36$, the model Young's modulus, E_m should be 833,000 psi.

In these cases, an elastic model, which properly takes into account the scale effects is designed and constructed. Sometimes it is not possible to build an elastic model due to lack of suitable material of proper modulus, small model, scaling difficulties, etc. In these cases, a segmented model can be built, in which individual segments are properly modeled. Usually, the segments are hinged together with rigid intermediate sections. The elasticity (such as, in bending) is introduced at the hinges. The number and stiffness of these hinges are selected to provide the scaled mode shapes of the model at its scaled natural frequencies.

In a torsional mode, the similitude condition similarly gives

$$(GI)_p = \lambda^5 (GI)_m \qquad (4.19)$$

where GI = torsional rigidity. Then, for a model with significant bending and torsional deflection, the model should additionally satisfy

$$G_p = \lambda G_m \qquad (4.20)$$

It is not an easy task to model bending and torsion simultaneously with one material. A case study on hydroelastic modeling will be provided in Chapter 9 to illustrate this difficulty.

4.4.6 Dynamic Similarity

The laws of dynamically similar fluid motions may be derived by three different methods: (1) Newtonian or integral method, (2) differential method and (3) dimensional method. In each case the selection of the physical quantities governing the flow are based on experience. The dimensional method has already been explained in section 4.4.1. In the Newtonian method, the ratios of the acceleration force and several of the impressed forces are equated, thus obtaining specific conditions for dynamic similarity. This has also been addressed previously. In the differential method, the differential equations for the two fluid motions, which are assumed to be dynamically similar, are written. Then the ratios of corresponding terms are equated to obtain the scaling laws.

In order to illustrate, consider an example of an open channel flow problem [Doodson (1949)]. For a uniform steady velocity U through a uniform cross-section A of a channel of width w, the continuity equation is written as

$$w \frac{\partial \eta}{\partial t} + \frac{\partial (UA)}{\partial x} = 0 \qquad (4.21)$$

where η is the elevation of the free surface. The momentum equation is expressed as

$$\frac{\partial U}{\partial t} + U \frac{\partial U}{\partial x} = -g \frac{\partial \eta}{\partial x} \pm \frac{CU^2}{R} \qquad (4.22)$$

Chapter 4 Experiments in Hydrodynamics and Vibration

in which R is the hydraulic radius of the channel and C is a constant. The relationship between the prototype and model variables involved in these two equations in terms of the respective scale factor is written as

$$v_p = \lambda_v v_m \tag{4.23}$$

where v denotes the variable under consideration. Note that the scale factor for cross-sectional area A is $\lambda_A = \lambda_x \lambda_\eta$. Then considering the above equations for the prototype, they may be written in terms of the model quantities as

$$\frac{\lambda_x \lambda_\eta}{\lambda_t} w_m \frac{\partial \eta_m}{\partial t_m} + \lambda_u \lambda_\eta \frac{\partial (U_m A_m)}{\partial x_m} = 0 \tag{4.24}$$

and

$$\frac{\lambda_x}{\lambda_t \lambda_U} \frac{\partial U_m}{\partial t_m} + U_m \frac{\partial U_m}{\partial x_m} = -g \frac{\lambda_\eta}{\lambda_U^2} \frac{\partial \eta_m}{\partial x_m} \pm \frac{\lambda_k \lambda_x}{\lambda_R} C_m \frac{U_m^2}{R_m} \tag{4.25}$$

For similarity to follow, Eq. 4.24 gives

$$\lambda_U = \frac{\lambda_x}{\lambda_t} \tag{4.26}$$

which is satisfied from the kinematic similarity. It follows from Eq. 4.25 that

$$\frac{\lambda_\eta}{\lambda_U^2} = 1 \tag{4.27}$$

and

$$\frac{\lambda_c \lambda_x}{\lambda_R} = 1 \tag{4.28}$$

The above analysis shows that, according to dynamic similarity between the model and prototype in this open channel flow, the identities in Eqs.

4.27-28 must be satisfied. The first of these equations (Eq. 4.27) may be shown to be a consequence of Froude's law.

The derived working formula from any of the three methods mentioned earlier contains dimensionless coefficients that are determined from model tests and applied to the prototype system operating under similar dynamical conditions.

4.5 Hydrodynamic Model

Hydrodynamic models are not just replicas of the prototype. They must also satisfy several properties of the prototype. Besides the geometrical properties, the dynamic properties of the model should be scaled so that the model responses duplicate the responses of the prototype. For hydroelastic testing, the material properties of the model should be satisfied as well. This section describes the requirements of these types of hydrodynamic models.

4.5.1 Modeling of Rigid Structure

The geometry of a submerged or semi-submerged structure is scaled dimensionally using the chosen scale factor. For a floating structure that is allowed to move or deflect in the flow of fluid, all the dynamic properties, e.g. displacement, moment of inertia, metacentric height (GM), and natural periods are properly scaled using Froude's law. Many of the details, e.g., appendages and small members, however, are often omitted. Even at a small scale (e.g. $\lambda = 100$ to 200), this modeling can provide reasonable results.

One of the most important elements of a successful vibration testing, besides the true model, is the test fixture that is used to establish the model properties or test the model in vibration. It is important to design the test fixture with the structure in mind. Similarly, the properties of the fixture must be carefully determined before the model is placed on it. If the fixture properties are inaccurate, then they will influence the model properties. Also, the size of the fixture should be comparable to the overall size of the model to be tested in the fixture. Oftentimes, the existing standard fixture at the test facility is used. It may need modification to fit the model being tested. If the fixture is too large or too small, then the possibility of error

Chapter 4 Experiments in Hydrodynamics and Vibration 149

introduced in the measurement and calculation of the model properties or response may be unacceptably large.

4.5.2 Modeling of Elastic Structure

If the size of a structure is such that the exciting forces allow it to deflect appreciably from its original configuration, then special care must be taken to model the elastic properties of the structure as well as its components. The hydrodynamic model is built of a suitable material, which reasonably satisfies the Cauchy law of similitude. The method of model building will depend on the requirements and scope of the testing and the results desired from such testing. The following methods that may be used in constructing such models are sequentially more complex:

- The structure is broken up into small segments. Each model segment is built as a rigid structure. The connecting element between adjoining segments provides the scaled axial, bending and torsional elasticity. This is a simple model that will be responsive to variable loads along its length. The limitation of this model is that the modes of vibration are not continuous, but are limited by the number of segments used in the model. Moreover, it is difficult to design connecting elements that properly account for the scaled elasticity and rigidity.

- For a long model, a segmented model is built with a beam element as a continuous backbone. The structure geometry is floated on this beam using a 'soft' material of minimal stiffness. The cross section of the beam is determined such that the bending stiffness of the prototype structure is modeled at the chosen scale. In this case the beam may be strain gauged to measure the structural bending moment. A laser positioning system may measure the mode shapes of the beam. This method will provide the required elastic model well. The limitation of the method is that the load distribution on the model, the model deflection and fluid slamming pressure on the hull are difficult to obtain.

- The entire shell of the structure model is built using a material of a reduced Young's modulus with a chosen thickness that satisfies the Cauchy similitude. The following are some of the metal and nonmetal materials that have been considered for building such a model: thin sheet aluminum with honeycomb design for the stiffeners, thin sheet copper with honeycomb design and foam for flotation, fiberglass of suitable thickness, Lexan sheets, foamed vinyl chloride, PVC sheets, polyethylene sheets, and polycarbonate with beam reinforcement. This is an expensive model to build and may experience detrimental properties from using plastic material.

In choosing the material, the following structural properties should be considered as well:

- Stress-strain relationship of the material is in the linear range over the test load range.

- Young's modulus is small so that the rigidity (material stiffness) may be properly modeled with workable thickness.

- The material is isotropic.

- It is easy to weld, glue or otherwise join the parts with sufficient strength and water tightness.

- Material is stable within the linear range with minimum creepage at the ambient temperature.

- The Poisson ratio is close to that of steel.

4.5.3 Modeling of Mooring Line

Floating structures in fluid are held in place with springs or mooring lines. Chains are often used to moor floating structures in water. The most important property of a chain is its weight (per unit length). The material and size of the model chain are chosen to model the weight at a small scale. The elasticity of the material should be verified to ensure the proper order of magnitude. The geometry of the chain is difficult to scale, which makes it nearly impossible to study its vortex-induced vibration at a small scale. It also introduces inaccuracy in simulating hydrodynamic damping of the chain generated from its own motion as well as from the wave and current action. In fact, hydrodynamic damping of the submerged mooring line has been shown to have a significant effect on the total response of a moored structure at its natural frequency.

There are three parameters that are important for the floating structure response in terms of the mooring line behavior. They are

- pretension in the spring or mooring line,
- stiffness of the spring, and
- load experienced by the structure at the line fairlead connection

For relatively large structures, if these quantities are modeled properly for a given environment, then the simulation of the structure response may be acceptably scaled, even if the spring lines are not physically modeled. The modeling of the outer geometry of mooring line becomes important, when the damping on the mooring line is considered important.

If the structure is comparatively small or the mooring system is long (e.g., in deep water mooring), then it may not be sufficient to model the stiffness characteristics alone. In these cases the environmental effect from the fluid on the lines may be significant compared to the environmental effect on the floating moored structure. Therefore, in these cases the replication of geometry or an equivalent representation of the line in model scale is important.

The difficulty of modeling and setup of a floating moored system in a test basin arises from several considerations, especially if the complete system is not modeled:

- the mooring stiffness is often nonlinear,
- the fairlead angle that the mooring line makes at the structure changes with time and loading for a given environment, and
- the initial mooring line angle requires change as different environmental conditions are simulated.

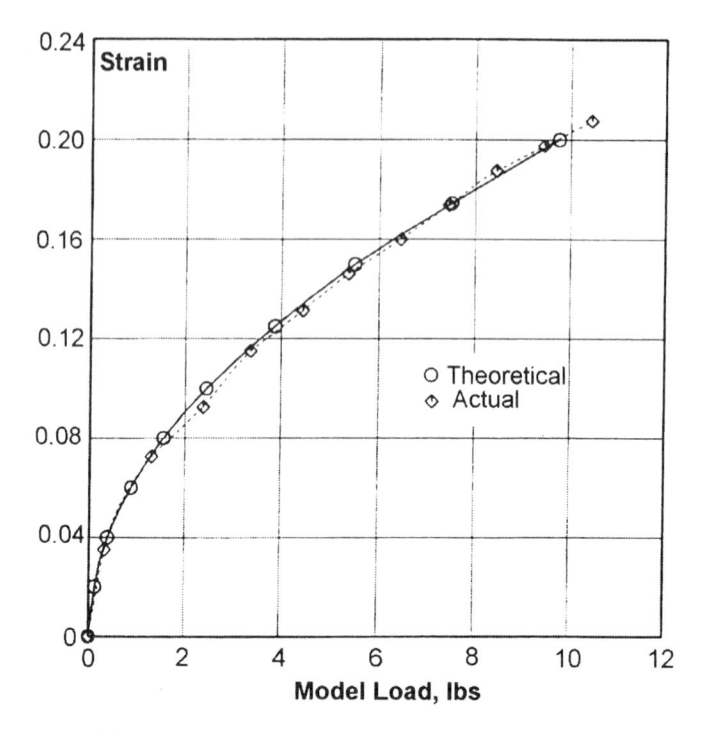

Fig. 4.9: Stiffness of a Typical Mooring Line

The mooring lines often take a catenary shape. A new type of steel riser in deep water is also placed in a catenary configuration from the structure to the ocean bottom. An example of the stiffness characteristics of a catenary mooring line for a floating structure is given in Fig. 4.9. This stiffness may be modeled in segments by a series of linear springs. If several spring components are used, for example, to simulate this nonlinear stiffness, these spring components acting in series simulate the lowest slope portion of the load-deflection curve. At the transition tension at which the lowest slope line changes to the next higher slope line, a cable or string is tied between the ends of the weakest spring. This prevents that spring from extending at higher tensions, resulting in a higher spring constant. This higher spring constant equals the slope of the next line segment of the load-deflection curve. Similarly, at the next transition tension, the second spring is tied off. This process is continued until the highest desired tension in model scale is achieved. The actual stiffness achieved for a model line is compared to the prototype (called theoretical) in Fig. 4.9.

Sometimes, the entire mooring line extending to the ocean bottom may not be modeled due to limited depth in the test facility. Distortion due to truncation in the mooring line length can be provided by additional springs. The lack of damping in the model chain due to this truncation may be corrected by external means, if deemed important.

If the mooring line dynamics are important, segmented plastic rods of suitable material of correct submerged unit weight connected by eye hooks will provide a uniform diameter for the mooring line (except for the small area of the hooks). It is easy to build and provides a reasonable estimate for the drag coefficient. One problem area for this model is with the segments in contact with the basin floor. Another alternative is to use an outer flexible tube of a diameter representative of the model diameter and a weighted cable inside. A thin-walled length of tygon or plastic tubing may be used for the lines. This method provides scaled damping effect, but may introduce significant bending stiffness in the lines.

For a taut mooring system, the elasticity of a cable/wire (tensile stiffness) is an important property that should be scaled with a suitable material at a small scale. For the success of small scale testing, it is important that the truncated taut mooring system simulation is kept simple and the mooring arrangement does not change with every environment. If the pretesioned line force is roughly linear with the line extension, the mooring line may be modeled with a set of linear springs. The initial fairlead angle of a truncated line is adjusted in order to match the computed design values. It is understood that this angle changes with loading from the environment. But for significant tension in the fairlead, the change of angle will be small. Since there is a large pretension in most of these taut lines, the error in the angle with varying load will be small. The initial tensions at the fairleads, which are monitored using a load cell located at the end of the mooring lines at the fairleads, are adjusted and maintained.

4.5.4 Modeling of Riser

Risers are composed of flexible tubes and buoyancy elements and often appear in a group or bundle. The riser system is subject to a constant tension, which may be modeled by a tensioned spring at the top. Individual risers may be modeled by metal or plastic tubing or other smooth surface material with sufficient rigidity and of the required diameter. Sometimes in a model, a riser bundle can be modeled by using an equivalent diameter single riser with the same overall drag area as the total riser array, and proper stiffness for the group of risers. Where the interference effect among the risers is an important area of investigation, all individual risers from the deck to the ocean floor should be modeled. This enables evaluation of the spacing among the risers under loading and any possible collisions between the risers.

Many submerged structures are held in place by tensioned members. The tension leg platform is moored by tensioned risers, which are called tendons. The tendons experience large vertical loads at the heave natural periods (known as springing) as well as higher frequency transient loads (known as ringing). The springing and ringing characteristics of the tendon can be simulated in a model test reasonably well. The tendon weight is not of special concern here. Tensile or axial stiffness is of primary importance for modeling the tendon. Proper choice of a plastic material makes it possible to simulate the stiffness at a small scale. However, many plastic materials exhibit creep, which should be handled properly. In a small-scale test of a TLP model to investigate springing load of the TLP, the steel tendons were modeled with delron plastic material. The creep present in the delron material was stabilized by pre-loading the tendons in the dry for several days before testing. Sometimes, additional springs are introduced in the tendon system to account for scaled stiffness. Bending stiffness of tendons may also be of some importance if the structure undergoes large horizontal excursions. But it is difficult to model both axial and bending stiffness simultaneously. Special techniques may be introduced by external means to approach the axial and bending stiffness simultaneously.

Consider an example of the modeling of a riser system connected between a floating platform and the ocean floor in deep water [LeMéhauté (1976)]. The motion of the platform under wave action obeys the Froude similitude provided the mass and volume of the platform are distributed accordingly. Because of the depth limitation in laboratory testing facilities, scaling of the riser in deep water often requires truncation of the overall length of the riser. This requires a special similitude law for the riser model. Because of the scale effect due to distortion, the motion of the deeper section of the riser will not be in similitude. On the other hand, the area of

Chapter 4 Experiments in Hydrodynamics and Vibration

intersection of the riser and the platform is of primary interest where the forces exerted on the riser will be the highest. The moment at the junction of the riser to the platform is given by the sum of the bending and hydrodynamic loads at this point:

$$M = \sqrt{TEI}\,\vartheta + M_H \qquad (4.29)$$

where T = tension in the riser, EI = riser stiffness, ϑ = angle of riser and M_H = moment due to the hydrodynamic force.

Inertial forces due to the fluid are in similitude as long as the outside diameter is scaled according to geometric scale. By necessity, the cross-section of the riser will probably be modeled by a solid rod. The mass of the riser is modeled such that the density of the model equals the average density of the prototype riser, including the steel annulus and the contained water:

$$\rho_m A_m = \rho_W A_W + \rho_R A_R \qquad (4.30)$$

where ρ = density, A = cross sectional area, and the subscripts W and R stand for the water and riser respectively. This will ensure proper inertia of the riser.

The drag force will generally not be in proper similitude due to the Reynolds effect. However, this effect is small compared to the inertia force near the water surface. The Cauchy similitude is ensured if the following relationships are satisfied. From the Froude similitude, one gets

$$M_p = \lambda^4 M_m \qquad (4.31)$$

$$\vartheta_p = \vartheta_m \qquad (4.32)$$

Therefore, the following relationship holds for the Cauchy similitude:

$$(TEI)_p = \lambda^8 (TEI)_m \qquad (4.33)$$

In order to determine the relationship for the elasticity between the model and the prototype riser, the expressions of T and I for the model and the prototype are examined. It is assumed that the hollow riser is modeled with a solid rod. The moment of inertia I for the model riser of radius R_m is

$$I_m = \frac{\pi R_m^4}{4} \tag{4.34}$$

The moment of inertia of the hollow prototype riser may be approximated as

$$I_p = \pi R_p^3 \Delta R_p \tag{4.35}$$

where R_p = outer radius of riser and ΔR_p = thickness of the annulus. Similarly, the tensions are expressed as

$$T_m = \pi R_m^2 E_m \frac{\Delta l_m}{l_m} \tag{4.36}$$

and

$$T_p = 2\pi R_p \Delta R_p E_p \frac{\Delta l_p}{l_p} \tag{4.37}$$

where $\frac{\Delta l}{l}$ is the relative elongation. Note that for similitude, $\frac{\Delta l_m}{l_m} = \frac{\Delta l_p}{l_p}$.

Substituting the above relationships:

$$E_p = \lambda \frac{R_p}{\Delta R_p} \frac{E_m}{2\sqrt{2}} \tag{4.38}$$

Note that if the riser tension models as a Froude model such that $T_p = \lambda^3 T_m$, then the modulus of elasticity between the model and prototype are related by

$$E_p = \lambda \frac{R_p}{2\Delta R_p} E_m \tag{4.39}$$

The first condition (Eq. 4.38) appears to be more appropriate for the reaction of the platform upon the riser. It also approximates well the motion of the riser in similitude. In this case, the quantities such as the inertia forces, bending forces and tension are in approximate similitude from the platform to the bottom.

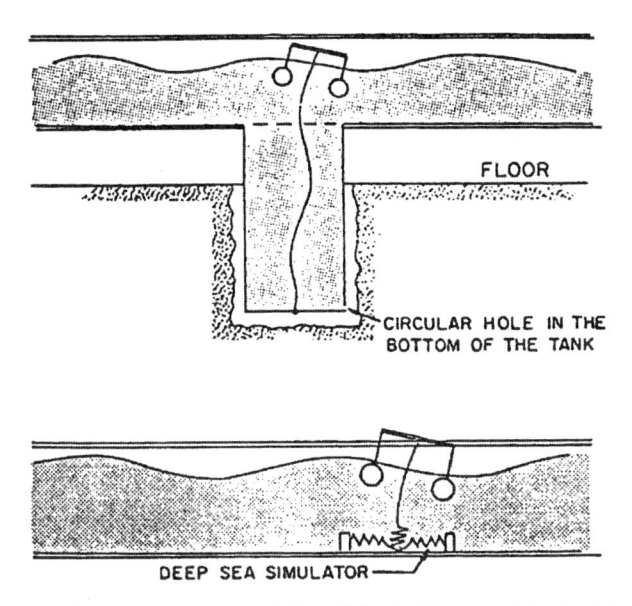

Fig. 4.10: Distortion of Deepwater Riser Model Due to Limited Basin Depth
[after LeMéhauté (1976)]

Consider a scale factor $\lambda = 200$ for the riser. At this scale for a steel riser, a wire of polytetra-flurethylene with a relative density of 2.2 and $E = 4 \times 10^5$ kPa (0.58×10^5 psi) is appropriate for the similitude requirements. Even at this scale the riser in a model basin may have to be distorted for very deep water (see Fig. 4.10). In Fig. 4.10 two such modeling schemes are shown. In the first case a deep section in the tank helps incorporating longer riser model, while in the second case, the truncated riser is modeled with additional springs. If the length of the riser is distorted so that

$$l_p = \lambda l_m \delta \tag{4.40}$$

where δ is the distortion factor. Then considering $\Delta l_p = \lambda \Delta l_m$, the modulus of elasticity is related by

$$E_p = \lambda \frac{R_p}{\Delta R_p} \frac{E_m}{2\sqrt{2}} \sqrt{\delta} \qquad (4.41)$$

For a given distortion factor, a model material of proper elasticity should be chosen to satisfy Eq. 4.41. This example illustrates the complexity of modeling small diameter long elastic members in limited water depth.

4.6 Vibration Instruments

In studying vibration of structures, one measurement is of utmost importance, namely, displacement of the structure, or its acceleration. Often the displacement is not measured directly, but is derived numerically (adjusting any offset) from the measured acceleration. In addition, it is important to know the excitation loads and the reaction forces for a system. These quantities allow one to assess the possibility of damage to the structure and take remedial measures.

The following standard instruments [see Goldstein (1983) for a detailed discussion] are often required during a model test to measure motions and vibrations of the structure model and the associated hydrodynamic loads:

- Linear and angular potentiometers
- Six D.O.F. mechanical motion system
- Optical motion tracking system
- Single axis strain-gauged load cells
- Three axis block load cells
- Accelerometers
- Pressure gauges
- Rotating dynamometer

These instruments are electronically connected to an automatic data acquisition system (DAS) so that the transducer signal may be automatically recorded. A simple schematic of a data acquisition system is shown in Fig.

Chapter 4 Experiments in Hydrodynamics and Vibration 159

4.11. The typical transducer signal is such that its output is given in microvolts. The low signal level is first amplified by an amplifier gain factor to a 0-5 volt or 0-10 volt limit. The signal is then conditioned, which may include analog filtering of noise and other unwanted signals (such as 60Hz electrical noise), and converted from the analog to digital form through an A/D converter. The digital signal is then stored into a computer memory for later viewing and analysis. Today these operations may be accomplished efficiently on a desktop personal computer. DAS cards are commercially available for insertion into the PC board slots.

```
Response → TRANSDUCER → AMPLIFIER & SIGNAL CONDITIONER → A/D CONVERTER → COMPUTER DATA BUS
```

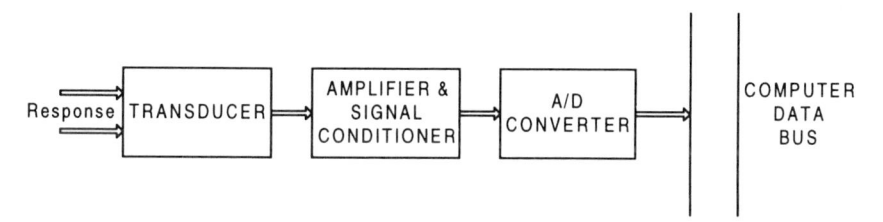

Fig. 4.11: Schematic Data Acquisition System

For a model subjected to a model environment, transducers receive a physical input from the model response such as displacement, acceleration, force, etc. and produce an equivalent electrical output. The transducer is designed so that this transformation from the measured response to volts is in the linear range for the level of response expected, which allows a single scale factor for conversion of the output to the required engineering unit. A few common means of measuring an input signal include a bonded strain gauge, a linear variable differential transformer (LVDT), a diaphragm pressure sensor, a sonic sensor from the reflecting surface and a resistance or capacitance probe. These components are placed on a finished transducer block, which is designed to measure an expected response in a model test.

For example, the strain gauge is glued strategically on a tension/compression member of a load cell designed for the desired load range. The load cell is attached between the model and a rigid mounting system. As the model is subjected to fluid flow, the load imposed by the fluid on the model is recorded by the load cell. After construction and waterproofing, these instruments are placed on a specially designed calibration stand and calibrated over the range of expected values. For example, the load cell is fixed on the calibration stand and known weights are hung in the direction of measurement from the load cell in increments, and the associated voltages are recorded. In the case of a capacitance wave probe, the calibration is achieved by placing it perpendicular to the water surface and moving it up and down in equal incremental steps in the water. The linearity of the instrument is verified and a scale factor in terms of the response unit per volt is generated. This factor is used to multiply the voltage output during the testing inside the DAS.

Testing of ultra deep-water structures require small scales. Instruments designed for the measurement of responses of a structure at a small scale may pose a problem due to their size compared to the model. However, many small precise and reliable instruments are available today. The measurement accuracy or instrument sensitivity at a small scale, even down to 1:200, is not a serious problem. The generally accepted overall measurement error in the complete setup at a reasonable scale of 1:50 to 1:100 is less than 5 percent. At a much smaller scale (smaller than 1:100), this measurement error may increase substantially in spite of the current level of precision in instrumentation. For small scale testing, this measurement error must be recognized and considered in the correlation and extrapolation of data to the prototype. Additional inaccuracies in the system may be introduced from superfluous physical phenomena not present in a larger scale model or the prototype. For example, amplified turbulence in the flow, or the capillary free-surface effect is exaggerated in smaller scaled testing.

4.6.1 Accelerometers

It is a common practice to measure the acceleration of various components of a vibrating system. Accelerometers are particularly suitable for high frequency vibrations. They are commercially available and relatively inexpensive. There are two types of accelerometers: closed loop type and open loop kind. Closed-loop systems appear to have levels of accuracy, stability and reliability that are several orders of magnitude greater than for open-loop accelerometers. These compact units come in single and multiple components and may be mounted directly on the model at the desired location. For underwater application they should be encased in a watertight box.

4.6.2 Motion Sensors

There are numerous methods of detecting and recording motion, but many have a tendency to affect the motion that they are meant to measure. Some methods measure displacement (e.g., linear transducers); other methods measure rotation (e.g., gyroscope). Many of these transducers are commercially available, with or without built-in amplifiers. In the past, the motions of a structure were often measured by attaching displacement transducers such as potentiometers directly to the body of the structure.

The mechanical system of motion measurement uses linkages and linear and angular potentiometers attached between the overhead carriage structure and the model. This system works well and is quite reliable for a large structure model. However, if this system is employed, it is important that its effect on the structure in terms of the overall inertia and damping introduced by the mechanical system be investigated.

In many cases the attachment of these sensors affects the motion of the structure, particularly when the structure is small. In particular, the friction in such a system may be detrimental to the motion of the model. In order to measure the motions of a model without physical connection to the body, an optoelectronic remote motion monitoring system is used. There are many such systems commercially available today. Examples are SELSPOT or OPTOPOS. The optoelectronic movement monitoring system uses small light-emitting diodes (LEDs) or similar sensors to identify selected points on the moving model. The LEDs emit infrared light sequentially, and the output from each camera is demultiplexed, sampled and stored to provide two analog voltage outputs per camera. These outputs represent the tangent of the vertical and horizontal displacement angles taken from the camera line of sight axis. The detector simultaneously measures the position of several sensors in sequence. The use of one camera will monitor positions in two dimensions, whereas, two cameras permit three-dimensional monitoring. Using two cameras and a minimum of three LEDs attached strategically to the structure, it is possible to obtain data that allows the calculation of the six motions of a model by software. A fourth sensor is generally added to the system so that the calculated motions at this sensor may be overlaid on the measured displacement for verification of the systems' functional accuracy. In operation, tiny infrared light emitting diodes are positioned and attached to a model in such a way that they are visible to two infrared sensitive cameras. In Fig. 4.12 four LEDs may be seen on the deck of a semisubmersible model. They are mounted on sticks of unequal lengths in order to avoid shielding.

Fig. 4.12: Typical Setup of an Optoelectronic Motion Measuring System

Typically, the accuracy is better than 2 percent. There are limitations to the use of this system. The LEDs work in the dry so that they must be placed above the free surface. The two cameras have a field of vision which, when combined, form a cubic region in space. If the sensors move out of this cubic region, the cameras can not record their position. When the sensing elements are placed farther away from the cameras, this region becomes larger, but reduces the measurement accuracy. Sunlight and reflections of the lights off the water or other reflective surfaces can create spurious readings. The sensors must be positioned on the structure such that they are never shielded from the camera by the movement of the structure or by one another. These are not serious drawbacks to the optoelectronic system in a controlled laboratory environment. The field of vision and the resulting accuracy are easily calculated from the calibration tables. The ambient light can be controlled, and all reflections may be negated by camera placement and shielding techniques. The location of each light is selected to maintain line of sight with both cameras. This system has been employed quite successfully in model basins. An example of a semisubmersible barge tested in waves with this system is shown in Fig. 4.12. It was moored in head seas by a matched pair of linear springs fore

Chapter 4 Experiments in Hydrodynamics and Vibration 163

and aft such that it had a natural period in surge of approximately 30 seconds. Figure 4.13 demonstrates typical results from the regular wave test. A ring gauge measured the bow mooring line load. A capacitance wave probe and the SELSPOT system measured the wave and surge motion, respectively. The mooring line load was used to verify the accuracy of the surge motion, since the linear spring constant of the springs in the lines were known.

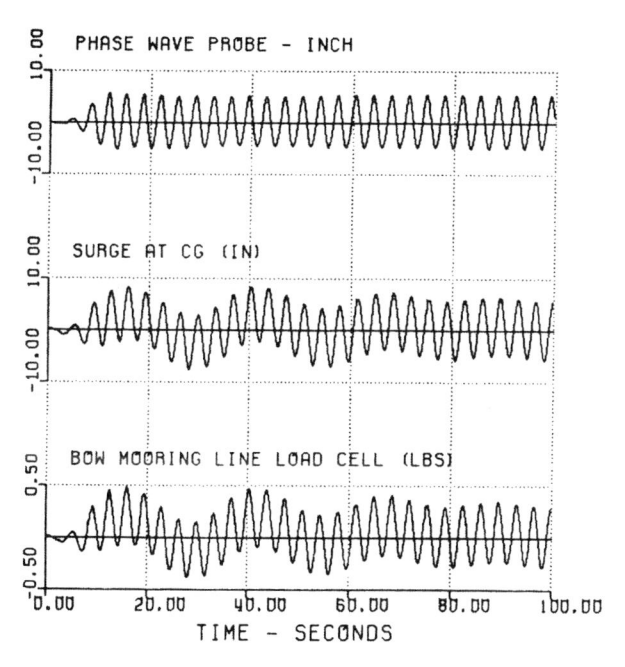

Fig. 4.13: Measured Time History by an Optoelectronic System

4.6.3 Load Measuring Device

In testing the vibration of a structure model subjected to fluid flow, it is often desirable to measure the environmental loads on the entire structure or a portion thereof. These measured loads are employed in the design and operational requirements of these structures. The type and design of these load measuring instruments depend on the model structure and the intended use of the instrument [Chakrabarti (1994)]. Submerged pipe and riser models in current are often directly strain-gauged on their surface. In fixed offshore structure testing, such as jackets and production platforms, the overall loads and overturning moments are measured by supporting the structure model on load cells specially designed for this purpose. For ship or marine structures as well as in transportation of offshore structures, the powering requirements are determined by towing the model at required speeds while attached to an instrumented towing staff, which measures the towing load. For testing floating moored structures, the mooring lines are instrumented with load cells. Normally, single component strain gauged ring type load cells are suitable for this measurement. For a submerged structure placed in a moving fluid, it is often desirable to measure local loads on an element of the structure. In this case, the element is suspended on the load cells to separate it from the remainder of the structure. In maneuvering tests of a submarine model, in which the model is moved in water by radio control, it is a common practice to measure loads on the planes and rudders as well as the propellers. The design of the dynamometers for these measurements is a difficult task due to high load requirements and propeller revolution coupled with limited available space.

There are two common types of load measuring devices. One uses strain gauges to measure the strain in the load cell element caused by the applied load. The other measures the relative displacement of two parts of the load cell element. In the strain-type load cell, the material of the active element is selected to have a linear stress-strain characteristic in its elastic range. The physical size of the element is designed to ensure that the material remains in its elastic range at the maximum expected loading.

For the strain type load cell, it is desirable to use bonded strain gauges. A bonded strain gauge consists of a continuous grid or filament of very small diameter wire or thin metallic foil mounted on paper or plastic. The filament has the property of linear variation of electrical resistance with strain. The bonded gauge is glued directly on the active member of the load cell. Sometimes, multiple gauges are used on one or several members of the load cell. The gauge or gauges become the electrical arms of a Wheatstone bridge.

This type of load cell requires elaborate waterproofing for the underwater use. Provisions should be made in the design for the application of this waterproofing material. This is particularly important if the space available for fitting these cells is limited, such as inside the stern planes or the propeller hub of a submarine model. Since the strain-gauged area is often reduced in size to make the cell more sensitive to the applied load, there is sufficient room for applying the necessary waterproofing on the gauges. In potting multi-component dynamometers, it must be ensured that the potting material does not increase the stiffness when spanning voids between flexures, substantially reducing its sensitivity.

In the displacement type load cell, one element of the load cell acts as a spring and is connected to a relatively fixed element through a position (or displacement type) transducer. It is ensured in this case that the deflection remains linear within the range of measured forces. One of the displacement type transducers is a Linear Variable Differential Transformer (LVDT). The LVDT is available commercially in a very compact form and is often hermetically sealed requiring no further waterproofing.

4.6.4 Design of a Load Cell

In testing small diameter members, e.g., risers, it is advantageous to be able to measure the nature and extent of stresses that exist at a particular point of interest in the member as a result of external forces. For the instantaneous measurement of stresses, it is convenient to use strain gauges. By suitable calibration of the strain gauges, the stress, which exists in the member at the location of the strain gauge, can be determined.

A simple strain gauge, however, does not distinguish between the various types of stresses, e.g., tensile, shear, or bending, which can exist in the member. For example, a strain gauge can not distinguish between a purely axial tensile load and a bending moment, each of which may produce a similar magnitude of deformation in the strain gauge. In order to permit a complete analysis of the stresses, which exist in the test member, it is desirable to distinguish among such stresses, and to measure selectively as well as instantaneously a particular type of stress.

A separate load cell may be independently designed for this purpose, and then inserted into a (cylindrical) member at a point in the structure where the stresses present in the member are to be measured. Such load cell (Fig. 4.14) will typically comprise of a pair of spaced parallel plates, each of which is adapted to be connected to a segment of the member. The spaced plates of the load cell are in turn interconnected by a plurality of spaced legs perpendicular to the plates and formed of a suitable rigid but elastic material. Base plates should be of sufficient thickness so that they do not deform appreciably under the design load. Legs are formed of a rigid but elastic material, preferably metal. The length and cross section of the legs should be adequate to permit the legs to withstand the forces applied to the load cell without being stressed beyond the elastic limit of the leg material. The legs are preferred to have a square section such that gauges may be applied to flat surfaces. It also insures uniform response irrespective of the direction of the applied force.

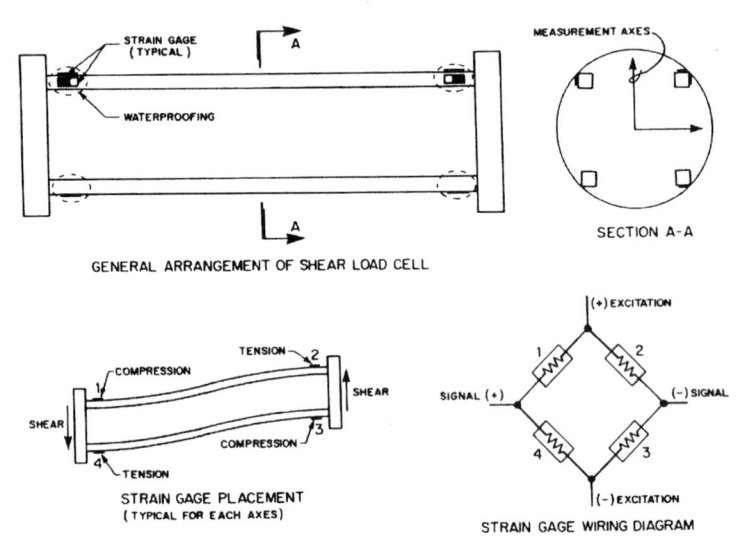

Fig. 4.14: Two-Component Shear-Moment Load Cell

Strain gauges are attached in interconnected diametrically opposite pairs at selected points along the length of two or more legs (Fig. 4.14), the number and location depending on the nature of the stress to be measured. For measuring shear stresses, strain gauges are placed near an end of the cell and on opposite sides of the leg. This compensates for the axial tension or compression in the leg, either from axial loads, weight or bending. By using the pair of strain gauges as one or two arms of a Wheatstone bridge (Fig. 4.14), the measured resistance of the circuit including the strain gauges is a measure of the stress applied to the load cell. Upon proper calibration, the load cell can be used to give an instantaneous measure of the magnitude and nature of the stresses created at the location of the load cell within the structural member. For a multicomponent load cell, it is desirable to design the cell that is equally responsive to shear and bending moment in any direction. The basis for design of this type of load cell is a fixed-end beam, with only end deflection (i.e., no end rotation).

To have equal resolution of shear and bending, the cell is designed with the stresses of the same magnitude. The shear and bending stresses [Timoshenko (1955)] are equated:

$$\frac{F}{8}(l-2x)\frac{6}{D^3} = \frac{M}{2sD^2} \qquad (4.42)$$

where F and M are the maximum shear load and bending moment respectively (Fig. 4.14), D is the size (dimension) of the cell beam which is usually of square section for multiple axis design. The quantities l and s are the length of beams and the spacing between them and x is the location of strain gauges from the end of beams for measuring shear loads. Assuming $x = D$, then

$$s = \frac{2MD}{3F(l-2D)} \qquad (4.43)$$

By assuming various combinations of beam size and length, an optimum value of the spacing s can be calculated from this equation.

Consider the example of a load cell, which was designed to measure bending moments up to about 270 inch-lb (30.5 Nm) and shear loads of up to about 8 lb (35.6 N). These magnitudes were determined from the computed expected response of the prototype and a chosen scale factor for the model. Of course, load cells are often designed with multiple future uses in mind. The design of the load cell yielded the following configuration. Circular base plates 4 inches (102mm) in diameter and 1/2 inch (13mm) thick were interconnected by four legs 2-3/8 inches (60mm) long symmetrically spaced on a 2-1/4 inch (57mm) diameter circle concentric with the base plate. The legs were square in cross section and 0.187 inches (4.7mm) in width. The load cell was machined from 6061-T6 aluminum. Commercially available strain gauges (Micro-measurement EA-13-062AQ-350 with option W) were attached in opposite pairs to the midpoint of each leg and also at a point spaced from an adjacent base plate by the width of the leg.

Static calibration of the load cell was accomplished by attaching the cell to an indexing table and applying various loads at a known distance from the cell. It is usually preferable to perform an initial calibration as a check after attaching the strain gauges and all required circuitry, but before applying the waterproofing. This allows for easier repair of any inconsistencies or defective areas without destroying the gauges or the cell.

After calibrating the load cell by the application of known torque, bending, shear and tensile loads as needed, the instrument is acceptable for measuring unknown loads within its range for a given accuracy. Interaction among different load directions, which is commonly known as cross-talk, should be within an acceptable limit.

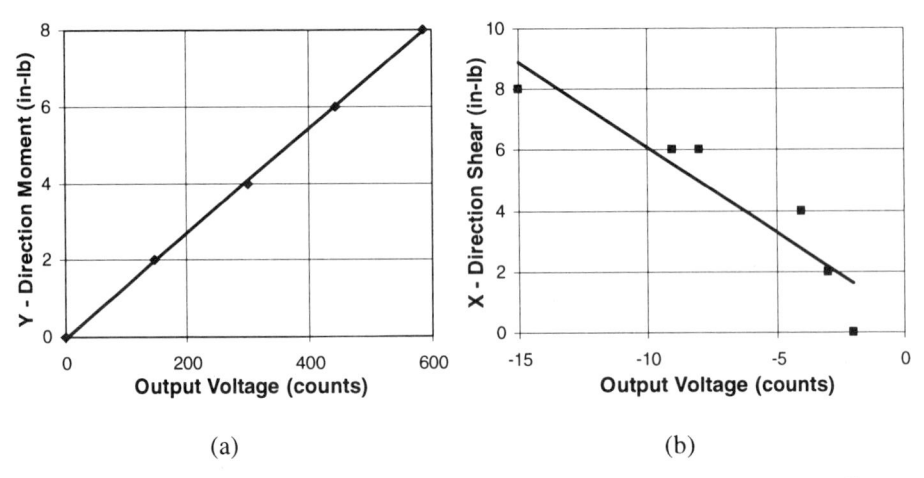

(a)

(b)

Fig. 4.15: Typical Calibration Curve and Cross Talk Due to Inline Loading

A typical calibration curve in bending is shown in Fig. 4.15. The y-axis is the applied bending moment in inch-lb (Fig 15a) and the x-axis represents the measured voltage. The interaction effect (Fig. 4.15b) in shear is on the order of 2 percent.

4.6.5 Design of a Propeller Dynamometer

The propeller dynamometer is designed to measure the hydrodynamic loads on a rotating propeller in a free-running radio controlled model of a submarine [Chakrabarti and Libby (1995)]. The dynamometer can be used to study the performance of any type of propeller or rotating equipment. The dynamometer must be designed to be very compact in shape and size so that it can be contained completely within the hub of the propeller. The propeller is supported entirely by the dynamometer so that all forces and moments acting on the propeller are transferred to the propeller shaft through the dynamometer. The dynamometer is instrumented with strain gauges. Six simultaneous voltage outputs are produced that can be used to determine the applied forces and moments on three orthogonal axes referenced to the propeller's centerline. The voltages are read from the dynamometer through sliprings (in order to maintain electrical contact during rotation) at the driven end of the propeller shaft.

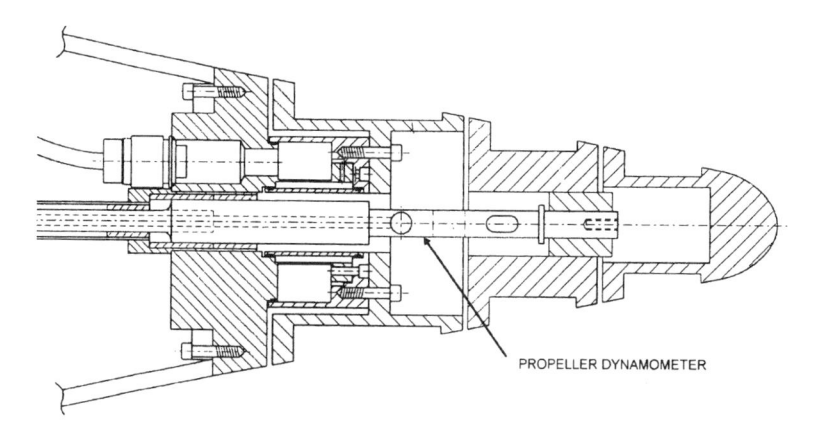

Fig. 4.16: Submarine Tail Assembly Showing Propeller Dynamometer

A dynamometer for the propeller vibrational loads of a submarine model was designed. The design details for the propeller dynamometer are shown in Fig. 4.16. During calibration of the propeller dynamometer small amounts of cross axis sensitivity among the six axes were measured. This cross-talk appeared to be linear for the larger values for which corrections could be made in the cross-axis direction. The six-by-six correction matrix generated during calibration of a six-axis propeller dynamometer is shown in Table 4.3. The diagonal terms represent scale factors close to one representing direct axis calibration. The off-diagonal terms are the cross axis corrections and are generally small. The larger off-diagonal calibration factors could be used during data reduction to reduce the error due to crosstalk.

Table 4.3: Correction Matrix for Propeller Dynamometer

IN/ OUT	X Force	Y Force	Z Force	X Moment	Y Moment	Z Moment
X Force	0.99996	0.019752	0.014351	-0.005378	0.032327	-0.050366
Y Force	-0.04238	0.999993	-0.00628	0	0	0.426764
Z Force	-0.10192	0.006559	0.999995	0	-1.458944	-0.007261
X Moment	0.189485	0.001987	-0.00525	0.999952	0.017058	-0.072731
Y Moment	0.072910	0.005428	-0.01586	0	0.999998	0.034852
Z Moment	0	0	0.003558	0	-0.012338	0.999992

4.6.6 Waterproofing of Transducers

The surface preparation, gauge bonding and waterproofing procedures are usually outlined in detailed step-by-step procedures by the strain gauge manufacturer.

Some types of waterproofing material used on a transducer are very corrosive to unprotected strain gauges before the material is fully cured. The manufacturer of strain gauges recommends that the gauges be protected by a layer of Teflon tape if corrosive materials are used. It is still possible for some corrosive vapor to come in contact with the gauge. It is sometimes advisable to coat the gauges with a layer of microcrystaline wax before the final waterproofing is applied.

An alternative to waterproofing, which avoids the preceding problem, is to seal the dynamometer in a pressure compensated housing filled with a non-conductive fluid such as oil. Submerged sliprings are often protected this way. The compensation mechanism is often arranged to make the internal pressure of this liquid slightly higher than ambient plus the dynamic so that small leakage is tolerable. However, any oil leaked in the water may cause a multitude of problems in the operation of the model test, e.g., contaminating other instruments, changing the characteristics of the fluid flow interaction, etc. Therefore, leakage properties of the dynamometer should be checked beforehand under pressure.

4.6.7 Excitation Level for Transducers

The magnitude of the output voltage of a strain-gauged transducer is directly related to its excitation voltage level. Usually, transducers are excited at a constant voltage on the order of 5 or 10 volts. Excessive excitation voltages can lead to instability (drift) of the transducer output, or even gauge failure. Strain gauges dissipate the power applied to excite them by producing heat. The selection of the proper excitation voltage level for a particular strain gauge involves several factors, including the size of the gauge's grid, the heat transfer characteristics of the material that is being gauged and the required output voltage level. In general, the excitation voltage is selected to be as high as possible while still meeting the gauge manufacturer's specification for the maximum allowable power dissipation for the particular application. Gauges used for long term measurements of static strains would typically be powered at a lower level than gauges being used to collect short term dynamic data like the submarine plane lift, drag and torque transducers. The lift and drag gauges used on the plane transducers are powered at the upper limit of the acceptable range so that they produce as high an output as possible. In this application, the excitation level is limited by the heat transfer characteristics of the thin web of transducer material where the gauges are adhered to the transducers. The gauge manufacturer's technical bulletins list long term drift in the gauge's output as a symptom of overpowering.

4.7 In-Situ Vibration Measurements

Often, components of prototype structures placed in service are instrumented to monitor their loading conditions. Two spar type production platforms producing oil and gas in the Gulf of Mexico having a number of production and export risers were equipped with instrumentation to assess their tension and bending loads. The instrumentation is expected to provide valuable information on the stability of the riser, its ability to resist maximum loading conditions and its fatigue life. One of them is a load measurement system described by McCarthy, et al. (2000). Similar systems are in service on five TLPs and a spar.

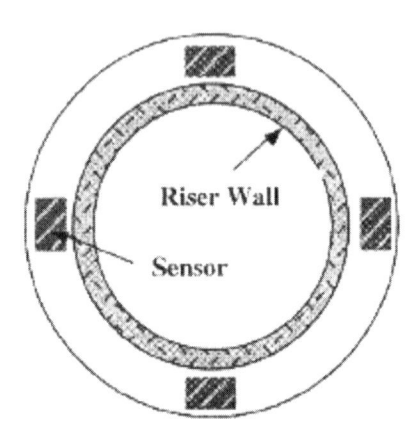

Fig. 4.17: Sensor Layout in the Load Measuring Unit [McCarthy, et al. (2000)]

The loads on the field risers are measured with load units placed in the risers. The riser tension measurement unit is part of a pup joint that is bolted into the riser. The sensors are mounted to a pair of rings that are welded to the exterior of the riser, as shown in Fig. 4.17. A concentric cylinder, welded to the riser at one end and to the lower sensor ring on the other end, is used to increase the gage length of the load cell. As the riser stretches under applied load, the sensors measure the extension between the rings. Four sensing elements measure the axial load, bending load, and the orientation of the bending load.

Using laboratory tests and in-service measurements, the tension measuring system configured with four sensors per load unit [McCarthy, et al. (2000)] can be used to measure dynamic bending moments at a sample rate of 10 Hz with sufficient accuracy to detect very low amplitude dynamic loads. With approximately 1/2 hour of data, a 0.5 ksi stress due to either dynamic tension or bending loads can be measured to approximately 2% accuracy, assuming that the vibration is stationary over the measurement interval. This accuracy estimate includes inaccuracies in the estimates of the axial and bending stiffness of the load cells.

4.8 Exercises

Exercise 1

Design a riser model for a 21-inch drilling riser. Assume that the riser is a uniform hollow steel pipe of 1-inch thickness. Choose a suitable scale factor and scale the axial stiffness in the model using Froude scaling.

Exercise 2

Consider testing a model of a submerged pipeline in a steady flowing current. The size of the pipe section to be modeled is 20m long by 1m diameter. The current speed is 0.5 m/s. The weight of the concrete coated pipe is 50 kg/m. Using Froude scaling and a scale factor of 1:20, design the model setup.

Exercise 3

Design a single component load cell that is capable of measuring tensile/compressive loads of up to 100 lbs. Limit the overall size of the cell to 6 inches. Compensate for any bending moment in the tension load measurement.

4.9 References

1. Chakrabarti, S.K., Offshore Structure Modeling, World Scientific Publishing Co., Singapore, 1994, 470 pages.

2. Chakrabarti, S.K., and Libby, A.R., "Design and Application of Force Dynamometers", Proceedings of American Towing Tank Conference, College Station, TX, 1995.

3. Doodson, A.T., "Tide Models", Dock and Harbour Authority, Vol. XXIX, No. 339, Jan. 1949.

4. Goldstein, R.J., <u>Fluid Mechanics Measurements</u>, Hemisphere Publishing Corporation, Washington, D.C. 1983.

5. LeMéhauté, B., "On Froude Cauchy Similitude," Proceedings on Specialty Conference on Coastal Engineering, Santa Barbara, CA, ASCE, Oct. 1965.

6. LeMéhauté, B., "Similitude in Coastal Engineering," Journal of the Waterways Harbors and Coastal Engineering Division, ASCE, Vol. 102, No. WW3, Aug., 1976, pp. 317-335.

7. Littlebury, K.H., "Wind Tunnel Model Testing Techniques for Offshore Gas/Oil Production Platforms", Proceedings of Thirteenth Annual Offshore Technology Conference, Houston, TX, 1981, pp. 99-103.

8. McCarthy, D.J., Madden, R., Coney, W.B., "Instrumentation for the Measurement of Tension and Bending Moments in Risers", Proceedings of ETCE/OMAE2000 Joint Conference, ASME, 2000.

9. Timoshenko, S.P., <u>Strength of Materials</u>, Third Edition, D. Van Nostrand Co., Inc., Princeton, N.J., 1955.

Chapter 5

Statistical Theory in Vibration

5.1 Introduction

So far, discussions have been limited to the deterministic aspects of vibration. In these cases, the excitation is simply a harmonic or finite multi-frequency function. Many examples have been cited where such an analysis is applicable. However, in practice, the excitation is often random. This is particularly true in areas such as mechanical vibration of machines, aerodynamics, and vibrations in fluid flow. Types of excitation such as those created by machines, impact loading, earthquakes, wind, or waves cannot be predicted deterministically. In fact, a random process is one whose instantaneous value is not predictable. Unlike the deterministic systems described earlier, there is no well-defined periodicity in a random process, and the value of a random variable at one instant of time is not related to that at any other instant, and cannot be predicted by simple means.

In these cases, one relies on the probability theory. Although the time variation of a random process cannot be predicted, the probability of the system experiencing a prescribed value being within a certain range is predictable on a statistical basis. Thus, there is no single maximum value of a process similar to a deterministic process, but rather, each extreme value is associated with a certain probability. The probability of 'one' provides an extreme value equal to infinity for any random system. However, for a probability of less than one, the extreme value of the response of a structure has finite amplitude. Thus, a structure is designed based on an extreme response value for a given probability level. The probability level is determined based on the prescribed design life of the structure.

There are many statistical theories that describe the probabilistic nature of a random process [see, for example, Chakrabarti (1990)]. It is not the intent here to cover all such theories. Instead, the applicable theories that are used in the analysis of the random vibration of a hydrodynamic system are discussed here. It is intended that this chapter will help the reader in understanding the subsequent chapter on Random Vibration without the requirement of a supplementary textbook. Thus, it is the intent of this chapter to make the book self-contained.

5.2 Random Variables and Processes

While the simple harmonic motion from a simple vibrational system is deterministic, the environment imposed on a physical system is quite often random in nature. This gives rise to the response of a system which is random. Random processes by nature are not deterministic and are functions of the time variable. They are often governed by probability laws that can be described mathematically. The random processes that can not be described by a probability law are considered chaos. In recent days, the so-called chaos is also being analyzed to find systematic ways to identify them, which is establishing itself as the chaos theory.

If one considers random samples of a system taken at different times, then the collection of these time histories describing the same phenomenon is called an ensemble. If one attempts to describe the random process by a statistical theory, then all of the time histories should follow this given statistical theory. There are two assumptions that are often imposed on a random process so that it can be described by a probability law. These are stationarity and ergodicity. The statistics of a stationary process are time invariant. This means that once the statistics are established, all random processes for the system may be described by the particular probability law. Thus, a random signal, $x(t)$, as a function of time t is considered stationary if the statistical properties of the variable x are independent of the origin of time measurement.

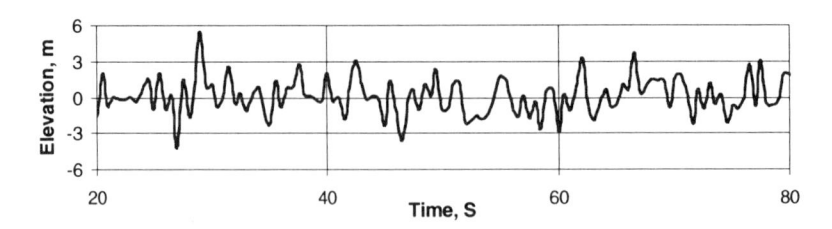

Fig. 5.1: Sample Record of a Random Process

Chapter 5 Statistical Theory in Vibration

Consider an ensemble of records given by the process $x(t)$. A sample of the record is shown in Fig. 5.1. The mean of the process $x(t)$ is defined as

$$\mu_x = E[x(t)] \tag{5.1}$$

where the notation $E[\,]$ denotes the (expected) average value of the quantity inside the brackets. For a sample $x(t)$ of record length T_s, the mean value is obtained from the integral:

$$\mu(x) = \lim_{T_s \to \infty} \left\{ \frac{1}{T_s} \int_0^{T_s} x(t)dt \right\} \tag{5.2}$$

which reduces to a summation for N samples x_i of a finite record length $T_s = N\Delta t$ at a discrete sampling rate of Δt:

$$\mu(x) = \frac{1}{N} \sum_1^N x_i \tag{5.3}$$

The standard deviation or the root mean square value of the process is defined as

$$\sigma_x = \sqrt{Var[x(t)]} \tag{5.4}$$

where *Var* represents the variance of $x(t)$. The standard deviation (i.e., the root mean square) of the variable $x(t)$ about the mean value μ_x is computed from the following integral:

$$\sigma(x) = \lim_{T_s \to \infty} \left\{ \frac{1}{T_s} \int_0^{T_s} [x(t) - \mu_x]^2 dt \right\}^{1/2} \tag{5.5}$$

in which the term inside the bracket is the variance *Var* about the mean value. For a record of finite length, this expression for an estimate of standard deviation reduces to

$$\sigma(x) = \left\{ \frac{1}{N} \sum_1^N (x(t) - \mu_x)^2 \right\}^{1/2} \tag{5.6}$$

where the mean value is obtained from Eq. 5.3. A coefficient of variation is a dimensionless measure of the variability of the record about the mean value, and it is defined as

$$Cov_x = \sigma_x / \mu_x \tag{5.7}$$

This quantity is a good measure of the scattering of experimental data about an expected value. The larger the value of Cov_x, the larger is the variation.

The auto-correlation function is obtained as

$$R(t_1, t_2) = E[x(t_1)x(t_2)] \tag{5.8}$$

The integral representation of the auto-correlation function becomes:

$$R(\tau) = \lim_{T_s \to \infty} \left\{ \frac{1}{T_s} \int_0^{T_s} x(t)x(t+\tau)dt \right\} \tag{5.9}$$

The process $x(t)$ is stationary if μ_x and σ_x are constant for all values of t, and R is a function only of $\tau = t_2 - t_1$.

If, in addition, the properties of $x(t)$ measured at different locations are invariant, it is also called homogeneous. In practice, the variable $x(t)$ may be stationary and homogeneous over a limited time and region. The variable $x(t)$ is considered ergodic if the measured realization of $x(t)$, say $x_1(t)$, is typical of all other possible realizations. Thus, a stationary process is ergodic if a time average of a single record in an ensemble is the same as the average across the ensemble. For an ergodic process, the sample mean, μ_x, from the ensemble approaches the mean and the sample variance, σ_x^2, approaches the variance of the process. These concepts of stationary, ergodic process are important in developing and applying short-term statistics of a system, even though such a hypothesis may not exist in reality. It allows one to derive the probability law of a process from the limited measured ensemble.

5.2.1 Probability Distribution Function

The probability distribution, or cumulative probability, as it is sometimes called, is defined as the fraction (or percentage) of time that a particular event is not exceeded. A probability distribution, $P(x)$, of a random

Chapter 5 Statistical Theory in Vibration 179

variable, $x(t)$, is a plot of the proportion of values less than a particular prescribed value (often considered the design value) versus the particular value of the variable x_0.

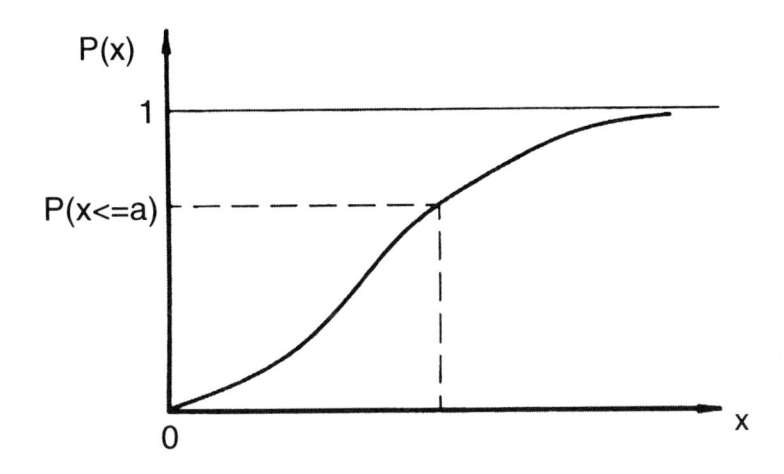

$$P(x) = prob[x(t) \leq x_0] \qquad (5.10)$$

Fig. 5.2: Probability Distribution Function

A typical probability distribution function of a random variable $x(t)$ is shown in Fig. 5.2. Note that it varies between the values zero and one and asymptotically approaches those limits.

5.2.2 Probability Density Function

The probability density function is defined as the fraction (or percentage) of time that a particular event is expected to occur. Thus, the probability density $p(x)$ is the rate of change of the distribution function and is obtained from the derivative of the distribution function:

$$p(x) = \frac{dP(x)}{dx} \qquad (5.11)$$

A typical density function is given in Fig. 5.3. The total probability of the event should be unity such that

$$\int_{-\infty}^{\infty} p(x)dx = 1 \qquad (5.12)$$

and the total probability for a particular value x is obtained as the integral

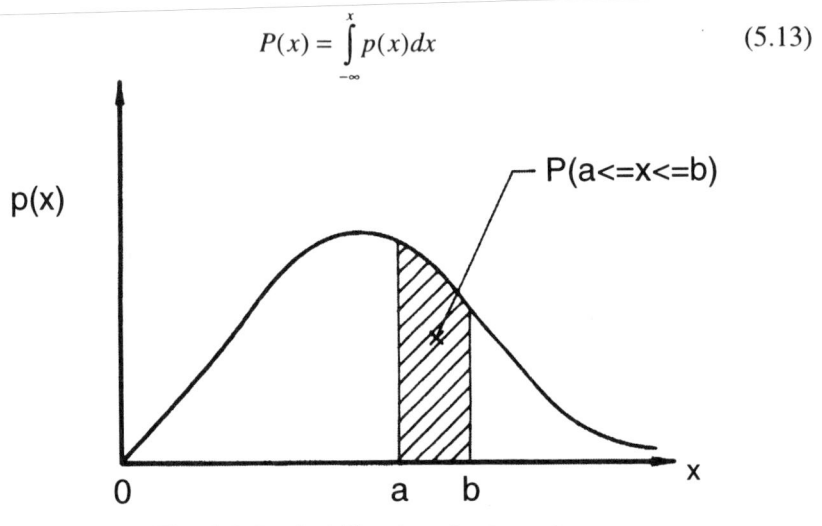

$$P(x) = \int_{-\infty}^{x} p(x)dx \qquad (5.13)$$

Fig. 5.3: Probability Density Function

The area between two prescribed values a and b under the probability density curve (Fig. 5.3) defines the probability that the result of the event will lie between the values a and b. Thus,

$$P(a \le x \le b) = \int_{a}^{b} p(x)dx \qquad (5.14)$$

In terms of the probability function, the mean value of the random variable is obtained by taking the moment of the density function about the origin, and dividing by the area of the density function (which is unity from Eq. 5.12):

$$\mu_x = E[x] = \int_{-\infty}^{x} xp(x)dx \tag{5.15}$$

Thus, μ_x may be considered as the distance to the centroid of the probability density function. Similarly, the variance of the random variable is computed from the density function as

$$Var(x) = \sigma_x^2 = E[(x-\mu_x)^2] = \int_{-\infty}^{x} (x-\mu_x)^2 p(x)dx \tag{5.16}$$

Therefore, the variance may be interpreted as the moment of inertia of the probability density function about its mean value. By expanding the right hand side of Eq. 5.16, and using Eqs. 5.12 and 5.15, one can show that the variance may be obtained from

$$\sigma_x^2 = E[(x^2)] - \mu_x^2 \tag{5.17}$$

5.2.3 Gaussian Distribution

One of the most common distribution functions that is often applicable in the random signal theory is the Gaussian distribution. The density function for this distribution is given by the formula

$$p(x) = \frac{1}{\sqrt{2\pi}\sigma_x} \exp\left[-\frac{(x-\mu_x)^2}{2\sigma_x^2}\right] \tag{5.18}$$

The Gaussian distribution and density functions are plotted in Fig. 5.4. Note that the density function is symmetric about the mean value μ_x of the random variable x. Thus, the probability of occurrence of a particular value of $x(t)$, namely, $\mu_x + \sigma_x$ is the same as the probability of a value at the same distance of the mean value on the opposite side, $\mu_x - \sigma_x$. The decay on both sides of the function is exponential, having a maximum value at the mean of $x(t)$. Thus, the density function represents a shape similar to a bell shape. A closed form expression of the distribution function is not available, and the computation of the distribution function is made using Eq. 5.13. Thus, the probability $P(x_1)$ of $x \leq x_1$ is given by the area of the density function between $-x_1$ and x_1 in Fig. 5.4.

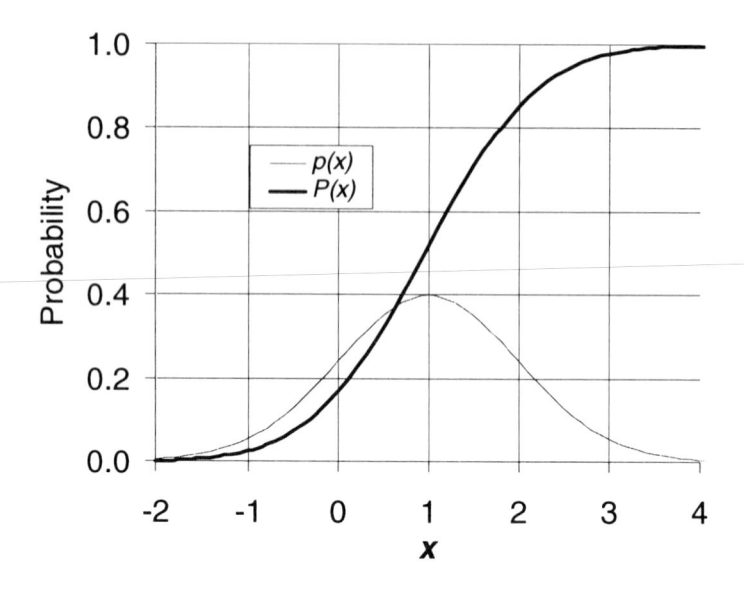

Fig. 5.4: Gaussian Probability Distribution and Density Function

This distribution is also called the normal distribution. From Eq. 5.18, it is seen that the normal distribution depends in general on the mean value and standard deviation of the variable $x(t)$. Therefore, a common notation for the normal distribution is

$$p(x) = N(\mu_x, \sigma_x) \tag{5.19}$$

When the mean value is zero and the standard deviation is unity, Eq. 5.18 reduces to

$$p(x) = \frac{1}{\sqrt{2\pi}} \exp\left[-\frac{x^2}{2}\right] \tag{5.20}$$

In this case the probability distribution function takes on the form of the well-known mathematical error function:

Chapter 5 Statistical Theory in Vibration

$$\Phi(x) = \int_{-\infty}^{x} \frac{1}{\sqrt{2\pi}} \exp\left[-\frac{\chi^2}{2}\right] d\chi \qquad (5.21)$$

In general, normalized $P(x)$ may be expressed as

$$P(x) = \Phi\left(\frac{x - \mu_x}{\sigma_x}\right) \qquad (5.22)$$

The values of $P(x)$ are tabulated for convenience for various values of mean and standard deviation of x and may be found in many statistical handbooks [see Lin (1967)].

5.2.4 Rayleigh Distribution

The probability density function for the Rayleigh distribution of a random variable $x(t)$ is written in terms of the mean value μ_x of x:

$$p(x) = \frac{\pi x}{2\mu_x^2} \exp\left[-\frac{\pi}{4}\left(\frac{x}{\mu_x}\right)^2\right], \qquad x > 0 \qquad (5.23)$$

The density function may be integrated in a closed form using Eq. 5.13 to obtain the expression for the distribution function, $P(x)$:

$$P(x) = 1 - \exp\left[-\frac{\pi}{4}\left(\frac{x}{\mu_x}\right)^2\right], \qquad x > 0 \qquad (5.24)$$

For the Rayleigh distribution, the standard deviation of the variable $x(t)$ is related to its mean value by the relation

$$\sigma_x = \sqrt{\frac{4-\pi}{\pi}}\mu_x = 0.523\mu_x \qquad (5.25)$$

Also, the mean value is $1.253 x_m$ where x_m corresponds to the variable at the maximum probability density. The density and distribution function for the Rayleigh distribution is shown in Fig. 5.5. For this plot, the mean value of $x(t)$ was taken as $\mu_x = 1$, meaning that the random variable $x(t)$ is normalized by μ_x. For the Rayleigh distribution to apply, the values of the random variable $x(t)$ are always positive.

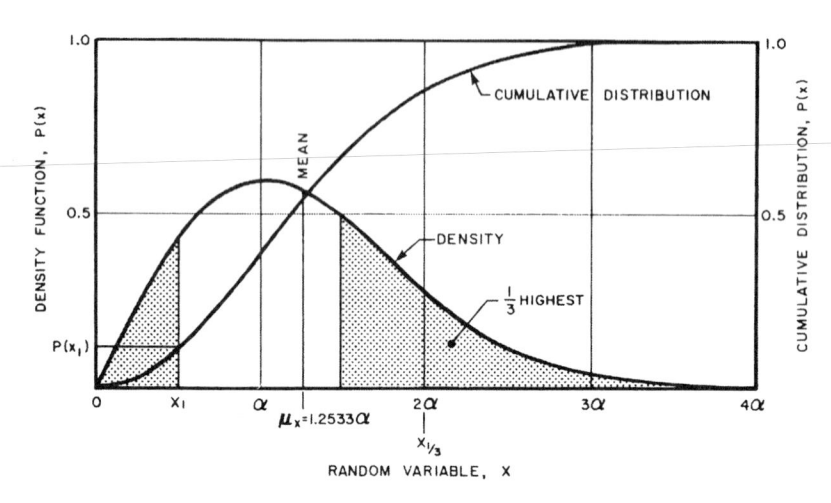

Fig. 5.5: Rayleigh Probability Distribution and Density Function

5.2.5 Weibull Distribution

The probability distribution function for the Weibull distribution of a random variable $x(t)$ is written in terms of constant parameters α and β:

$$P(x) = 1 - \exp\left[-\left(\frac{x}{\beta}\right)^\alpha\right], \qquad x > 0 \qquad (5.26)$$

This is commonly known as a two-parameter Weibull distribution. The probability density function may be obtained by differentiating Eq. 5.26:

$$p(x) = \frac{\alpha}{\beta}\left(\frac{x}{\beta}\right)^{\alpha-1} \exp\left[-\left(\frac{x}{\beta}\right)^\alpha\right], \qquad x > 0 \qquad (5.27)$$

The mean value of the distribution function may be found from the relation:

Chapter 5 Statistical Theory in Vibration

$$\mu_x = \beta \Gamma\left(\frac{1}{\alpha}+1\right) \tag{5.28}$$

where Γ is the Gamma function [Abramowitz and Stegun (1964)].

In order to fit a random process to the Weibull distribution, the probability level, P of the partitioned random data, $x(t)$ is established first. The values of the parameters α and β may then be evaluated by a least square technique using the relationship of a straight line for the Weibull distribution:

$$\ln\left[\ln\frac{1}{1-P}\right] = \alpha \ln x - \alpha \ln \beta \tag{5.29}$$

There are other types of Weibull distribution involving more parameters. However, the two-parameter Weibull distribution is more commonly used in the vibration analysis.

5.2.6 Gumbel Distribution

The Gumbel distribution [Gumbel 1958)] is an exponential type distribution that is considered an extreme value distribution of maxima. It is unbounded and grows in a logarithmic scale. It is given in terms of two parameters, α and β:

$$P(x) = \exp[-\exp(-\alpha(x-\beta))], \qquad x > 0 \tag{5.30}$$

The probability density function is the derivative of the distribution function:

$$p(x) = \alpha \exp(-\alpha(x-\beta))\exp[-\exp(-\alpha(x-\beta))], \qquad x > 0 \tag{5.31}$$

If the random variable x is normalized by σ, $\bar{x} = x/\sigma$ and the relationship between α and β with σ and μ are used, the probability density function may be written in terms of constants $a_1 = 1.2825$ and $a_2 = 0.45005$, and the correlation coefficient $Cov = \sigma/\mu$:

$$p(\bar{x}) = a_1 \exp(-a_1(\bar{x}-1/Cov+a_2))\exp[-\exp(-a_1(\bar{x}-1/Cov+a_2))], \quad \bar{x} > 0 \tag{5.32}$$

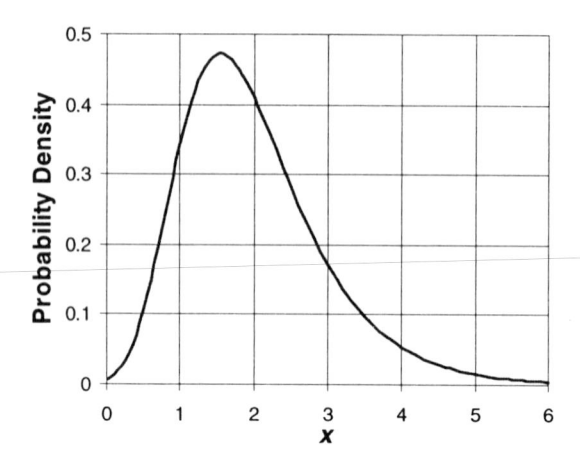

Fig. 5.6: Gumbel Probability Density Function

The probability density function is plotted for $Cov = 0.5$ in Fig. 5.6. The Gumbel distribution is used when investigating the extreme value distribution of the maxima of a random process.

5.3 Power Spectrum and Statistics

A study of the power spectrum and its statistical properties for a vibrating system is often useful in helping understand the physical mechanism that creates the vibration of the system. It describes the frequency content in the vibration signal and the magnitude of energy associated with each of these frequencies. In this regard, the vibration need not necessarily be random. Deterministic systems may also be analyzed by this method.

Consider the stationary random data represented by a finite record length of sample $x(t)$. Assume that the time interval of the record is T_s, being a short-term description of the random variable. The Fourier transform of the variable $x(t)$ of finite length, commonly known as finite Fourier transform, is written as (an extension of the Fourier series):

$$X(\omega, T_s) = \int_0^{T_s} x(t) \exp(-i\omega t) dt \qquad (5.33)$$

Assuming that the record $x(t)$ is sampled at a time interval Δt and N is the sample size, the record length $T_s = N\Delta t$. For a finite record length, there exists a Nyquist (upper limit) frequency given by

$$\omega_c = \frac{\pi}{\Delta t} \qquad (5.34)$$

This is then the cut-off frequency for such calculation for the frequency description of the variable $x(t)$. Since T_s is the record length, the fundamental frequency and the incremental frequency of this computation are $2\pi/T_s$.

The power spectral density of the sample may be obtained in two different ways. The first method used in the past is through an auto-correlation function of $x(t)$ of the form of Eq. 5.8. This process is slow and is seldom used in present day computations [see Bendat & Piersol (1980) for details]. The second method is more straightforward and efficient and is obtained by the Fourier transform of the original data. The extra speed is particularly possible through the use of a Fast Fourier transform (FFT). The Fourier transform $X_k(\omega, T_s)$ of the variable $x(t)$ as a function of frequency ω over a finite record k of length T_s is given by

$$X_k(\omega, T_s) = \int_0^{T_s} x_k(t) \exp(-\omega t) dt \qquad (5.35)$$

In practice, an average estimate of the spectral density is made over a number of records. The average estimate of the spectral density function $S_{xx}(\omega)$ is obtained as a function of frequency from the relation:

$$S_{xx}(\omega) = \lim_{T_s \to \infty} \frac{2}{T_s} E[|X_k(\omega, T_s)|^2] \qquad (5.36)$$

in which the operator E denotes the average expected value over the index k ($k = 1,...n_s$, the collection of records). In this case, the total amount of data analyzed is $n_s T_s$.

Note that since this density function is an estimate, it is important that the value of the index n_s is large in order to reduce the estimate error. Since the analog time history is converted to N samples at an equally spaced time interval of Δt, no significant loss of information occurs as long as

$$N \geq 2BT_s \tag{5.37}$$

where B is the cyclical frequency (Hz) bandwidth of the data. Aliasing effects may be avoided as long as the data has no significant mean square value in the frequency range above the cut off frequency, ω_c.

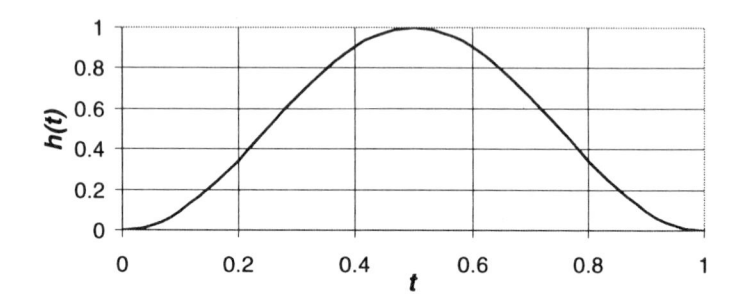

Fig. 5.7: Hanning Time Window

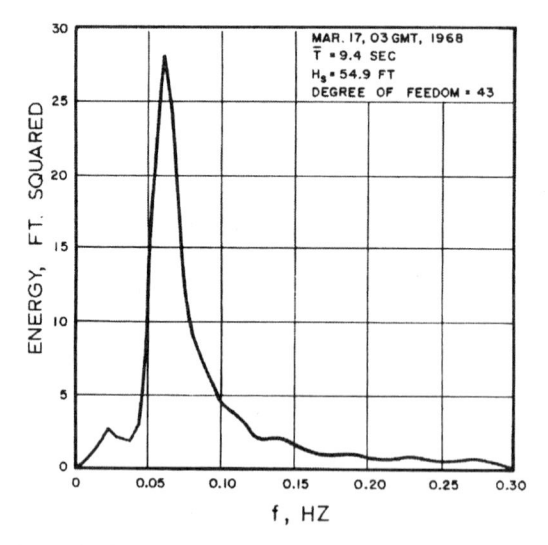

Fig. 5.8: Sample Ocean Wave Spectral Energy Density Function

Chapter 5 Statistical Theory in Vibration

Because the estimate is based on a finite length of record representing an infinite time history, it may be viewed as the product of the infinite record length times a rectangular window having a value of unity within the finite sample, and a value of zero outside this window. Thus, the estimate introduces leakage in the spectral density values as side lobes outside of the frequency range. This is corrected by the use of a time window, such as Hanning window (Fig. 5.7). The time window redistributes the energy of the estimate so that the side lobes disappear and the energy is brought back to the frequency range present.

An example of the spectral energy density function as a function of cyclic frequency (Hz) is given in Fig. 5.8. The energy density represents the estimate by the above method of an ocean wave recorded in the North Atlantic. The peak period and significant wave height are shown in the figure.

5.4 Extreme Values and Envelope Functions

In a design one is interested in the maximum design value of a response. It is, therefore, important to know the extreme maxima of random samples of a structure response when subjected to random excitation. The probability distribution of maxima of such samples will be derived here. Methods of deriving short-term extreme design values will be illustrated.

5.4.1 Extreme Value Function

Consider random samples X_i (I = 1,2,...N), each of which is of size k, and assume that $X_1, X_2,...X_N$ are arranged in the ascending order of their values. Thus, X_1 is a random sample of size k of the smallest value, while X_N is a random sample of size k of the largest value. It is also logical to assume that the samples X_i are independent.

The probability distribution of the largest value $X(N)$, is written as

$$P(X_{(N)} \leq x) = P(X_1 \leq x, X_2 \leq x, X_3 \leq x,..., X_N \leq x,) \qquad (5.38)$$

But since each sample X_i is independent, the resultant probability is the product of the independent samples:

$$P(X_{(N)} \leq x) = P(X_1 \leq x)P(X_2 \leq x)P(X_3 \leq x)...P(X_N \leq x,)$$
$$= P(x)P(x)P(x)...P(x)$$
$$= [P(x)]^N$$

(5.39)

Similarly, the distribution of the smallest $X(1)$ can be expressed as

$$P(X_{(1)} > x) = P(X_1 \leq x, X_2 \leq x, X_3 \leq x..., X_N \leq x)$$
$$= Q(X_1 > x)Q(X_2 > x)...Q(X_N > x)$$

(5.40)

where Q is the probability of exceedence. Note that the probability of exceedence is related to the probability distribution by the relation:

$$Q(X_1 > x) = 1 - P(X_1 \leq x) = 1 - P(x)$$

(5.41)

Then

$$P(X_1 > x) = [1 - P(x)]^N$$

(5.42)

5.4.2 Design Extreme Value

One of the objectives of a model test is to derive design parameters for a vibrating system. These parameters may then be calibrated with analysis techniques. The calibrated analysis technique may then be used efficiently to optimize the design variables in order to deduce the value for the extreme responses, such as maximum motion, tension, and stresses in members of a structure and its components. This is illustrated by the following example.

One of the random processes of interest is the motion response of a moored floating structure. Such structures experience response in random waves, which are composed of wave frequencies and low frequencies corresponding to the difference frequencies within the wave spectrum. The prediction of extreme responses of a moored floating structure is critically dependent upon both the low frequency and wave frequency responses. It has been customary in offshore engineering to assume that stable statistics can be obtained from 3-hour sea states for the wave frequency responses,

Chapter 5 Statistical Theory in Vibration 191

which have an average of 1000 zero upcrossings. When it comes to the low frequency responses whose periods are approximately 100 - 200 sec., the 3-hour sea states are simply inadequate in determining stable statistics. This is because the number of zero upcrossings for the low frequency oscillation is only about 100 to 150 (corresponding to a 100-sec period), which is simply too small a sample size.

Dyer and Ahilan (2000) computed the confidence intervals for the mean and standard deviation of a normal process. The mean and standard deviations for various sample sizes were taken as 1000 and 400 respectively and the confidence level was chosen as 95%. The number of peaks within a storm was increased from 100 to 1000. For the given confidence level, the standard deviation of a normal process was found to increase by a factor of 3 or 4 when the number of samples was reduced from 1000 to 100. This clearly shows that even the standard deviation of the low frequency process remains unstable over a 3-hour period. As the wave frequency response "rides" upon these low frequency responses, neither motions nor line tensions would be confidently predicted from 3-hour tests. This may be a major shortcoming of model tests of this kind. Much longer test duration, therefore, is desired if one wants more reliable extreme design values from the low frequency response of floating structures.

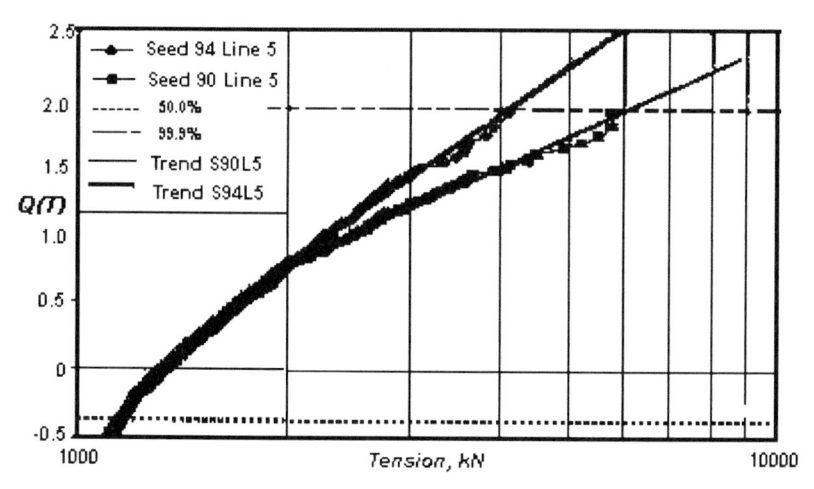

Fig. 5.9: Weibull Analysis of Two Independent Identical Simulations [Dyer and Ahilan (2000)]

192   *Theory and Practice of Vibration and Hydrodynamics*

The inadequacy of short simulations is further demonstrated by the example taken from Dyer and Ahilan (2000) and shown in Fig. 5.9. The results represent probability of exceedence of two 3-hour simulations on tension of a mooring line. The two simulations have identical input parameters except for the phases of wave and wind components. The 50 and 99.9% confidence levels are shown in the figure. Note in the figure that the two most probable maxima (MPM) differ by more than 35%. This figure also illustrates that the 3-hour simulation is inadequate in predicting the extreme response.

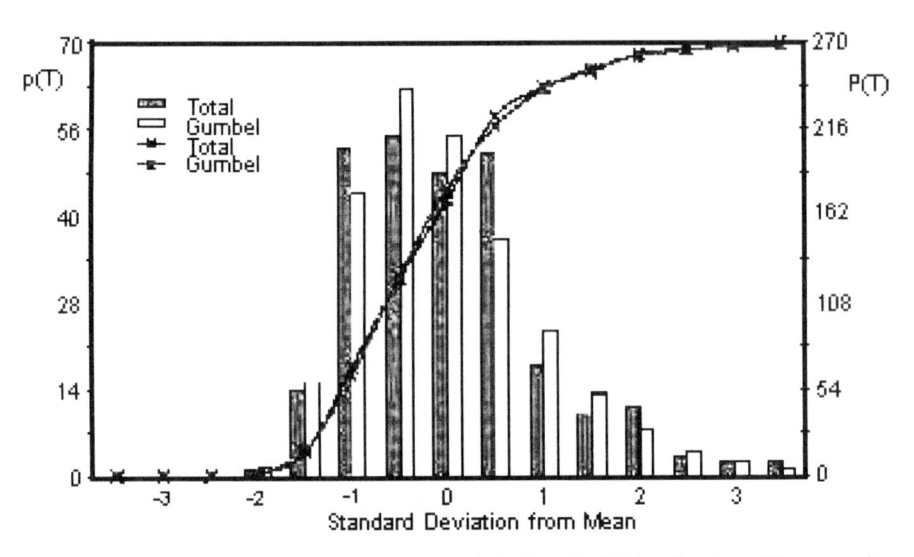

Fig. 5.10: Fitting of Simulation with Gumbel Distribution [Dyer and Ahilan (2000)]

Because the probability of exceedence plot is not linear, the most probable maxima is predicted by fitting a straight line to the upper part of the extreme response probability curve. The choice of the fraction of the peaks varies considerably from one application to another, such as the top 50 to 100 peaks. Each of these methods will predict different 3-hour extremes. Based on this example, the question is what is the most appropriate extreme value to be used by a designer, who is typically seeking to identify the most probable maximum value of the extreme response. This can be identified with considerably greater confidence, if considerations are given with regard to model test duration. Accordingly, tests should be performed for approximately 1000 times the longest natural period of the floating structure. This will typically mean performing model tests for the full-scale equivalent time period of about 18 hours. However, this length of test is impractical and will invariably cause wave reflections to build up near the model. In order to avoid this problem, multiple (of the order of 6) independent 3-hour tests could be conducted. These tests are then combined to compute the MPM value.

An alternative approach to the long test duration is direct numerical simulation of the response in a manner similar to the way in which results have been produced for the test duration. Extreme value theory suggests that the peaks from different three-hour storms of the same magnitude would be Gumbel distributed. Figure 5.9 shows a histogram of tension peaks of the mooring line. The cumulative distribution of peaks is plotted with the Gumbel distribution. It is seen that the probability distribution of the tension peaks obtained from about 250 3-hour simulations is well modeled by the Gumbel distribution.

The MPM value of a process distributed as Gumbel is given by the following equation:

$$\text{MPM} = \mu - 0.45\sigma \qquad (5.43)$$

where μ and σ are the mean and standard deviation of the peaks. Thus, by performing multiple simulations, the need to determine the MPM value by fitting a straight line to a Weibull plot can be completely circumvented. As the number of simulation increases, the mean and standard deviation of the peaks stabilize to produce a robust most probable maximum value.

5.5 Response Spectra

For a linear vibrating system, the response of the system varies as the amplitude of a harmonic excitation and, hence, is a function of only the frequency of excitation. One of the major goals of most analyses or model tests is to develop the Response Amplitude Operator (RAO) for a response of a vibrating system. The RAO is the magnitude of the linear transfer function between the forcing function imposed on a structure and the resulting response function of the structure. If these two quantities are recorded as a function of time in a model test with a random excitation, then the RAO may be computed from these two measurements. The time histories of the measured excitation and structure response are first converted to the frequency domain from which the spectral densities are computed. A simple calculation of the RAO from the spectral densities is shown below. The method is illustrated with an example in the following section.

5.5.1 Transfer Function

Consider an input signal of the sinusoidal form having the frequency ω.

$$x(t) = X \exp(-i\omega t) \tag{5.44}$$

If the response due to this input is given by the output signal of the form:

$$y(t) = Y \exp(-i\omega t) \tag{5.45}$$

then the ratio

$$H = Y / X \tag{5.46}$$

is the transfer function of the system having the frequency ω. If the transfer function H is unique for the frequency ω, then the system is called a linear system.

This concept is extended to a random signal having infinite frequency content, resulting in the following frequency domain relationship:

$$Y(\omega) = H(\omega) X(\omega) \tag{5.47}$$

where $Y(\omega)$ = the Fourier transform of the random response $y(t)$, $X(\omega)$ = the Fourier transform of the random excitation $x(t)$, and $H(\omega)$ = the transfer function for a linear system.

Given the density spectrum of an input random signal and a transfer function, the response spectrum may be computed by a simple formulation. The method works only if the system is linear and there is indeed a unique transfer function for the response of the system. Consider that $S_{xx}(\omega)$ is the autospectral energy density spectrum of the input signal and $S_{yy}(\omega)$ is the

Chapter 5 Statistical Theory in Vibration

corresponding energy density spectrum of a measured response. The autospectrum of the signal is computed from

$$S_{xx}(\omega) = \lim_{T_s \to \infty} \frac{1}{T_s} E[X(\omega, T_s) X^*(\omega, T_s)] \tag{5.48}$$

while the response autospectrum becomes

$$S_{yy}(\omega) = \lim_{T_s \to \infty} \frac{1}{T_s} E[Y(\omega, T_s) Y^*(\omega, T_s)] \tag{5.49}$$

The variables X and Y are defined in Eq. 5.44 and 5.45. Keep in mind that H, Y, and X here are all complex quantities. The superscript asterisk implies the complex conjugate. The expression for the response spectrum may be computed from the generalized transfer function, H and the imposed (input) spectrum S_{xx} by the formula:

$$S_{yy}(\omega) = H^2(\omega) S_{xx}(\omega) \tag{5.50}$$

An example of this method is illustrated in Fig. 5.11. The upper curve in the figure is an input wave spectrum. The middle curve is a force RAO, which may be measured or computed. The bottom curve is the response force spectrum, which in this case is computed using Eq. 5.50.

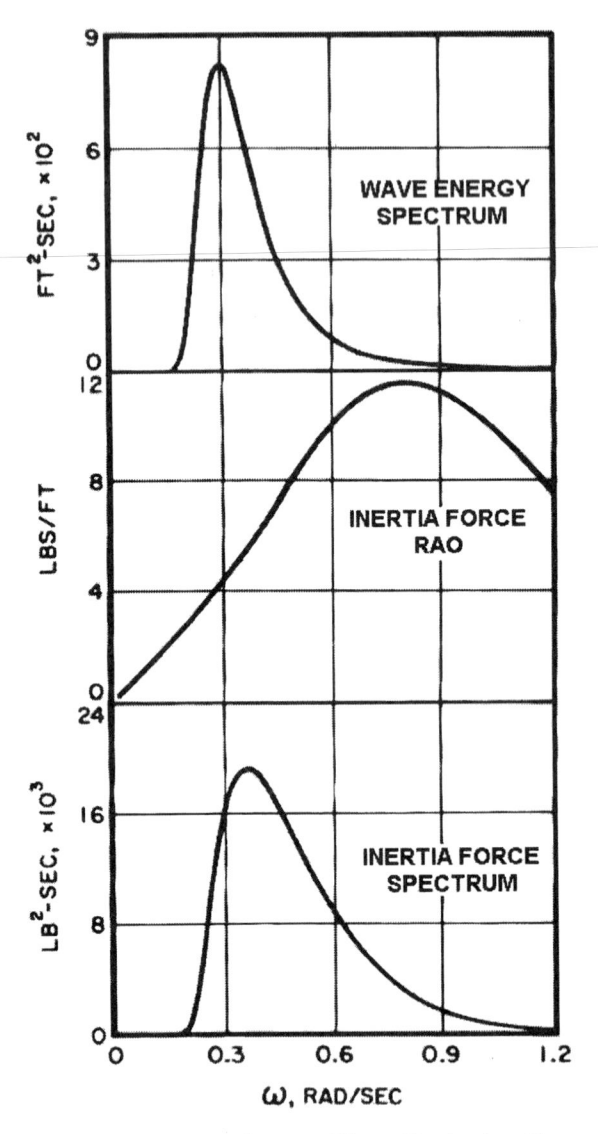

Fig. 5.11: Response Spectrum from a Given Excitation Spectrum and an RAO

Chapter 5 Statistical Theory in Vibration 197

Alternately, the relationship in Eq. 5.50 may be written as

$$H(\omega) = \left[\frac{S_{yy}(\omega)}{S_{xx}(\omega)}\right]^{1/2} \quad (5.51)$$

Thus, if the input random signal and the corresponding output signal are known (e.g., from experimental measurements) and the system is linear, then the transfer function may be computed from Eq. 5.51. This is a direct method of computing the response amplitude operator for the system.

While this is a very straightforward method, it has some drawbacks. This method may produce uncorrelated signal noise. In other words, the RAO gives a value from the response where the input signal is so small that it is hidden in the noise level and the response and input signal are not correlated. Additionally, the information on the phase relationship between the response amplitude and wave amplitude is lost in this method of computation. In order to avoid these problems, a cross-spectral density analysis method [Bendat and Piersol (1980)] may be employed.

The phase relationship between the response and the input signal may be restored by the cross-spectrum method. The cross spectrum between the input signal and response is given by the quantity S_{xy}, which is computed from the input and output signals by the relation:

$$S_{xy}(\omega) = \lim_{T_s \to \infty} \frac{1}{T_s} E[X(\omega, T_s) Y^*(\omega, T_s)] \quad (5.52)$$

In this case, both the input and output signals are combined to compute the co- and quadrature spectra. In complex form, the co- and quad spectra, C and Q, respectively, are related to the density spectra, S_{xx} and S_{xy} by the relation

$$C(\omega) + iQ(\omega) = \frac{S_{xy}(\omega)}{S_{xx}(\omega)} \quad (5.53)$$

In this case, the amplitude transfer function is computed as

$$H(\omega) = \left[C^2(\omega) + Q^2(\omega)\right]^{1/2} \quad (5.54)$$

whereas the phase transfer function is obtained from

$$\varepsilon(\omega) = \tan^{-1}\left[\frac{Q(\omega)}{C(\omega)}\right] \quad (5.55)$$

Also, the accuracy of the computation may be determined from the coherence function, which is calculated from

$$\gamma^2(\omega) = \frac{S_{xy}(\omega)}{[S_{xx}(\omega)S_{yy}(\omega)]^{1/2}} \qquad (5.56)$$

If the value of the coherence function is close to one, then the output and input signals are correlated and the transfer function is valid and accurate in this region. When this value is near zero, they are uncorrelated and the transfer function is inaccurate. This then provides the confidence on the transfer function and the region where the transfer function is valid. In general, a coherence value greater than 0.6 is sought for a reasonably good correlation.

5.5.2 Cross Spectral Analysis

As shown in the earlier section, the transfer function under the cross-spectral theory is a complex quantity and retains the amplitude and phase relationship between the forcing function and the response function. The following presentation documents one of the example problems to verify the computation method and the possible calculation errors.

For a linear dynamic single degree of freedom system (SDOFS) the equation of motion for the response is given by

$$m\ddot{y}(t) + c\dot{y}(t) + ky(t) = x(t) \qquad (5.57)$$

where y is the response function corresponding to the forcing function x. The forcing function $x(t)$ can be the time history of a deterministic or a random function.

The transfer function $H(\omega)$ may be expressed in terms of the system properties of the SDOFS. By assuming a complex forcing function of unity and solving the equation of motion for $H(\omega)$, the system property description [Clough and Penzien (1975)] of $H(\omega)$ i.e. the exact transfer function becomes:

$$H(\omega) = \frac{1}{-\omega^2 m + i\omega c + k} = \frac{(k - \omega^2 m) - i\omega c}{(k - \omega^2 m)^2 + (\omega c)^2} = A(k - \omega^2 m) - iA(\omega c)$$

$$(5.58)$$

in which the quantity A is the reciprocal of the denominator. The magnitude of the RAO is obtained from Eq. 5.58:

$$RAO(\omega) = |H(\omega)| = \frac{1}{\left[(k-\omega^2 m)^2 + (\omega c)^2\right]^{1/2}} \quad (5.59)$$

Thus, if the system properties (i.e. c, k, & m) are known one can find the RAO from Eq. 5.59. If the system properties are not known, but the excitation time history and the response time history are measured, the RAO may be calculated from the spectral relations of Eq. 5.51.

The ideal situation for applying this technique for response spectrum is to have time records of infinite length from which to compute the Fourier transforms. In practice, the time record length T_s must be finite, and for this reason the Fourier transforms are functions of T_s as well as ω [Bendat and Piersol (1980)]. The "real life" situation also dictates that the Fourier transforms are the average transforms that are obtained by averaging the transforms of many subrecords. These subrecords are obtained by dividing one long time record into a finite number of sections (or subrecords). If the long time record depicts a stationary and ergodic process, increased averaging will insure that the measured Fourier transform will approach the true transform. In other words, increased averaging means less random error, which means less dependence of the estimate of the Fourier Transform on the short time record T_s. The variables S_{xx} and S_{yy} shown in Eqs. 5.48 – 5.49 are the autospectra obtained from the average Fourier transforms the excitation and response respectively and S_{xy} in Eq. 5.52 is their cross-spectrum obtained from the average Fourier transform.

5.5.3 Error Analysis

Since discrete frequency increments are used for the Fourier transform instead of the ideal continuous case, negative bias errors will result. This means that all of the measured density spectra will be slightly smaller than the actual spectrum. This problem will always occur for "density" functions, such as the ones used here [Bendat and Piersol (1980), pg. 41]. The bias error normalized with respect to the quantity being measured may be associated with any density spectrum as follows:

$$\varepsilon_b = \frac{1}{3}\left(\frac{\Delta f}{BW}\right)^2 *100 \qquad (5.60)$$

where ε_b = "maximum" normalized bias error in percent, Δf = frequency increment, and BW = spectrum half-power bandwidth. This maximum is measured by locating the largest peak in the spectrum and calculating the bias error associated with that peak using Eq. 5.60. Bias errors less than 2% are considered negligible. The inability to take infinite time records introduces random errors. The normalized random error formulas are given in Table 5.1.

Table 5.1: Random Errors in Single Input/ Single Output Problems

Description	Quantity	Random Error
Autospectrum	$S_{xx}(f), S_{yy}(f)$	$\dfrac{1}{\sqrt{n_a}}$
Cross-spectrum modulus	$\|S_{xy}(f)\|$	$\dfrac{1}{\gamma_{xy}(f)\sqrt{n_a}}$
Coherence function	$\gamma_{xy}(f)$	$\dfrac{1-\gamma_{xy}^{2}(f)}{\gamma_{xy}(f)\sqrt{2n_a}}$
Cross-spectral RAO	$\|H_c(f)\|$	$\dfrac{\left[1-\gamma_{xy}^{2}(f)\right]^{1/2}}{\gamma_{xy}(f)\sqrt{2n_a}}$
Autospectral RAO	$\|H_a(f)\|$	$\dfrac{\left[1-\gamma_{xy}^{2}(f)\right]^{1/2}}{\gamma_{xy}^{2}(f)\sqrt{2n_a}}$
n_a = the number of averages (i.e. subrecords or sections).		

The coherence function (γ) indicates the amount of nonlinearities, noise and spectral bias errors present in the measurements. A completely noise-free linear system (containing no bias errors in its spectral measurements) will have a coherence of one for all frequencies. A system whose output is completely unrelated to its input will have a coherence of zero for all frequencies. A coherence of 0.6 or above is considered acceptable. It has also been shown (Eq. 5.56) that the coherence represents the ratio of the cross-spectral RAO to the autospectral RAO; thus, for any system exhibiting coherence less than one, the autospectral RAO will always be larger than the cross-spectral RAO.

Assuming that the coherence will be some value less than one, the table shows that the autospectral RAO will always possess more random error than the cross-spectral RAO. The RAO random error formulas also show that as long as the coherence is nonzero and the bias errors are not excessive, the RAO may be estimated to any degree of accuracy one desires simply by increasing the number of averages.

The coherence function for a system can fall below the value of one due to any (or combination) of the following reasons:

1. Extraneous noise is present in the measurements,
2. Resolution bias errors are present in the spectral estimates,
3. The system relating $y(t)$ to $x(t)$ is not linear, and
4. The output $y(t)$ depends on other inputs besides $x(t)$.

Assuming that the time record being analyzed is of a given duration, increasing the number of averages means decreasing the subrecord length T_s. Whenever the subrecord length is decreased, the frequency resolution Δf becomes coarser increasing the bias error. It is evident that there are conflicting requirements between the bias error and the random error when working with a time limited set of data. If it is impossible to increase the duration of the test run, one must find the appropriate number of averages that will give a reasonable random error and the appropriate subrecord length to give an acceptable bias error. If the duration of the test run is not time limited, one must determine the subrecord length, which yields an acceptable bias error first and then multiply that subrecord length by the required number of averages (to reduce the random error) to obtain the necessary test run duration. To determine the necessary subrecord length, one must estimate the half-power bandwidth of the data being analyzed and then solve the bias error formula for Δf.

The following guidelines are utilized to analyze the estimates of the transfer function $|H(\omega)|$ and phase $\varepsilon(\omega)$ in order to determine the existence of random and bias errors:

- If $\gamma_{xy}(\omega)$ falls broadly over a frequency range where $|H(\omega)|$ is relatively small, this might indicate measurement noise included in the output and/or contributions of other uncorrelated inputs.

- If $\gamma_{xy}(\omega)$ falls broadly over a frequency range where $|H(\omega)|$ is not near a minimum value and $\gamma_{xx}(\omega)$ is relatively small, then measurement noise included at the input should be suspected.

- If $|H(\omega)|$ peaks sharply at system resonance frequencies and $\gamma_{xy}(\omega)$ does not, then system nonlinearities might be suspected, as well as resolution bias errors in the spectral estimates. To distinguish between resolution problems and system nonlinearities, the spectral estimates may be repeated utilizing a sampling window.

5.6 Exercises

Exercise 1

Consider a floating structure that is subjected to a one degree of freedom motion. The equation of motion is given in Eq. 5.57. If the forcing function is known as a function of time, Eq. 5.58 may be numerically solved in time by using a piecewise linear acceleration scheme [Clough and Penzien (1975)]. In this example, it is assumed that the forcing function was a measured irregular wave time history. The linear acceleration scheme was used to solve for the response numerically, and the response time history was created. The parameter values used for the example SDOFS are:

$k = 0.80$ lb/ft, $m = 0.075$ slugs, $c = 0.0583$ slugs/sec, $f_n = 0.52$ Hz

The exact transfer function is as follows:

$$H(\omega) = A[0.80 - 0.075\omega^2] - iA[0.0583\omega] \qquad (5.61)$$

and

Chapter 5 Statistical Theory in Vibration 203

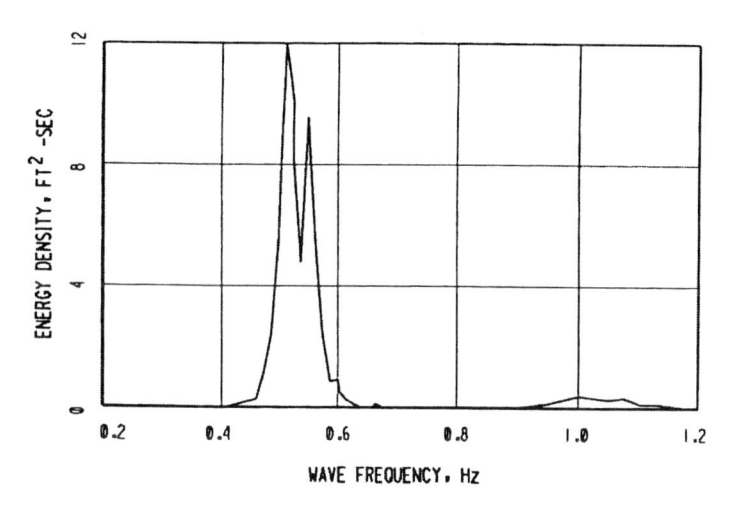

$$A = \frac{1}{\left[(0.80 - 0.075\omega^2)^2 + (0.0583\omega)^2\right]^{1/2}} \quad (5.62)$$

Fig. 5.12: Measured Wave Spectrum

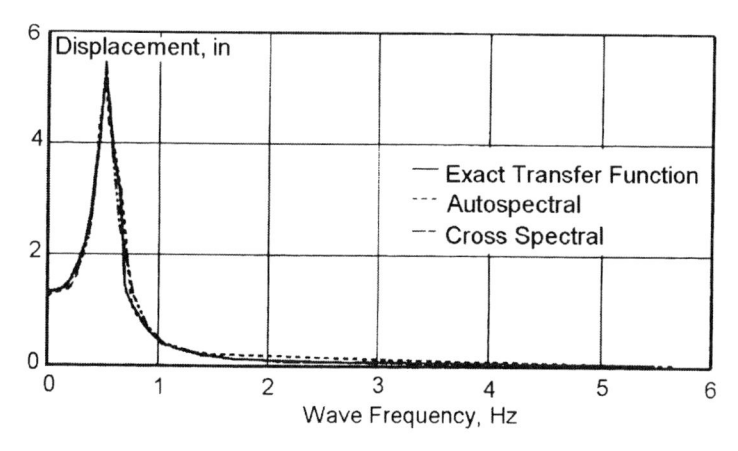

Fig. 5.13: Example SDOF System RAO

The wave was measured in a wave basin using a 5 Hz low-pass Butterworth filter. A spectrum of the measured waveform (Fig. 5.12) shows the peak at approximately 0.5 Hz. One will note that there is relatively little energy between 2 and 5 Hz. Since the Butterworth filter removes frequencies above 5 Hz, it is safe to say that absolutely no energy exists in the recorded wave data above 7 Hz. Any energy above 7 Hz that is present in the spectrum can only be due to numerical leakage [Brigham (1974), pg.140]. Figures 5.13 and 5.14 show the exact RAO and phase functions for this system. These exact functions are plotted as solid lines on the corresponding function estimate plots. Figure 5.15 presents the response function energy density spectrum generated by Eq. 5.49.

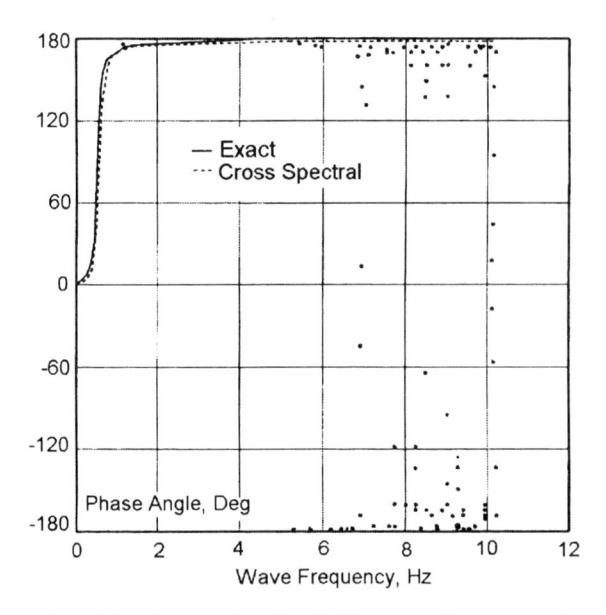

Fig. 5.14: Example SDOF System Phase Angle

For the case of Hanning smoothing, the spectral analysis estimates of the RAO (Fig. 5.13) are very close to the true RAO function. Figure 5.14 shows an excellent estimation of the phase function below 5.5 Hz. The coherence function (Fig. 5.15) has a value close to unity at frequencies below 5.5 Hz. Above 5.5 Hz, the coherence falls far below unity.

Chapter 5 Statistical Theory in Vibration 205

Since the data being analyzed is contrived, causes 1,3, and 4 for low coherence stated above (section 5.5.3) will not pose a problem. This leaves cause 2, resolution bias errors as the only possible reason that could cause the coherence to drop. Large bias errors will occur in spectral estimates at frequencies where the frequency resolution is not fine enough to pick up all of the peaks and troughs of the spectrum. This can typically occur at frequencies where the spectrum is extremely peaked, or jagged. Although the magnitude of the calculated spectrum is very small above 5.5 Hz, an expanded scale view of the spectrum reveals a very jagged spectrum above 5.5 Hz; hence, large bias errors are very likely to be present in this area.

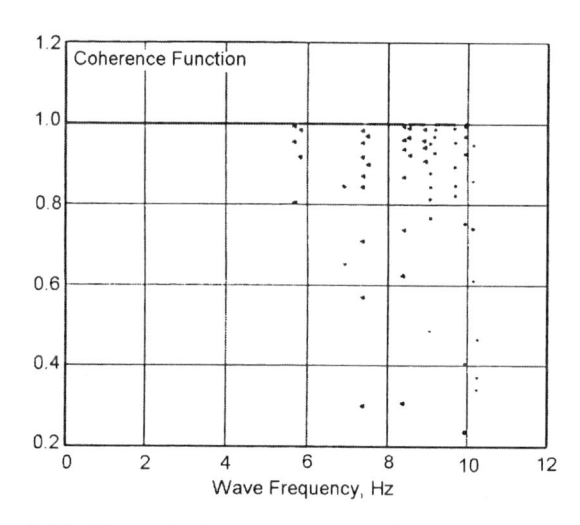

Fig. 5.15: Example SDOF System Coherence Function

Exercise 2

Find the total probability of an event between the values of 0.3 and 0.6 of normalized variable x/μ_x using the normal distribution function.

Exercise 3

Show that the total probability of a normal distribution for one standard deviation about the mean is 0.674.

Exercise 4

Derive the expression for the bias and random error for an autospectral density function. Compute the bias and random error for a record having a length of 180 sec and a collection of samples n_a of 10.

5.7 References

1. Abramowitz, M. and Stegun, I.A., <u>Handbook of Mathematical Functions</u>, Government Printing Office, Washington, DC, 1964.

2. Bendat, J.S., and Piersol, A.G., <u>Engineering Applications of Correlation and Spectral Analysis</u>, John Wiley & Sons, New York, 1980.

3. Brigham, E.O., <u>The Fast Fourier Transform</u>, Prentice-Hall, Inc., New Jersey, 1974, pg.140.

4. Chakrabarti, S.K., <u>Nonlinear Methods in Offshore Engineering</u>, Elsevier, Amsterdam, 1990.

5. Clough, R.W., and Penzien, J., <u>Dynamics of Structures</u>, McGraw-Hill, New York, New York, 1975.

6. Dyer, R. C. and Ahilan, R. V., "The Place of Physical and Hydrodynamic Models in Concept Design, Analysis and System Validation of Moored Floating Structures", Proceedings of ETCE/OMAE2000 Joint Conference, ASME, 2000.

7. Gumbel, E.J., <u>Statistics of Extremes</u>, Columbia University Press, New York, 1958.

8. Lin, Y.K., <u>Probability Theory of Structural Dynamics</u>, McGraw-Hill, New York, 1967.

Chapter 6

Random Vibration

6.1 Introduction

The vibration of a structure may be characterized as one of the following: (1) periodic, having one or more definite frequencies, (2) stationary random, having a fixed set of statistical parameters or (3) non-stationary random, which has changing statistical properties. The hydrodynamics of structures give rise to each of these types of vibration characteristics, depending on the excitation experienced by the structure. The first kind of periodic vibration has already been covered earlier. This chapter will mainly discuss the second kind of random vibration of structures.

6.2 Random Single Degree of Freedom Linear System

The vibration of structures may be periodic, but is not often harmonic. This means that the non-harmonic periodic motion may not be described by a single frequency of vibration, as is the case for the simple harmonic motion.

If a linear single degree of freedom system (SDOF) is subjected to an excitation composed of several frequencies, then the response of the system is expected to include the same frequencies and will not be harmonic any more. The response amplitudes will not be in proportion to the amplitudes of the individual amplitudes of the excitation frequencies but will depend on the transfer function. If one of the frequencies approach the natural frequency of the SDOF, then the response at that frequency will very likely maximize. As the number of components of this group of excitation frequencies is increased, the response will appear to be random. For a discrete number of frequencies, however, a response of the SDOF may be obtained by superposition as long as the system is linear. The system in this case is said to undergo a non-harmonic response.

6.2.1 Non-Harmonic Vibration

A general expression for the non-harmonic periodic motion is derived with the use of Fourier frequency analysis. The Fourier series analysis describes the accumulated motion in terms of a series of harmonic motions with related frequencies. In particular, the frequencies in a Fourier series are multiples of the fundamental frequency, which provides the longest period of structure motion. Mathematically, the motion may be represented by

$$x(t) = \sum_{n=0}^{N} X_n \sin(n\omega t + \varepsilon_n) \qquad (6.1)$$

where the quantities X_n are the amplitudes and ε_n are the corresponding phases and N is the number of components. An alternate representation of Eq 6.1 is

$$x(t) = \sum_{n=0}^{N} (a_n \cos n\omega t + b_n \sin n\omega t) \qquad (6.2)$$

The relationship between the quantities a_n, b_n and X_n, ε_n may be obtained by equating the right hand side of Eqs. 6.1 and 6.2:

$$a_n = X_n \sin \varepsilon_n \qquad (6.3)$$

$$b_n = X_n \cos \varepsilon_n \qquad (6.4)$$

so that the amplitudes are

$$X_n = \sqrt{a_n^2 + b_n^2} \qquad (6.5)$$

and the phases can be expressed by

$$\varepsilon_n = \tan^{-1}(a_n / b_n) \qquad (6.6)$$

The summation is finite having a total number of terms of $N+1$. The first term X_0 is independent of the time (static) and represents as offset from zero. The frequencies for the subsequent terms are multiplier of ω up to $N\omega$. The period of the fundamental frequency ω is

Chapter 6 Random Vibration

$$T = \frac{2\pi}{\omega} \qquad (6.7)$$

whereas the periods of the higher frequencies are a fraction of the fundamental frequency. Thus, the overall period of the non-harmonic motion is T, regardless of the value of N. Moreover, the higher the value of N, the motion appears more random. The smaller the value of ω, the repeat period becomes longer as well. Thus, it is possible to represent a very complex motion with the Fourier series function having a close resemblance to the original time history of motion. This will be illustrated with an example.

An example of the non-harmonic periodic motion is given in Fig. 6.1. In this case, only three related frequencies are chosen, i.e. $N = 4$. Moreover, since most vibrations take place about the equilibrium position, the static offset value of X_0 is taken as zero in this example. The fundamental period is chosen as $T = 10$ seconds. The three displacements are taken as $X_1 = 1$, $X_2 = 0.5$ and $X_3 = 0.25$.

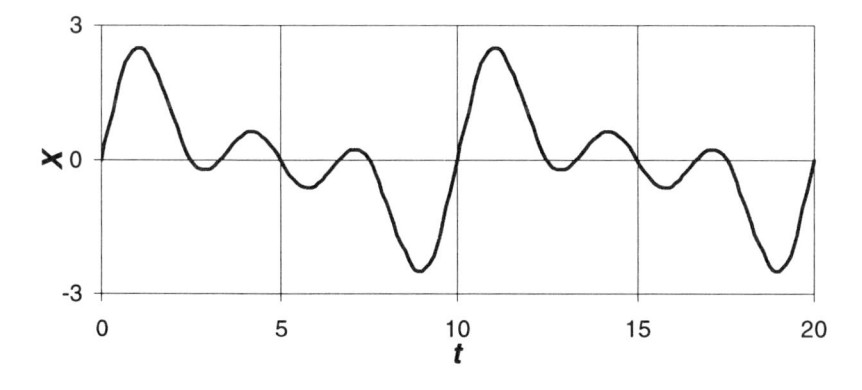

Fig. 6.1: Non-Harmonic Periodic Motion

Assume that the complex motion $x(t)$ may not be represented exactly by the finite Fourier series representation in Eq. 6.1. In this case, Eq. 6.1 may be considered an approximate representation. The expressions of the coefficients a_n and b_n are computed by multiplying both sides of Eq. 6.2 by $\sin n\omega t$ and $\cos n\omega t$ in sequence.

Then the coefficients a_n and b_n are determined by the integration of Eq. 6.2. For example, the first integral equation becomes

$$\int_0^T f(t)\sin n\omega t\, dt = \sum_{n=0}^{N} \int_0^T (a_n \cos n\omega t + b_n \sin n\omega t)\sin n\omega t\, dt \tag{6.8}$$

It can be shown that the first integral on the right hand side within the parenthesis is an odd function between the interval 0 and T (i.e. antisymmetric about $T/2$) so that the integral over one cycle becomes zero. Therefore, Eq. 6.8 determines b_n. Carrying out the integration on the right hand side, the expressions for a_n and b_n may be similarly shown to be

$$a_n = \frac{2}{T}\int_0^T f(t)\cos n\omega t\, dt \tag{6.9}$$

$$b_n = \frac{2}{T}\int_0^T f(t)\sin n\omega t\, dt \tag{6.10}$$

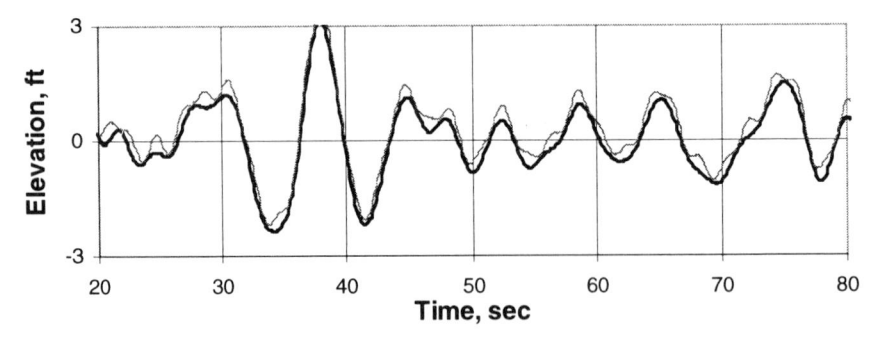

Fig. 6.2: Pseudo-Random Motion with Fourier Fit

Consider an example of the time history of a pseudo random vibration shown in Fig. 6.2. The random signal shown as the thin line was obtained from the time history simulation of a given power spectrum. A Fourier series was fitted to this motion history using a fundamental frequency corresponding to the length of the time history, $\omega = 2\pi/T_s$ (where T_s is the length of the record) and the number of components $N = 25$. The result for a

Chapter 6 Random Vibration 211

short time span is shown as the thick line in Fig. 6.2. While some of the details are missing, the match between the two traces is quite good. Thus, it is possible to represent a random vibration signal with an equivalent Fourier representation. This is an important tool since it permits treatment of a random signal in terms of known harmonic oscillations. As long as the response is linear, each component may be treated separately and the final result may be obtained by linear superposition (simple summation of the individual responses). However, this is possible only if the response is linear.

6.2.2 Random Vibration

Unlike the non-harmonic vibration, random vibration includes a band of frequencies and should be described as a continuous function of frequencies. Therefore, the analysis of a random signal should be carried out through Fourier transform type of function rather than the Fourier series of discrete frequencies. The Fourier integral allows expressing the signal as a continuous function of frequencies. For a linear system the relationship between an input signal and the corresponding random output signal may be stated by the integral

$$y(t) = \int_0^\infty x(t-\tau)h(\tau)d\tau \qquad (6.11)$$

in which x(t) and y(t) are the input and output random signals. The function h(t) is the response due to an unit impulse function $\delta(t)$ and is assumed known. This function will be discussed in further detail in a subsequent section. Assume that the input random signal of finite length may be represented reasonably well by a Fourier series of the form given by Eq. 6.1. However, since the signal is random the summation is extended from $-\infty$ to ∞. Then the response for a linear system may be written by the relation:

$$y(t) = \int_0^\infty \sum_{n=-\infty}^\infty X_n \exp[in\omega(t-\tau)]h(\tau)d\tau \qquad (6.12)$$

in which use has been made of the complex algebra where the phase angle has been absorbed. This expression may be simplified as follows:

$$y(t) = \sum_{n=-\infty}^{\infty} X_n \exp(in\omega t) \int_0^{\infty} \exp(-in\omega\tau) h(\tau) d\tau \qquad (6.13)$$

Note that the last integral in Eq. 6.13 is the frequency representation of the response function. The response time history will retain the statistical properties of a random signal as long as a large number (e.g., 1000) of frequencies are used in this summation.

6.3 Response of SDOF System to Random Input

In practice, random vibrations are a common occurrence. There are many examples in which the system responds to excitations that are random. In fluid structure interaction problems, such cases may be described as structures responding to wind effect or wave action. In these cases the structure undergoes a random motion. Such responses cannot be described by the simple means used in the earlier sections. There are two different methods that are used in describing the responses of a system subjected to a random input. The first is a statistical description of the responses similar to those introduced in Chapter 5. The other is a spectral description in which the power spectrum of the response is given in the frequency domain, as developed in Chapter 5.

Consider a lightly damped structure. The random excitation will produce a random response of this structure. If an auto-spectrum of the random response is computed, the spectrum will tend to maximize at the peak of the excitation spectrum. However, it will also peak at the response frequencies at which the response peaks. If the structure responds to multiple frequencies, the response may have several peaks corresponding to these response frequencies, which correspond to the normal modes or resonant frequencies. Therefore, the response spectrum reveals the modes of vibration of a structure when subjected to a broad band spectrum.

6.3.1 Frequency Domain Analysis

The frequency domain analysis is an efficient method in obtaining response of a system. By the nature of the analysis, it is easier to handle linear

Chapter 6 Random Vibration

systems accurately by this technique. However, certain nonlinear systems may also be considered in a frequency domain method.

6.3.1.1 Frequency domain linear system

It has been shown that the Fourier transform of the response of a system X is the product of the Fourier transform of the random excitation F_0 and the transfer function or the frequency response function H.

$$X(\omega) = H(\omega)F_0(\omega) \tag{6.14}$$

The transfer function of a single degree of freedom system may be calculated in the following way. The equation of motion is written as

$$m\ddot{x} + c\dot{x} + kx = F_0 \exp(i\omega t) \tag{6.15}$$

The complex form is a simpler mathematical description of the equation in which the amplitude and phase of the response are included in the real and imaginary terms. It has already been shown in the linear response case that the transfer function has the form

$$H(\omega) = \frac{1/k}{1 - \left(\dfrac{\omega}{\omega_n}\right)^2 + i2\zeta\left(\dfrac{\omega}{\omega_n}\right)} \tag{6.16}$$

The damping factor ζ determines the level of damping in the linear system. An approximate measure of the damping factor may be made by the width of the response curve $H(\omega)$ at the half power points. This is illustrated in Fig. 6.3. The half power points are defined as points in the curve where the energy is half the energy obtained at the natural frequency:

$$|H(\omega)|^2 = \frac{1}{2}|H(\omega_n)|^2 \tag{6.17}$$

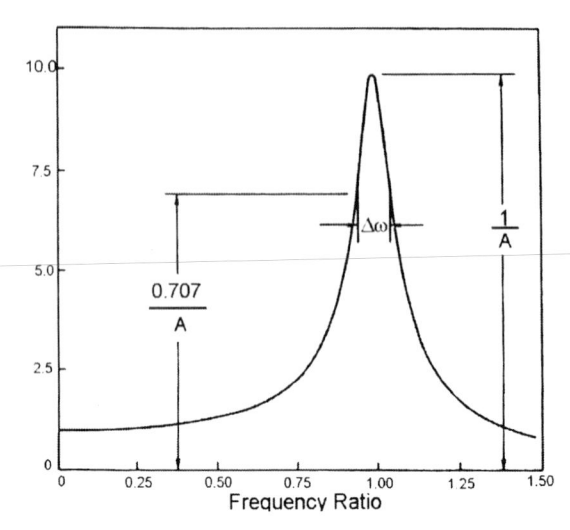

Fig. 6.3: Example of Resonance Curve Showing Half Power Point

If this width between the half power points is given by $\Delta\omega$, then the measure of the damping factor is obtained from the following relation:

$$\zeta = 2\frac{\Delta\omega}{\omega_n} \qquad (6.18)$$

The frequency difference between the half-power points is often referred to as the bandwidth of the system.

6.3.1.2 Frequency domain nonlinear system

For linear systems, the structure response due to random waves is straightforward to compute as long as the linear transfer function is known. Floating structures are also susceptible to vibrations at their natural frequencies outside the range of frequencies (i.e., higher or lower) present in random waves. These are caused by nonlinear forces experienced by the floating structures. Large moored floating structures in the ocean typically have long natural periods in surge since the mass is large and springs in the mooring lines are relatively soft. The motions of these structures are

Chapter 6 Random Vibration

sensitive to the low frequency second-order forces and are excited by them. The second-order forces arise from the combination of neighboring frequencies in the random waves and are proportional to the square of the wave amplitude. The square of the wave envelope is, therefore, important when considering this low frequency behavior since the slowly-varying second-order response of the structure is significantly affected by grouping patterns present in the time history of waves. This "groupiness" function is also called the second-order spectrum and is defined in terms of the difference frequency and spectral energy density as

$$S_{xx}^{(2)}(\omega) = 8\int_0^\infty S_{xx}(\mu)S_{xx}(\omega+\mu)d\omega = \frac{16}{\pi}\int_0^\infty R_{xx}^2(\mu)\cos(\omega\mu)d\mu \qquad (6.19)$$

in which the subscripts xx are introduced for the spectral estimate to indicate that it is an auto-spectrum (i.e., from a single signal). The second equality in Eq. 6.19 is written in terms of the autocorrelation function $R_{xx}(\mu)$. Figure 6.4 shows an example of the groupiness function of the (JONSWAP) wave spectrum. In the generation of random waves in a wave model basin, the generated wave grouping must be accurate when compared to the theoretical grouping. The match should be particularly good at the small values of ω. The measured groupiness function of a random wave generated in a wave tank is compared with the computed function in Fig 6.4. The trend and correlation shown in the plot are typical. While this example is given in terms of a random wave, it is equally applicable to other types of random excitation as long as the structure is responsive to low frequency oscillation (i.e., have a long natural period and low damping).

When higher order responses are expected, such as the low frequency response of a moored floating system, then the second-order spectrum of the wave profile is needed in the computation of the response. Assuming $H_2(\mu, \omega+\mu)$ as the quadratic transfer function (QTF) between the excitation and response in which μ and $\omega+\mu$ are a pair of frequencies, the total response spectrum is computed from

$$S_{yy}(\omega) = |H_1(\omega)|^2 S_{xx}(\omega) + 8\int_0^\infty |H_2(\mu,\omega+\mu)|^2 S_{xx}(\mu)S_{xx}(\omega+\mu)d\mu \qquad (6.20)$$

The first term is the linear response spectrum, while the second term is the contribution from the second order response.

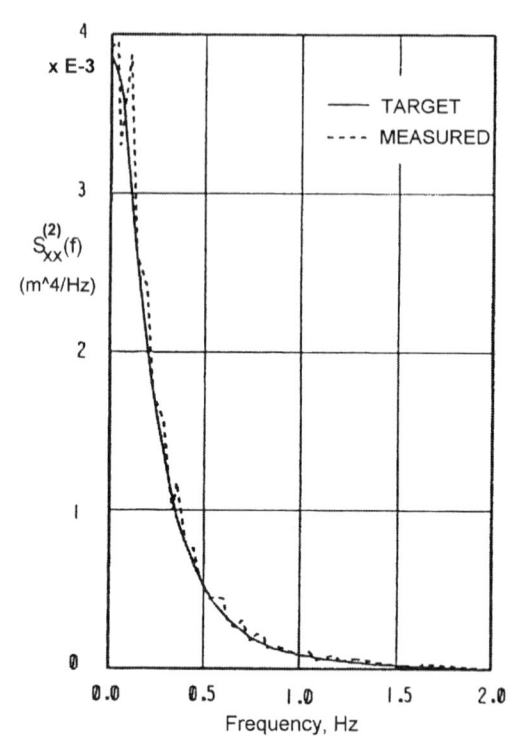

Fig. 6.4: Groupiness Function for a JONSWAP Spectrum

6.3.2 Time Domain Analysis

For random excitation, frequency domain analysis is possible only if the system is linear or linearized such that it can essentially be treated as a linear system. On the other hand, the analysis by the pseudo frequency domain design may become inadequate when strong nonlinearities are present. If the system has nonlinear contribution from the damping or restoring forces, then an accurate analysis of response should maintain these nonlinearities for which a time domain solution should be sought. Moreover, it may be

Chapter 6 Random Vibration 217

preferable to seek a time domain solution to a random excitation even for a linear system. The time history of the response allows direct application of the statistical analysis to the generated response.

6.3.2.1 Time domain linear system

Consider the equation of motion for a linear dynamic single degree of freedom system introduced earlier:

$$m\ddot{x}(t) + c\dot{x}(t) + kx(t) = F(t) \tag{6.21}$$

in which $F(t)$ is a random forcing function and $x(t)$ is the corresponding motion for the system. This relationship may be written in a convolution integral form known as Duhamel integral:

$$x(t) = \int_{-\infty}^{\infty} F(\tau)h(t-\tau)d\tau \tag{6.22}$$

The function h is known as the unit impulse response function or weighting function for the system [see Bendat and Piersol (1980)] and was introduced earlier. The Fourier transform of both sides yield the relationship in the frequency domain:

$$H(\omega) = \int_0^{\infty} h(\tau)\exp(-i\omega\tau)d\tau \tag{6.23}$$

Equation 6.23 implies that the weighting function h is the Fourier transform of the frequency response function H of the system. Thus given a well-defined input function $F(t)$, one can obtain the output function $x(t)$ through the weighting function $h(t)$. Note that the transfer function is computed from the cross spectrum $S_{Fx}(\omega)$ as

$$H(\omega) = \frac{S_{Fx}(\omega)}{S_{FF}(\omega)} \tag{6.24}$$

Then the weighting function is computed from the integral

$$h(t) = \int_{-\infty}^{\infty} \frac{S_{Fx}(\omega)}{S_{FF}(\omega)} \exp(-i\omega t)d\omega \tag{6.25}$$

The unit response impulse function is similar to the cross-correlation function if the input spectral density $S_{FF}(\omega)$ is uniformly unity. The integral in Eq. 6.25 may be solved numerically after the expression for $h(t)$ has been substituted. It is applicable to a linear system.

6.3.2.2 Time domain nonlinear system

If the nonlinearity in a system is strong, then a time domain approach becomes more viable. One such system is a submerged moored system subjected to ocean wave. Ocean mooring systems are characterized by a nonlinear restoring force, a structural damping force and a coupled velocity-dependent exciting force. The restoring force includes material discontinuities and geometric nonlinearities associated with large amplitude motion. The exciting force includes nonlinear effects and periodic components governed by steady (e.g., current-induced) and unsteady (e.g., wave-induced) viscous drag, radiation damping and convective inertia effects [see, for example, Chakrabarti(1990)]. The response of these nonlinear systems to external and parametric excitation may yield periodic and aperiodic motion. This latter part may be amplitude modulated quasi-periodic and chaotic. For floating systems, the coupling of multiple degrees of motion of the structure and separation of fluid flow from the submerged surface of the structure further complicate the system behavior. The system dissipative force includes coupled quadratic wave-induced damping governed by the relative motion between the fluid particle and the structure itself. A time domain solution approach is most appropriate for such a system.

In order to develop the nonlinear equation of motion, a small, submerged buoyant body is considered moored to the ocean bottom by a soft mooring line of linear stiffness k. The system is described in Fig. 6.5. The body is taken as a fully submerged sphere [see Gottlieb (1996)]. The system is subjected to a finite amplitude wave described by the second-order Stokes wave, and is free to respond in the vertical XY plane as shown in the figure. The lift effect in the normal direction to the XY plane is not considered in this analysis. The equation of motion is written as a two degree of freedom system – one along X and the other along Y. In the X-direction, the equation of motion that includes the above-mentioned nonlinearities has the form

Chapter 6 Random Vibration

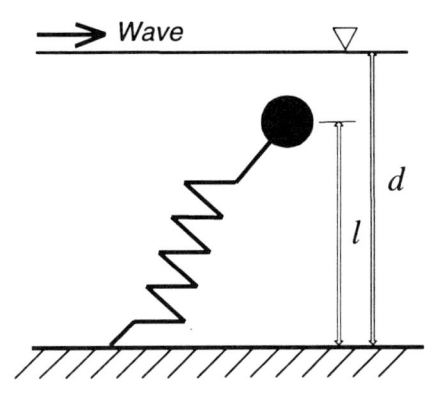

Fig. 6.5: A Moored Submerged Sphere Subjected to Wave

$$(m + \rho V C_A)\ddot{x} + c\dot{x} + \frac{1}{2}\rho C_D A v_{rel}(\dot{x}-u) + C(x,y)x = \rho V(1+C_A)\dot{u}_{nl} \quad (6.26)$$

in which m is the mass of the sphere, A its projected area and V its displacement. The quantity C is the restoring force per unit displacement in the X-direction involving instantaneous location of the sphere center and spring stiffness k. The relative velocity is given by

$$v_{rel} = \sqrt{(\dot{x}-u)^2 + (\dot{y}-v)^2} \quad (6.27)$$

and the nonlinear wave acceleration component is

$$\dot{u}_{nl} = \left[\dot{u} + u\frac{\partial u}{\partial x} + v\frac{\partial v}{\partial y}\right] \quad (6.28)$$

In Eq. 6.26 the first and second terms are the structure and added mass inertia. The third and fourth terms are the linear damping and relative-velocity quadratic damping terms. The fifth term is the restoring force of the moored system, which is dependent on the position of the body in reference to its equilibrium position. The right hand term is the exciting force based on the water particle acceleration, which includes the convective inertia terms given by Eq. 6.28. Note that the convective inertia term arises from the Navier-Stokes equation, and is included when a nonlinear wave theory is applied. The quantities u and v are the water particle velocities along X and Y by the nonlinear Stokes wave theory. For a sphere, the added mass coefficient C_A may be taken as 0.5. An equation similar to Eq. 6.26 is written for the vertical (Y) direction. Equation 6.26 is normalized by dividing by the mass coefficient, $(m + \rho V C_A)$. Then, the equation of motion in the X-direction is written in terms of nondimensional system parameters denoted by β, γ, δ, and μ.

$$\ddot{x} + \beta \dot{x} + \delta v_{rel}(\dot{x} - u) + \gamma x = \mu \dot{u}_{nl} \tag{6.29}$$

The forms of the nondimensional parameters are easy to derive. For example, the form of β is

$$\beta = \frac{B}{m + \rho V C_A} \tag{6.30}$$

For a sphere, the expression for γ is given by

$$\gamma = 1 - \frac{1-\alpha}{\sqrt{x^2 + (1+y^2)}} \tag{6.31}$$

where

$$\alpha = \left(\frac{\rho V - m}{k}\right)\frac{g}{l} \tag{6.32}$$

Also,

$$\delta = \frac{0.5 \rho C_D A l}{m + \rho V C_A} \tag{6.33}$$

$$\mu = \frac{\rho V (1 + C_A)}{m + \rho V C_A} \tag{6.34}$$

Similarly, in the vertical direction, the equation of motion becomes

$$\ddot{y} + \beta \dot{y} + \delta v_{rel}(\dot{y} - v) + \gamma(1 + y) = \mu \dot{v}_{nl} + \alpha \tag{6.35}$$

Chapter 6 Random Vibration 221

in which $\dot{v}_{nl}$ is the vertical acceleration including the convective term. Note that the unforced (i.e., $\mu = 0$), undamped ($\beta = \delta = 0$) system is governed by a single parameter α, that is bounded from above and below ($0 \le \alpha \le 1$), by the spring stiffness and the system buoyancy.

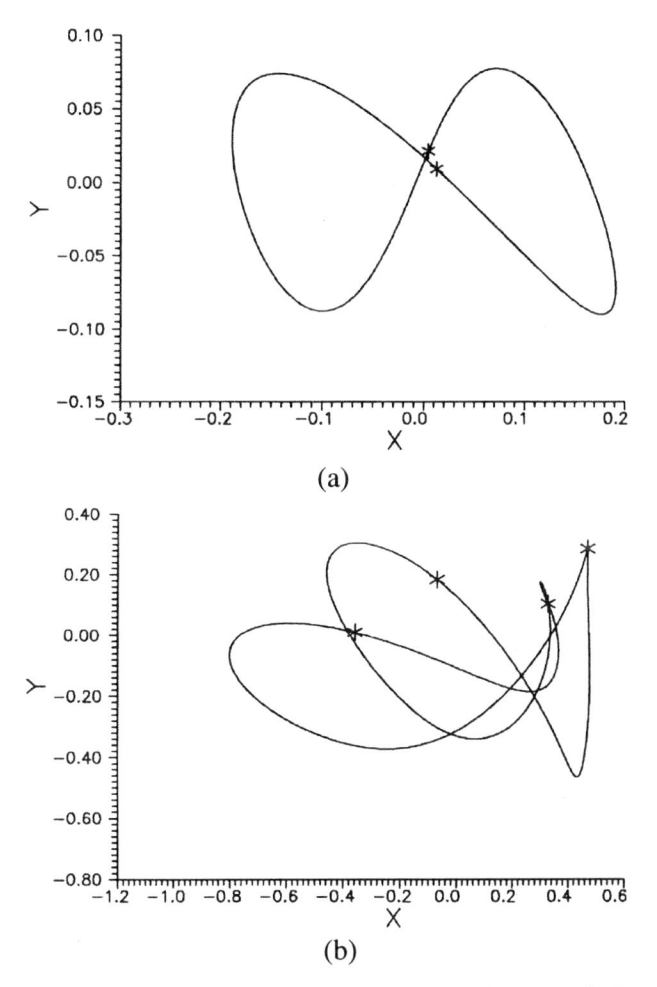

Fig. 6.6: Phase-Plane Diagram for Periodic and Subharmonic Response of a Submerged Sphere in Waves [Gottlieb (1996)]

In the dynamic equations the chaotic response of the system evolves from the nonlinearity in the convective terms. The threshold of the system stability, on the other hand, is controlled by the energy dissipation mechanism of the system. Equations 6.29 and 6.35 are solved for x and y in the time domain. The plot of x vs. y describes the XY plane motion of the sphere and is called the phase-plane diagram. The stability of the motion and any bifurcation from the periodicity becomes clear in such a diagram. At relatively low amplitude waves the motion of the sphere is quite stable. However, as the amplitudes are increased the convective terms take over and the motion becomes nonperiodic with subharmonics in them. At even larger amplitudes, the response amplitude becomes high and the periodic motion bifurcates showing higher order subharmonics. Two examples are shown in Fig. 6.6 for Stokes second order wave using $\alpha = 0.025$ for which the natural frequency in the vertical direction is twice the horizontal frequency. The damping is constant in both examples with values of $\beta = 0.05$ and $\delta = 0.1$. The first one (Fig. 6.6a) corresponds to a low wave amplitude showing periodic solution for the above values. In the second example (Fig. 6.6b), the response is due to a high wave showing quadruple periodic subharmonic solution. The loss of symmetry generates from the nonlinear relative-velocity damping of the form used in the analysis. The convective term enhanced the bifurcation sequence.

6.3.3 Random Decrement Technique

Hydrodynamic damping for a floating structure has been found to be dependent on the excitation frequency. In other words, damping in the presence of external excitation is higher than the free oscillation damping. In a single frequency response, the initial response of the structure will consist of impulse response at the resonant frequency as well as the excitation frequency response. In this case it is easy to separate the two by simply filtering out the excitation frequency response, as long as the excitation frequency is separated from the natural frequency by a measurable amount. This has been demonstrated earlier.

Since the response from a random excitation has a frequency band that may cover the system natural frequency, the standard technique for a decayed free oscillation may not be used to determine damping. In this case, the damping present in the random response may be determined by the random decrement technique [Yang, et al. (1985)]. The random decrement technique helps one to obtain curves of extinction from the measured response time history in random excitation. Once the decay curve corresponding to the resonant extinction is derived, the damping coefficient can be determined by the same method outlined in Chapter 3.

The basic concept of the random decrement method is based on the fact that the random response of the structure due to random excitation may be viewed to consist of a deterministic component and a random component. The deterministic component corresponds to the impulse and/or step function. The random component may be assumed to have a stationary average. By averaging several samples of the same random response, the random part of the response will be averaged out, leaving only the deterministic component of the response. The deterministic part is similar to the free-decay response from which the added mass and damping may be computed as shown earlier. The following procedure may be used to obtain this decayed oscillation.

Consider a structure having a linear response, when subjected to a random excitation. Assume the force to be a known, zero-mean, stationary, Gaussian random process. Hence, the response will also be a zero-mean, stationary, Gaussian process. The random decrement technique computes the average of a large number of segments of response $y(t)$ in the following manner. The oscillation time history is divided into an ensemble of segments of equal lengths. The starting time t_i of each segment is selected such that $\{y(t_i)\}$ is a constant and the slope has alternating positive and negative values. These segments are then ensemble averaged, giving a signature starting at the chosen level. The initial slope is zero because an equal number of positive and negative slopes have been chosen in the ensemble. The choice of the constant level is arbitrary and can be given as a percentage of the zero-mean rms value of the random time history. What remains after averaging is a decayed oscillation of the signature whose length depends on the total record length and the number of averages performed.

An example of extraction of the random decrement signature from response $\{y(t)\}$ is shown in Fig. 6.7. The top trace is the response of a floating structure measured in high frequency waves in a wave tank. The slow oscillation response corresponds to the natural frequency response of the floating system. It is desired to obtain the damping present in the system. If the deterministic high frequency portion of the response is filtered out, the second trace shows the irregular low frequency response. This filtered data is then subjected to the random decrement technique using the procedure described above. The final result is shown in the bottom trace, which yields the decayed oscillation after the average. The damping coefficient of this decayed trace may be determined by the logarithmic decrement method described in Chapter 3.

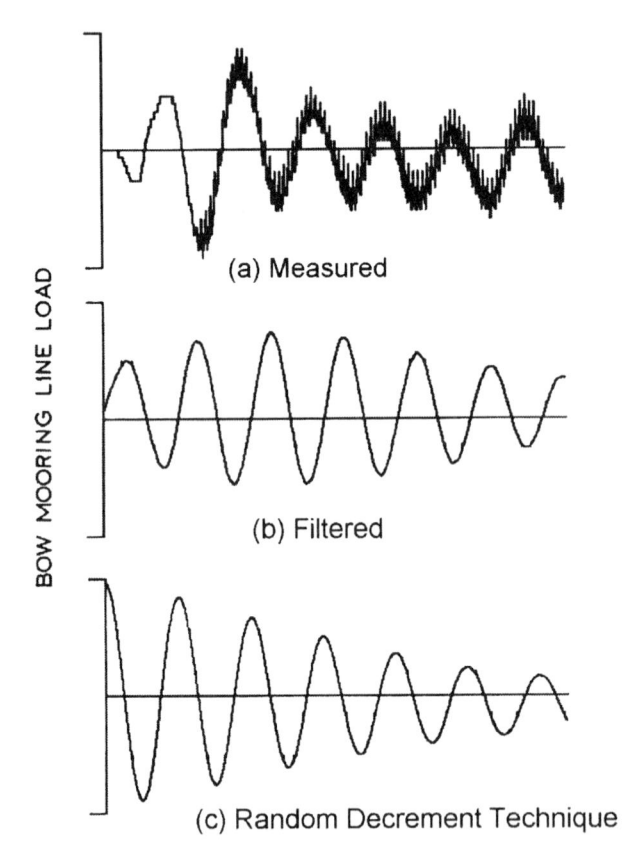

Fig. 6.7: Random Decrement Technique for a Random Wave Response

6.4 Random Response of Multi-Degree of Freedom System

The response of a multi-degree of freedom system due to harmonic excitation was discussed in Chapter 3. Since only one frequency was involved in that case, the values of the hydrodynamic coefficients corresponding to this frequency were applied. The equation of motion for a random wave is in principle similar. However, the frequencies are mixed and the added mass and damping values change from one frequency to the next. For a frequency domain solution, the analysis may be restricted to constant added mass and damping values on the assumption that the added mass and damping do not vary appreciably over a range of frequencies. However, this is an overly simplified assumption since the hydrodynamic coefficients are often strongly dependent on the excitation frequency. In order to avoid this deficiency, a time domain simulation of motions under random excitation should be considered. In this case, convolution integrals are needed for the inertia and damping terms. At each time step of numerical analysis, these integrals are computed over all frequencies of interest to include the added inertia and damping quantities varying with frequency.

6.4.1 Multi-Moduled Time-Domain Solution

Floating structures in an offshore operation often appear in a group. Figure 6.8 shows such a multi-moduled system in which a tanker is moored to a buoy, which, in turn, is spread moored to the ocean floor by multiple mooring lines. The tanker acts as a production platform to which a shuttle tanker is moored in tandem. A floating vessel held by several mooring lines at angles to its principal axes experiences nonlinear line loads, as well as coupling among its modes of motions. The single point mooring of a vessel to a buoy or tower in oblique seas will involve large angles of yaw and nonlinear wave loads, in addition to the nonlinear mooring line loads. Coupled motions of two moored floating structures in the vicinity of each other involve a 12 degrees of freedom system. Multiple floating structures will increase the order of DOF accordingly. Moreover, the exciting forces and hydrodynamic added mass and damping will include interaction among the modules.

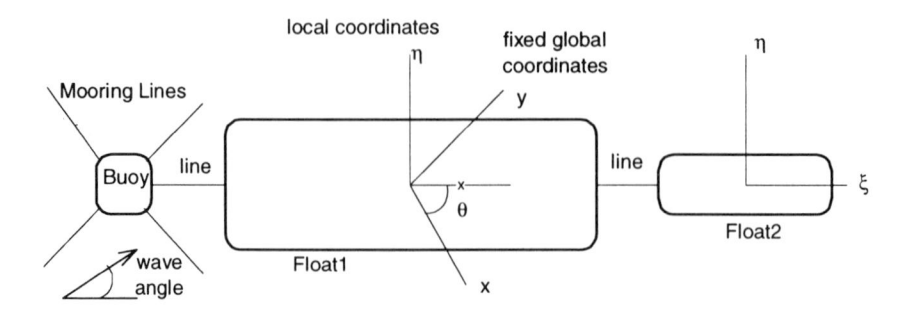

Fig. 6.8: Definition Sketch for a Multibody Moored System

Because most mooring systems are three dimensional and nonlinear, they are difficult to treat analytically in the frequency domain. As a result, a numerical time domain analysis is needed to facilitate the description of such systems and to evaluate their dynamic behavior. The analysis may include a vessel, which is maintained on station by mooring lines, or by a buoy or an articulated tower for a single point mooring system. The vessel's angular displacements are assumed small (with the exception of yaw). The coordinate system is chosen to permit large angular displacements in yaw. A formulation will be presented for the dynamics of a floating vessel subject to random excitation moored to an articulated tower as an example of this time domain approach.

This approach permits the nonlinearity in the excitation and reaction forces to be maintained in the analysis, and provides explicit results (i.e., responses as functions of time). At each time step, a set of second-order differential equations is solved to obtain the accelerations of the system. The solution is initiated with prescribed values for the displacements and velocities, and these values are calculated for the next time step from the derived accelerations by a forward integration scheme.

To describe the motion of the moored vessel, two sets of Cartesian coordinates are selected (Fig. 6.8). The first global set, viz., x, y, and z, is fixed in space. The positive x axis is chosen in the longitudinal direction; the y-axis is in the vertical direction, and the z-axis is chosen to complete the right-handed system. The second local sets, ξ, η, ζ, coincide with the principal axes of the individual vessels through their center of gravity (CG),

Chapter 6 Random Vibration   227

and partake in their motions. The hydrodynamic properties of the vessels are described in these coordinate systems. The equilibrium position of one vessel CG may be selected as the origin of the fixed coordinate system. These coordinates then describe the movement of the vessel's CG relative to its initial position.

The motion of each vessel about its CG is best described in terms of the Eulerian angles, θ, ψ, and ϕ. The yaw, θ, is measured about the vertical axis, i.e., the translated y axis; the pitch, ψ, is measured about the rotated transverse axis, and the roll, ϕ, is measured about the rotated longitudinal axis of the vessel, viz., the x axis. When the motions of the vessel are small, these Eulerian angles will coincide with the angular displacements about the space fixed axes (see Chapter 3, Eq. 3.137). The ξ, η, and ζ axes are related to the x, y, and z axes by their respective unit vectors.

If an articulated tower is attached to the floating vessels, then the instantaneous position of the tower is described by the Eulerian angles Ω and ε. The precession angle ε is measured from the vertical axis of the tower passing through its pin, and the oscillation angle Ω is measured from the tower's transverse axis through the pin.

The expression for the system's kinetic energy [Chakrabarti and Cotter (1989)] is given by:

$$\begin{aligned}T =& \sum_{i=1}^{M} \frac{1}{2}[m^i(\dot{x}^{i^2} + \dot{y}^{i^2} + \dot{z}^{i^2}) + I^i{}_\xi (\dot{\phi}^i - \dot{\theta}^i \sin\psi^i)^2 \\&+ I^i{}_\eta (\dot{\psi}^{i^2} + \dot{\theta}^{i^2} \cos^2\psi^i) \\&+ (I^i{}_\zeta - I^i{}_\eta)(\dot{\psi}^i \sin\phi^i - \dot{\theta}^i \cos\psi^i \cos\phi^i)^2] \\&+ [I_t(\dot{\Omega}^2 + \dot{\varepsilon}^2 \sin^2\Omega) + I_l\dot{\varepsilon}^2 \cos^2\Omega]\end{aligned} \quad (6.36)$$

where M is the number of modules, and the last two terms arise from the presence of the articulated tower. The equations of motion for the system are obtained from the generalized Lagrange's equations, which equate to the excitations:

$$\frac{\partial}{\partial t}\left(\frac{\partial T}{\partial \dot{x}^i}\right) - \frac{\partial T}{\partial x^i} = F^i \quad (6.37)$$

in which the right hand side is the sum of all the forces acting on the system. The translational equations of motion for the individual modules i in the structure in surge, sway and heave ($k = 1,2,3$ respectively) are written as

$$m_k^i \ddot{\eta}^i_k = f_k^i; k = 1,2,3 \tag{6.38}$$

The three rotational equations for roll, pitch and yaw respectively are written as

$$\frac{d}{dt}[I'_\xi(\dot{\phi}^i - \dot{\theta}^i \sin\psi^i)$$
$$- (I'_\zeta - I'_\eta)(\dot{\psi}^i \cos\phi^i + \dot{\theta}^i \cos\psi^i \sin\phi^i) \tag{6.39}$$
$$(\dot{\psi}^i \sin\phi^i - \dot{\theta}^i \cos\psi^i \cos\phi^i) = M_\phi^i$$

$$\frac{d}{dt}[I'_\eta \dot{\psi}^i + (I'_\zeta - I'_\eta)[(\dot{\psi}^i \sin\phi^i - \dot{\theta}^i \cos\psi^i \cos\phi^i)\sin\phi^i]$$
$$+ I_\xi(\dot{\phi}^i - \dot{\theta}^i \sin\phi^i)\dot{\theta}\cos\psi \tag{6.40}$$
$$+ I_\eta^i \dot{\theta}^2 \sin\psi^i \cos\psi^i - (I'_\zeta - I'_\eta)(\dot{\psi}^i \sin\phi^i - \dot{\theta}^i \cos\psi^i \cos\phi^i)$$
$$\dot{\theta}^i \sin\psi^i \cos\phi^i = M_\phi^i$$

$$\frac{d}{dt}[I'_\eta \dot{\theta}^i \cos^2\psi^i - I'_\xi(\dot{\phi}^i - \dot{\theta}^i \sin\psi^i)\sin\psi^i \tag{6.41}$$
$$- (I'_\zeta - I'_\eta)(\dot{\psi}^i \cos\phi^i - \dot{\theta}^i \cos\psi^i \sin\phi^i)\dot{\theta}^i \cos\psi^i \sin\phi^i = M_\theta^i$$

The tower oscillation and precession angles are described by

$$I_t \ddot{\Omega} - (I_t - I_l)\dot{\varepsilon}^2 \sin\Omega \cos\Omega = M_\Omega \tag{6.42}$$

$$I_t(\sin^2\Omega + I_l\cos^2\Omega)\ddot{\varepsilon} + (I_t - I_l)\dot{\Omega}\dot{\varepsilon}\sin 2\Omega = M_\varepsilon \tag{6.43}$$

In the above equations, the left-hand sides represent the inertia terms with the structure mass and moment of inertia in the respective directions. The moments due to excitation on the vessel are given in terms of the principal axes ξ, η, and ζ. The resultant loads on the structure system are due to the mooring lines, buoyancy, wave, wind, and current action. The right-hand side represents all the resultant forces and moments on the vessels and the tower in the respective coordinate directions. The internal loads are the buoyancy and hydrostatic loads while the external loads consist of the wind, wave and current loads, mooring line and connector loads and hydrodynamic added mass, damping and nonlinear drag loads (and moments). The steady drift loads and oscillating drift loads (and moments) may also be included in the right hand side. The buoyant forces and moments on the vessel are due to the hydrostatic pressure and are functions of y, ϕ, ψ, and θ with the dependence on the polar angle θ being only directional. The nonlinear damping forces are included as the relative-velocity drag on the right hand side in terms of drag coefficients, C_D in six directions. The wave forces and the hydrodynamic coefficients are introduced from the linear diffraction & radiation theory (see Chapter 7) as functions of frequency. The interaction among adjacent floating vessels is introduced from the multibody interaction effect in the diffraction problem. The shadowing effect of vessels due to large yaw angle is taken into account by interpolating the exciting forces at each time step based on the horizontal positions of the vessels. In general, the exciting forces will be functions of all the displacements of the vessel including yaw. However, generally these other displacements are small, such that their effect on the exciting wave force may be neglected without losing accuracy. The motion time history of the vessels lends itself to a finite difference scheme.

Consider an example of an SPM system having a tanker and a tower. The mooring line may become slack during a wave cycle and has highly nonlinear stiffness characteristics (see Fig. 4.9). The motion time history of the SPM for a random wave of significant height of 20 ft (6.1 m) is computed by the above system of equations and is shown in Fig. 6.9. In this case, a buoyant tower is attached to a tanker with a nonlinear mooring line. The analysis includes a steady uniform current, a steady drift force and an oscillating drift force. The steady-state time history of the tower oscillation and the tanker surge are plotted along with the mooring line load. The line load shows that the line goes slack during part of the cycle. Such results have typically been observed in small-scale model tests of an SPM system. The tower oscillates in the inline direction with the frequencies of the random wave. The tanker, additionally, responds at its natural frequency in surge, which is long. The mooring line becomes slack during part of the wave cycle.

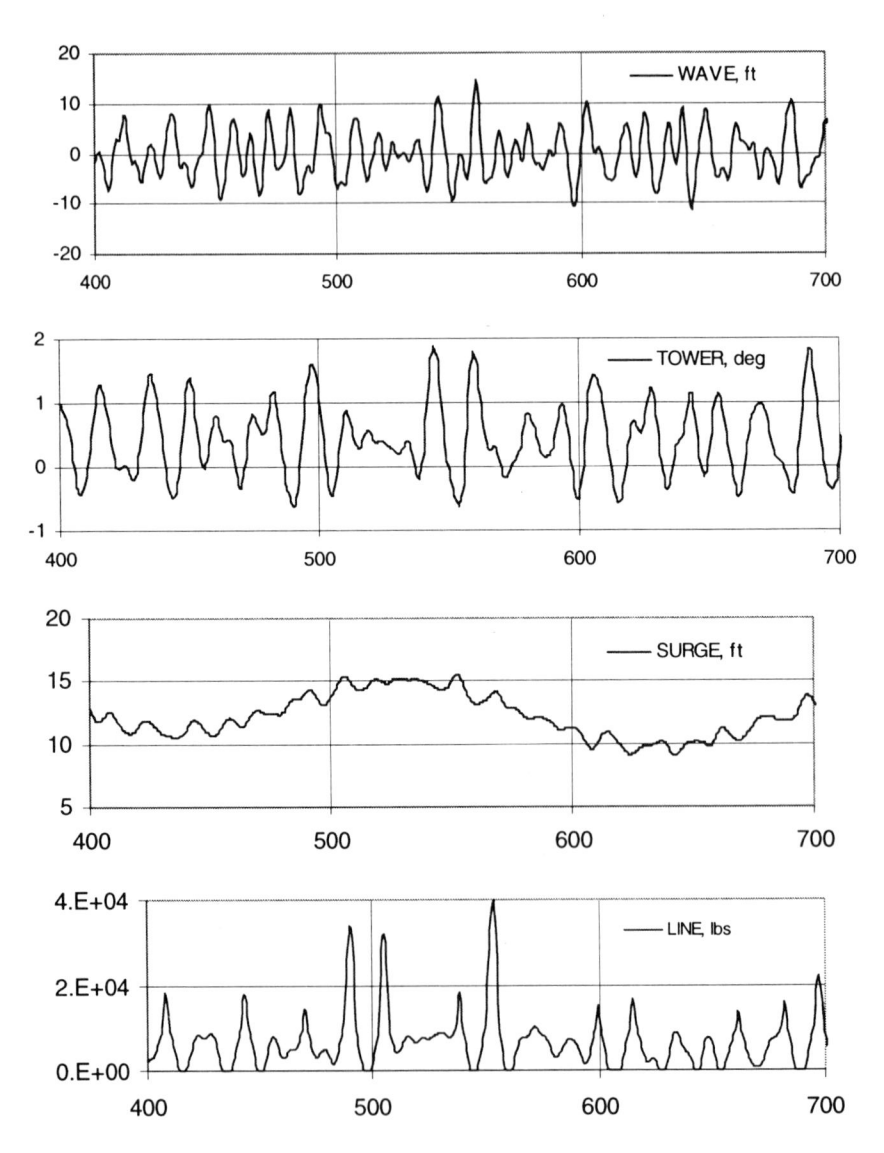

Fig. 6.9: SPM Ship & Tower Motions and Line Loads in Random Waves

6.5 Response of Nonlinear Systems to Random Input

Many structures undergo nonlinear random vibrations. In Chapter 3, several nonlinear structural responses from harmonic excitation were discussed. The same type of nonlinearity is also possible for structures subjected to random excitation. It is difficult to handle such nonlinearity in exact mathematical form for random input without resorting to the time domain method. Several methods are available to approximate these nonlinearities for random cases. One of them is called the stochastic linearization.

6.5.1 Stochastic Linearization of Force

One of the nonlinear forces encountered by submerged structures in oscillatory flows is the nonlinear drag force. For a simple harmonic flow, the oscillatory forces on a small structural member has been shown (in Section 2.7.5) to be the sum of an inertia force and a drag force, expressed by the Morison equation. The same formula has been extended to random oscillatory flow (e.g., ocean waves) in terms of the random flow velocities and flow accelerations. Thus the form of the random force per unit length in time domain is written as

$$f(t) = \rho C_M \frac{\pi D^2}{4} \dot{u}(t) + \frac{1}{2}\rho C_D D |u(t)|u(t) \tag{6.44}$$

in which the velocity $u(t)$ and acceleration $\dot{u}(t)$ are random functions. In this expression the hydrodynamic coefficients are succinctly assumed constant. The first term on the right hand side is linear, and computation of the power spectrum of this force is straightforward, as has been shown earlier. The difficulty in computing the force spectrum for Eq. 6.44 is with the second term, which is nonlinear. In this case, a stochastic linearization technique is employed [Borgman (1969)]. The linearization of the velocity squared term takes the form

$$|u(t)|u(t) = \sqrt{\frac{8}{\pi}} \sigma_u u(t) \tag{6.45}$$

in which σ_u is the rms value of the random velocity signal $u(t)$. Once this approximation is introduced in Eq. 6.44, the force spectrum is easy to compute as long as the hydrodynamic coefficients are considered invariant:

$$S_f(\omega) = \left(\rho C_M \frac{\pi D^2}{4}\right)^2 S_{\ddot{u}}(\omega) + \frac{8}{\pi}\left(\frac{1}{2}\rho C_D \sigma_u\right)^2 S_u(\omega) \qquad (6.46)$$

in which $S_u(\omega)$ and $S_{\ddot{u}}(\omega)$ are the velocity and acceleration spectra, respectively. Since the rms value of the random velocity time history may be computed from the time history itself, the computation of the force spectrum on the structure is straightforward. This was carried out for the force measured on an instrumented section of a vertical cylinder in random waves in a wave tank. The comparison of the computed force spectrum with the measured force spectrum is shown in Fig. 6.10.

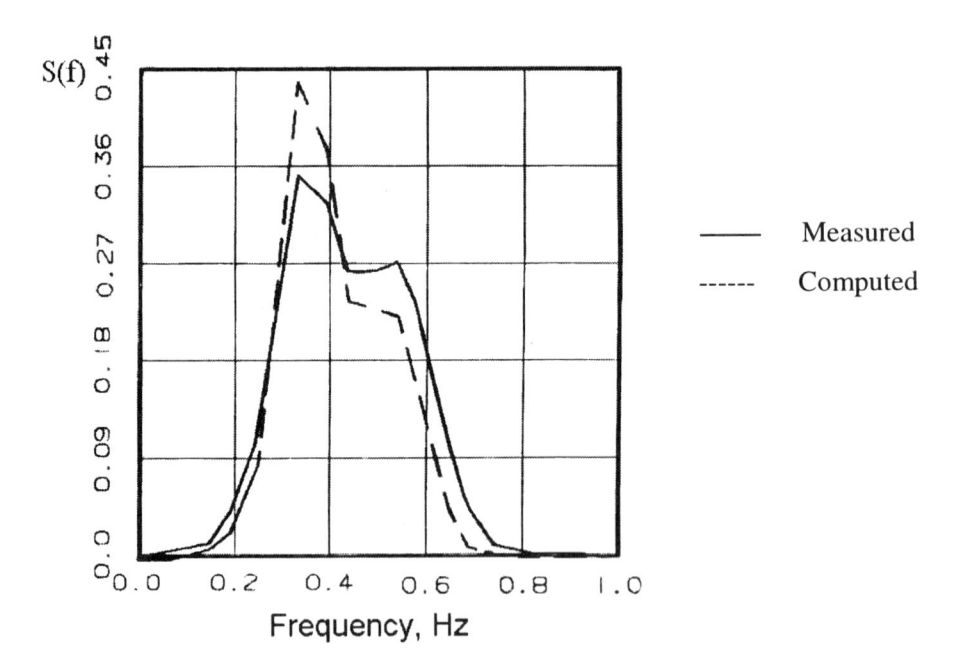

Fig. 6.10: Computed vs. Measured Force Spectra on a Vertical Cylinder

Note that the first peak of the force spectrum (which is mainly contributed from the linear inertia force term in Eq. 6.46) corresponds to the peak in the wave spectrum. However, the second peak at the higher frequency appears from the nonlinear drag force acting on the vertical cylinder. The computation by the approximate linearized formula in Eq. 6.46 matches the measurement well.

6.5.2 Stochastic Linearization of Response

In a random oscillatory flow, a floating structure will undergo a random oscillation. The motions of these structures in harmonic flows have been addressed in Chapter 3. Consider a floating structure whose single degree of freedom motion is described by the dynamic equation

$$m\ddot{x}(t) + c_1 \dot{x}(t) + c_2 |\dot{x}(t)| \dot{x}(t) + kx = F(t) \tag{6.47}$$

in which the force $F(t)$ is a random excitation, and the structure experiences a nonlinear damping with the coefficient c_2. This equation was addressed for harmonic excitation in Chapter 3. For the computation of the response spectrum due to random excitation, the nonlinear equation is first linearized. The method of stochastic linearization is similar to the above section. The nonlinear damping term is approximated using the rms value of the structure velocity $\sigma_{\dot{x}}$ by the relation:

$$|\dot{x}(t)| \dot{x}(t) = \sqrt{\frac{8}{\pi}} \sigma_{\dot{x}} \dot{x}(t) \tag{6.48}$$

This approximation is substituted in Eq. 6.47, which makes the equation linear in the time function. The spectral relationship of this equation of motion may be written as

$$S_x(\omega) = H^2(\omega) S_f(\omega) \tag{6.49}$$

in which the transfer function is derived as

$$H(\omega) = \frac{1}{\left([k - m\omega^2]^2 + [(c_1 + \frac{8}{\pi}c_2\sigma_{\dot{x}})\omega]^2\right)^{1/2}} \qquad (6.50)$$

Therefore, if the flow velocity (hence, acceleration) time history is known, its effect on the floating structure and the corresponding structure motion may be computed in the following way. First the force spectrum is computed, for example, by the Morison equation as shown above. Equations 6.48 and 6.49 are applied to derive the motion spectrum. This is illustrated in the example in Fig. 6.11. In the example, the results are shown including linear and linearized nonlinear damping. Near the resonance frequency, the nonlinear damping provides a significant effect on the motion response of the structure.

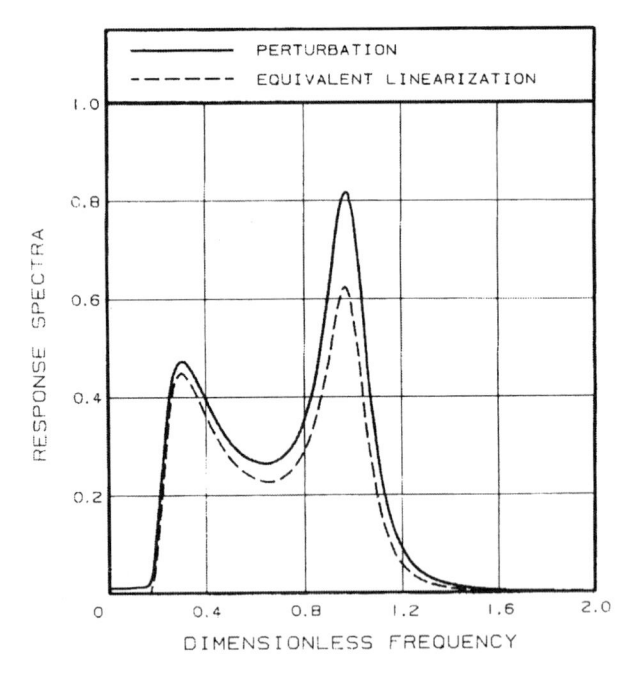

Fig. 6.11: Motion Response Spectrum for a Nonlinearly Damped SDOF System

6.6 Numerical Methods in Stochastic Dynamics

The systems described in the earlier section are rigid body systems. In these cases, the structure modules may be treated as having a single mass and undergoing six degrees of freedom. Therefore, even in the presence of multiple modules in the system, the elasticity enters only in the mooring system and the problem is essentially solved as a multiple spring-mass system. However, if the structure itself undergoes large deformation, then such an approach is not appropriate. In these cases, the structure is represented by many mass elements which all experience different displacements, so that the structure takes on different shapes as the excitation varies with time. The numerical analysis methods that should be chosen for such systems are described briefly in this section. One of the most popular techniques for analyzing these systems is the finite element method. While this method generally applies to deterministic dynamics, the technique may be equally applicable to stochastic inputs for linear and certain nonlinear systems.

6.6.1 Eigenvalues and Eigenvectors

A linearly damped dynamic flexible system subjected to an excitation may be represented by a finite number of elements. Then the dynamic equation of motion for the system may be written in a matrix form as

$$[M]\{\ddot{X}\} + [C]\{\dot{X}\} + [K]\{X\} = \{F(t)\} \qquad (6.51)$$

in which $[M]$ = mass matrix, $[C]$ = damping matrix, $[K]$ = stiffness matrix, and $\{F(t)\}$ = external force vector as a function of time. The vector $\{X\}$ represents N degrees of freedom response of the system. The N degrees of freedom correspond to N nodes representing the continuous system. In this case each node undergoes a deflection or displacement. The quantities $[M]$, $[C]$ and $[K]$ are N x N mass, damping and stiffness matrices. The mass matrix is generally symmetric and banded and could even be diagonal. It includes the added mass effect from the fluid. The matrix $[K]$ is symmetric and banded. Both these elements are positive. The matrix $[C]$ is the damping matrix and is also symmetric in most cases. The forcing function is a function of time, which provides the response of the system as a function of time as well. Consider for now that the force is absent. Then the undamped free vibration of the system may be described by the familiar form

$$[M]\{\ddot{x}\} + [K]\{x\} = 0 \qquad (6.52)$$

If the N nodes of the system are displaced from the original position and released, then the motion of the nodes will be described by Eq. 6.52. Assuming that the structure is supported at prescribed locations, the nodes can only move in harmonic oscillations depending on the natural modes of vibration of the system. Since no energy is dissipated in the absence of damping, the motion will theoretically be perpetual. The solution may be written in the familiar form:

$$\{x\} = \{X \exp(i\omega t + \varepsilon)\} \qquad (6.53)$$

where ω is the natural frequency and ε are the phase angles corresponding to the response amplitudes X. On substituting this expression in Eq. 6.52 and writing $\lambda = \omega^2$, one obtains,

$$([K] - [\lambda M])\{X\} = 0 \qquad (6.54)$$

Equation 6.54 is a homogeneous matrix equation, which will have nontrivial solutions (i.e. the vector $X \neq 0$) only if

$$|K - \lambda M| = 0 \qquad (6.55)$$

In other words, the determinant of the characteristic equation (as Eq. 6.55 is called) must vanish. This equation represents an Nth order polynomial in λ. The roots of the polynomial are called the characteristic values or eigenvalues. Thus, in general, for the N degrees of freedom, there are N independent eigenvalues (λ_i, $i=1,2,3...N$). For each value of λ_i, one would expect a nontrivial solution x_i. These solutions are called the eigenvectors. The eigenvalue and the corresponding eigenvector are called an eigenpair.

Consider a simple example for N=2. The mass and stiffness matrix are given by

$$[M] = \begin{bmatrix} 3 & -1 \\ -1 & 2 \end{bmatrix} \text{ and } [K] = \begin{bmatrix} 5 & 1 \\ 1 & 4 \end{bmatrix} \qquad (6.56)$$

Substituting these matrices in Eq. 6.55, the eigenvalues may be computed to be $\lambda_1 = 1$ and $\lambda_2 = 19/5$. Similarly, the values of the eigenvectors are obtained from Eq.6.54 as

$$\{x_1\} = \begin{bmatrix} 1 \\ -1 \end{bmatrix} \text{ and } \{x_2\} = \begin{bmatrix} 3/4 \\ 1 \end{bmatrix} \qquad (6.57)$$

The frequencies of the free vibration may be computed simply from

$$\omega_i = \sqrt{\lambda_i}; \quad i = 1,2,3...N \qquad (6.58)$$

If the eigenvalues are arranged from the lowest to the highest, then the first frequency ω_1 is called the fundamental mode. For each of these frequencies, the system assumes a mode shape, and the shape corresponding to ω_1 is called the fundamental mode shape.

The first three mode shapes of a flexible pipe with pinned-pinned joints are shown in Fig. 6.12. Note that the modes correspond to the bending of the pipe.

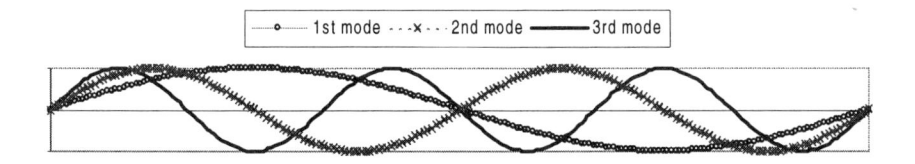

Fig. 6.12: Bending Modes of a Flexible Pipe with Pin Connections

6.6.2 Structure Finite Element Method

The most common finite element technique in solving the vibrations of a continuous system is the matrix displacement or stiffness method [Broch (1980)]. In this method, the displacements are the unknowns. The continuum is divided into a finite but large number of elements connected at their nodal points. Each discrete element is idealized. The displacements of these elements are assigned assumed values and the complete solution is obtained by combining these displacements so that the system satisfies the force equilibrium and displacement compatibility at the nodal points. The ensuing matrix equation is solved numerically by inversion of the stiffness matrix at each time step. The resulting solution provides the displaced shape of the structure at each time increment. Knowing the displacement, other desired quantities are calculated.

The matrix equation for the static analysis of the structure displacement due to time invariant loads is written as

$$[F] = [K][X] \qquad (6.59)$$

in which $[F]$ = force vector, $[X]$ = displacement vector and $[K]$ = stiffness matrix for the system. For a one-spring element the two forces are related to the two displacements as

$$F_1 = k(x_1 - x_2) \qquad (6.60)$$

$$F_2 = k(x_2 - x_1) \qquad (6.61)$$

which can be express in the matrix form:

$$\begin{bmatrix} F_1 \\ F_2 \end{bmatrix} = \begin{bmatrix} k & -k \\ -k & k \end{bmatrix} \begin{bmatrix} x_1 \\ x_2 \end{bmatrix} \qquad (6.62)$$

where F_i are the forces and x_i are the displacements. For a two-spring system, the equations become

$$F_1 = k_1(x_1 - x_2) \qquad (6.63)$$

$$\begin{aligned} F_2 &= k_1(x_2 - x_1) + k_2(x_2 - x_3) \\ &= -k_1 x_1 + (k_1 + k_2)x_2 - k_2 x_3 \end{aligned} \qquad (6.64)$$

$$F_3 = k_2(x_3 - x_2) \qquad (6.65)$$

or in matrix form:

$$\begin{bmatrix} F_1 \\ F_2 \\ F_3 \end{bmatrix} = \begin{bmatrix} k_1 & -k_1 & 0 \\ -k_1 & k_1+k_2 & -k_2 \\ 0 & -k_2 & k_2 \end{bmatrix} \begin{bmatrix} x_1 \\ x_2 \\ x_3 \end{bmatrix} \qquad (6.66)$$

Returning to the dynamic equation of motion in Eq. 6.51, the mass matrix is generated for the elements based on how the structure is divided into elements. The forces are computed at the node of these elements based on the environmental effects on the geometry of the particular element. Similarly, the damping, which is considered linear here, represents the damping value corresponding to that element. The dynamic equation of motion may be solved in a number of different ways, such as frequency response method, normal mode method, statistical technique, etc.

Various shapes can be used in describing the finite elements depending on the complexity of the structure. The most common and easiest are the beam elements. But there are other elements, such as shear panels, shell and ring elements, and solid elements. It is beyond the scope of this book to go beyond this treatment of a continuous system. The reader is referred to the book by Zienkiewicz (1977) and others on this subject for further details.

6.7 Response Statistics in Random Excitation

In nature, the environment produces an excitation that is often random. One of the fundamental tasks in structure design is to estimate the extreme load and structure response. Reliable extreme value predictions are vital in order to avoid compromising the safety and economy of a particular structure development.

Present engineering practice regarding extreme value predictions is not uniform. One direct approach is to base the predictions on maxima recorded in model experiments when a large number of maxima are sampled. However, in practice, this method is difficult to apply because not enough observations are generally available from model tests. Moreover, the scaling of the model data to the real structure is often questionable. A more common and accepted approach is to assume that the response process is Gaussian, or nearly Gaussian, and to base the predictions on well-established formulas.

In particular, in this approach the maxima may be derived from the Rayleigh distribution function (See Chapter 5). This method is applicable to systems that are linear or may be linearized without appreciable errors.. A floating structure may respond to a first order and a second order excitation, such as slow drift of the vessel at its natural period, in which case the Rayleigh distribution is not appropriate. In this case, a superposition of the two responses is applicable. It may be assumed that the total response consists of two components, a Gaussian, wave frequency part and a second-order, slowly-varying part near the natural period of the system. Various formulas, e.g., square root of sum of the squares that combine these two responses are used to estimate the total extreme from component extreme response estimates.

Table 6.1: Characteristics of Example TLP [Teigen & Naess (2000)]

Item	Value
Draft	28.50m
Column diameter	8.75m
Column separation distance	28.50m
Pontoon height & width	6.25m
Surge & sway natural period	133.5s
Yaw natural period	121.0s

This section will describe how such results may be obtained for a particular case. The method is generally applicable to a wide class of moored structures. The example of a small Tension Leg Platform (TLP) is chosen here from Teigen & Naess (2000). A TLP is a floating structure vertically moored in the ocean environment. The major dimensions of the example TLP are shown in Table 6.1. The reason for choosing a TLP is because it includes both the linear and nonlinear horizontal response in the coupled motion of the moored structure.

A direct approach in the extreme value prediction may be based on solving an associated eigenvalue problem for the first- and second order motion response. The probability distribution function (pdf) for the component and the combined process is computed. This method can be used to obtain asymptotic extreme values. Another method is conceptually simpler, but more involved in practice. The dynamic system is simulated in the time domain, and a large number of realizations of the same underlying process are generated. The resulting time histories are subsequently analyzed statistically, and the probability density and extreme value distributions for the component and total response processes are determined.

The excitation is chosen as a short-crested sea whose directional spectral density is given by $S(\omega,\theta)$. A short-crested wave is a random wave that varies not only with frequency, but also with direction. It is usually assumed that these two dependence may be uncoupled as

$$S(\omega,\theta) = S(\omega)D(\theta) \qquad (6.67)$$

where $S(\omega)$ is a wave spectrum (such as a PM or JONSWAP) and $D(\theta) = D_0\cos^{2s}(\theta - \theta_0)$ is a directional function where D_0 is a normalizing constant, θ_0 is the main direction of wave travel and $2s$ is the wave directional spreading parameter.

It is assumed that the total horizontal wave-induced response of the TLP structure subjected to a random, short crested sea can be written as a sum of a linear and a quadratic response component. Specifically, let $X(t)$ denote the total surge response process of the TLP. Then

$$X(t) = X_1(t) + X_2(t) \qquad (6.68)$$

where the subscript 1 refers to the linear component while subscript 2 is the quadratic part. The linear response time history is written in terms of the linear transfer function H_1 (similar to unidirectional waves) as follows:

$$X_1(t) = \sum_{l=-L}^{L}\sum_{n=1}^{N} H_1(\omega_n,\theta_l)a_{nl}B_{nl}\exp(-i\omega_n t) \qquad (6.69)$$

in which the wave components run from 1 to N, the spreading function varies from $-L$ to L and the function $H_1(\omega_n,\theta_l)$ denotes a directional, linear transfer function, which depends on frequency and wave direction. The wave amplitude is based on a small section ($\Delta\omega$, $\Delta\theta$) of the directional spectrum and is given by

$$a_{nl} = \sqrt{S(\omega_n, \theta_l) \Delta\omega\Delta\theta} \qquad (6.70)$$

In the above equation an equidistant discretization is assumed for simplicity, which is not mandatory. It is easy to modify the above equations to cover the situation of nonequidistant discretization. The nonlinear response time history is obtained from the second-order transfer function H_2

$$X_2(t) = \sum_{l=-L}^{L}\sum_{k=1}^{L} \sum_{n=1}^{N}\sum_{m=1}^{N} H_2(\omega_{n-m}, \theta_l, \theta_k) b_{nmlk} B_{nl} B^*_{mk} \exp[-i(\omega_n - \omega_m)t] \qquad (6.71)$$

in which the quantity $\{B_{nl}\}$ is a set of independent, complex Gaussian $\Phi(0, 1)$ variables with independent, identically distributed real and imaginary parts and * signifies complex conjugate. The quantity b_{nmlk} is the square of the wave amplitude given by

$$b_{nmkl} = [S(\omega_n, \theta_l) S(\omega_m, \theta_k)]^{1/2} \Delta\omega\Delta\theta \qquad (6.72)$$

The function H_2 is the quadratic transfer function based on bichromatic waves and two associated directions. Both H_1 and H_2 are obtained from the wave diffraction theory using the FEM or BEM. Note that the surge response of a TLP has important contributions both from the first order wave response and from the second order slow drift response based on the difference frequency. This is assumed in Eqs. 6.68, 6.69 and 6.71.

The combined extreme response incorporates the coupling between the linear and the quadratic responses. Let X_m be the combined extreme value of the surge response $X(t)$ over a time period T_s. Then the extreme value distribution $P(x)$ of the surge response is calculated by the formula

$$P_{X_m}(x) = \exp(-\nu T_s) \qquad (6.73)$$

where ν is the average level upcrossing rate of the total surge response. The average upcrossings rate of the mean value level is used for a Gaussian process. The total level upcrossing rate is computed based on the individual rates for the wave frequency and slow drift components as follows:

$$\nu(x_{ref}) = \left(\frac{(\nu_{X^{(1)}}(0))^2 m_0^{(1)} + (\nu_{X^{(2)}}(x_{ref}))^2 m_0^{(2)}}{m_0^{(1)} + m_0^{(2)}} \right)^{1/2} \qquad (6.74)$$

where m_0 refers to the zeroth moment (variance) of the respective spectrum, $v_X^{(1)}(0)$ = mean zero upcrossing rate of the Gaussian component X_1, and $v_X^{(2)}(x_{ref})$ = the mean zero upcrossing rate of the slow-drift component X_2. Here $v^{(2)}(x_{ref}) = 1/T_n$, where T_n denotes the surge slow-drift period.

The principal advantage of the time simulation is that all major aspects of the physical model can be included. In other words, nonlinearities associated with damping/mooring etc. are an integral part of the mathematical formulation, and that these terms do not have to be simplified. The most important consequence is that time simulation is more appropriate for large amplitude motions where the nonlinearities tend to take effect, and for which the maxima will occur. Assume the process to be ergodic. The record length should be long in order to obtain estimates within a certain confidence band. Within the design community, a 4-hr period is often considered to be "sufficiently long". Table 6.2 gives an overview of the fluctuations in the low order cumulants from 50 independent 4 hr. realizations. Table 6.2 refers to one particular sea state with H_s = 10 m, T_p = 10 s and with a narrow band $\cos^{32}(\theta - \theta_0)$ spreading function. Subscripts 1 and 2 refer to first-order and slow drift component processes.

Table 6.2: Fluctuations in TLP Time Simulation Response Statistics
[from Teigen & Naess (2000)]

Item	Mean	St. Dev	COV	Range
μ	3.871	0.128	0.033	3.659 – 4.209
σ_1	1.150	0.023	0.020	1.107 – 1.215
σ_2	5.831	0.484	0.083	4.801 – 7.305
σ	5.946	0.475	0.080	4.932 – 7.398

COV = Coefficient of Variation (i.e. mean / standard deviation)

The variation in these quantities in Table 6.2 is considerable, and shows the danger in basing all statistical predictions on a single realization. As one would expect the slow drift standard deviation has the largest scatter. An accurate estimate of the "true" cumulants (e.g. mean (μ) or standard deviation (σ)) has to be extracted from the distribution of sampled cumulants. The discrete distributions of the mean and standard distributions of the TLP surge obtained from the 50 individual samples when compared to a corresponding normal distribution were in general agreement.

Note that the individual realizations are all "drawn" independently from the same underlying sample space. Hence, if the distribution for a single sample point has mean value μ and standard deviation σ, the arithmetic mean of N sample points will have a mean value μ_N and standard deviation σ_N given by:

$$\mu_N = \mu; \qquad \sigma_N = \sigma/\sqrt{N} \tag{6.75}$$

In view of the Central Limit Theorem, the distribution for the arithmetic mean of N sample points should asymptotically approach a normal distribution for a large value of N. Note that the normal distribution is completely defined by the mean and the standard deviation (Chapter 5). Therefore, by making use of the results from Table 6.2 and the foregoing analysis for the arithmetic mean of N samples, one can determine the number of realizations that are needed in order to obtain results within a certain confidence band.

Consider the estimate of the standard deviation for the slow drift response (σ_2) which has the largest fluctuations. Assuming a 95% probability, this estimate should stay within 3% of the "true" value. For a normal distribution, approximately 95% of the distribution falls within $\mu \pm 1.96\sigma$. The mean value and standard deviations are 5.831 and 0.484 respectively (Table 6.2). Consequently, assuming that these are the "true" estimates, the number of realizations N needed to get an estimate within the desired accuracy is given by the inequality

$$1.96 \times 0.484/\sqrt{N} \leq 0.03 \times 5.831 \tag{6.76}$$

which gives $N \geq 29.42$, so that 30 realizations are required. If instead, the estimate is needed to be within 2% of the target value with 95% probability, the number of realizations increases to 67.

Although these results are obtained within a specific context with chosen environmental and structural parameters, the methodology may be considered general. Furthermore, the example given above is at least indicative of the level of robustness for statistical estimates obtained by the time simulation approach.

6.8 Exercises

Exercise 1

Compute the surge motion response spectrum of a floating barge moored with linear springs in waves. Assume the barge to be a block of length 250 ft (76.2 m), beam 20 ft (6.1 m), and draft 10 ft (3.05 m) in a water depth of 200 ft (61 m). The natural period of the spring mass system in surge is 10 s. Assume a linear damping factor of 10 percent ($\zeta = 0.1$) and an added mass equal to the displaced mass of the barge. The barge is subjected to an inertia force given by $F_0\eta(t)$ where the force amplitude is related to an average acceleration of the displaced fluid

$$F_0 = m\frac{g}{D}\frac{\sinh kd - \sinh ks}{\cosh kd}\frac{\sin kl}{kl} \qquad (6.77)$$

where m = mass, g = gravity, d = water depth, D = draft, $s = d - D$ and l = length of the barge. The quantity k is the wave number, which is related to the wave frequency ω by Eq. 2.45.

The random wave profile is given by $\eta(t)$ whose frequency spectrum representation is given by a Pierson-Moskowitz (PM) spectrum model having a significant wave height H_s of 53.7 ft (16.4 m) written (in an alternate form to Eq. 4.1) as

$$S(\omega) = \frac{5}{16}H_s^2\frac{\omega^{-5}}{\omega_0^{-4}}\exp[-1.25(\omega/\omega_0)^{-4}] \qquad (6.78)$$

and the peak frequency of the wave spectrum

$$\omega_0^2 = 0.161 g/H_s^2 \qquad (6.79)$$

Note that for a single frequency the motion response may be written in terms of the time function as

$$x(t) = \frac{F_0 m/\omega_n^2}{[(\omega_n^2 - \omega^2)^2 + (2\zeta\omega_n\omega)^2]^{1/2}}\cos(\omega t + \varepsilon) \qquad (6.80)$$

where ε is the phase angle between the motion and the incident wave.

Compute the input wave spectrum from Eq. 6.78. Using the frequency domain approach, the transfer function between the motion and input wave is written as

$$H(\omega) = \frac{\dot{u}_0}{[(\omega_n^2 - \omega^2)^2 + (2\zeta\omega_n\omega)^2]^{1/2}} \qquad (6.81)$$

Show the transfer function for the barge in surge in a plot and demonstrate that there are several peaks in the transfer function, especially at the high frequency end. These appear from the last term in Eq. 6.77 caused by the barge length being a multiple of the wave lengths where the forces on the barge disappear (zero average acceleration). Calculate the response spectrum from the input spectrum and the RAO. Show that most of the wave energy is concentrated at the low end of the frequency band near the natural frequency. The additional peaks in the response spectrum appear from the multiple peaks in the RAO.

Exercise 2

Compute the second order group spectrum for a PM spectrum model using a significant wave height of H_s = 20 m. Repeat the computation for the same wave using JONSWAP spectrum for γ = 3.3. Comment on the low frequency contents of the two.

Exercise 3

Derive Eqs. 6.39 and 6.42 from Eq. 6.36 using Lagrange's relationship. Use the excitation moment term on the right-hand side. Show the sources of the excitation term and discuss the possible nonlinear terms in the equation of motion.

Exercise 4

Compute the force spectrum for a unit length of a vertical cylinder assuming the PM wave spectrum model as the excitation. Use C_M = 2.0 and C_D =0.7 and stochastic linearization for the drag force. Plot the inertia and drag contribution separately and comment on the effect of the nonlinear term in the force spectrum.

6.9 References

1. Bendat, J.S., and Piersol, A.G., Engineering Applications of Correlation and Spectral Analysis, John Wiley & Sons, New York, 1980.

2. Borgman, L.E., "Ocean Wave Simulation for Engineering Design", Journal of Waterway, and Harbor Division, ASCE, Vol. 95, No. WW4, Nov, 1969, pp 557-583.

3. Broch, J.T., Mechanical Vibration and Shock Measurements, Bruel & Kjaer, 1980.

4. Chakrabarti, S.K., Nonlinear Methods in Offshore Engineering, Elsevier, Netherlands, 1990.

5. Chakrabarti, S. K. and Cotter, D. C., "Motions of Articulated Towers and Moored Floating Structures", Journal of Offshore Mechanics and Arctic Engineering, ASME, Vol. 111, No. 3, Aug. 1989, pp. 233-241.

6. Gottlieb, O., "Bifurcations of a Nonlinear Small-Body Ocean Mooring System Excited by Finite Amplitude Waves", International Conference on Offshore Mechanics and Arctic Engineering, ASME, 1996.

7. Teigen, P., and Naess A., "Response Statistics of a Moored Structure in Short-Crested Waves", International Conference on Offshore Mechanics and Arctic Engineering, OMAE 6073, ASME, 2000.

8. Yang, J.C.S., et al., "Determination of Fluid Damping Using Random Excitation," Journal of Energy Resources Technology, Vol. 107, June 1985, pp. 220-225.

9. Zienkiewicz, O.C., The Finite Element Method in Engineering Science, 3rd edition, McGraw-Hill, 1977.

Chapter 7

Fluid Structure Interaction

7.1 Introduction

Fluid dynamics is a major topic of applied mathematics and physics, and many of the more relevant theories of these sciences have contributed to the study of fluid flow problems. Fluid dynamics is also one of the most challenging areas of computational mechanics. The simulation of fluid dynamics problems is always a rigorous test of any numerical method, particularly when dealing with nonlinear transient problems.

For structures placed in the path of flowing fluids, fluid structure interaction plays an important role in their operational behavior. The motion of the fluid may be steady or oscillatory. It has been shown in Chapter 2 that the fluid flow and the associated fluid field are altered due to the presence of the structure. In addition, if the structure is long or tall, whether land-based or ocean-based, it exhibits some degree of compliance to react to the effects of the environmental force. The dynamics of the structures is important, as demonstrated by the compliant nature of these structures, through their capacity to deform significantly due to the environmental forces. For the safe and cost-effective design and operation, it is therefore crucial to ensure their operability and their strength, stability and fatigue life in their design. For example, two complaint offshore towers installed in the Gulf of Mexico in over 1600 ft (500m) of water are two of the tallest towers of their kind in the world built to date. Due to the global flexibility of the supporting towers, the upper parts of such structures experience significant fluid-structure interaction and move significantly under the wind, wave and current loads. In this case, the topside and the upper buoyancy tanks may be treated as rigid bodies for the motion analysis. However, the longer supporting structures beneath them will be elastic and will experience global bending and torsional modes.

This chapter will examine the effect of fluid flow on structures, and will discuss the fluid-structure interaction due to fluid flow. The discussion will start with the structures placed in steady flow. The steady flow can be in the form of current, which may be uniform with depth or may have a vertical shear. Unlike steady flows, structures in waves experience large oscillating force in the direction of flow. The fluid structure interaction is more complex in this case. The method of computation of forces on submerged structures and the corresponding motions of floating structures in waves will be discussed. Large floating structures are considered rigid and their motions in six degrees of freedom are computed as a rigid body analysis. When multiple structures are present in close proximity, the interactions among them are important. This is demonstrated with the example of a two-body interaction problem. Many long slender structures in steady or oscillating flows experience large deformations in some of their elements. The analysis method in these cases should additionally consider the stiffness of the structural material. This flexibility of the structure and the associated fluid structure interaction are analyzed using examples of a long slender riser and flexible mooring line.

7.2 Structures in Steady Flow

When structures are placed within a flowing fluid (liquid or gaseous) medium, the fluid flow pattern is altered and restraints are required to maintain the position of the structure within the fluid medium. The loads on the structure are normally characterized as drag (in-line with the direction of flow) and lift (transverse to the direction of flow). In the inline direction, the load can be described as being composed of a steady component and a smaller higher frequency component, which are both functions of the shape of the structure and existing flow conditions. In the transverse direction, the load is mainly oscillatory having multiple frequencies.

Downstream of the structure, the fluid flow regime will eventually return to its unaltered upstream condition due to fluid viscosity and damping. The region of altered flow directly behind the structure is defined as the "wake", and investigators have demonstrated that a one-to-one relationship exists between the extent of wake and the restraint loads. The fluid-structure interaction has been shown to be determinant, i.e., for the

Chapter 7 Fluid Structure Interaction

same body and flow conditions, the restraint loads will have the same steady and higher frequency components. The wake, although seemingly chaotic, will have a similar width and will have the same frequency content as the restraint loads. In this regard, the body is defined by its shape, surface roughness and orientation. The fluid is defined by its viscosity. The flow is defined by its velocity, vorticity and turbulence. The regime can be described by its boundaries.

Table 7.1: Flow Regimes in a Uniform Flow Past a Cylinder

FLOW REGION	*Re* RANGE	FLOW CONDITION	FORCE ON CYLINDER
Laminar	0 - 40	No separation of flow from the body. Continuous stream lines.	In-line forces only.
Subcritical	$40 - 5 \times 10^5$	Broken stream lines. Well defined periodic vortex shedding.	Strouhal number dependent lift force frequency. Steady in-line force. Smaller oscillating in-line force at twice the lift force frequency.
Supercritical	$5 \times 10^5 - 7 \times 10^5$	Ill-defined vortices.	Rapid decrease in drag force. Lift and drag forces at higher frequencies.
Transcritical	$> 7 \times 10^5$	Vortices are less well defined and persistent. Turbulence due to fluid viscosity produces a randomness.	Fluid-structure interaction similar to subcritical range.

Under the conditions of a smooth circular cylinder of infinite length that has been fixed with its axis perpendicular to the direction of uniform flow without turbulence and with no boundary effects, the fluid-structure interaction can be uniquely defined by the Reynolds number, Re (Eq. 2.16). Under these stringent conditions, the fluid-structure interaction can be subdivided into four distinct flow regimes: laminar, subcritical, supercritical and transcritical, which are described in Table 7.1 [see LeMehaute (1976)]. The laminar flow regime exists at a very low Reynolds number where the flow stays attached to the body causing only an inline force. As the Reynolds number increases, the flow detaches from the body and regular vortices are formed and shed. When the Reynolds number is further increased into the supercritical regime, the vortex shedding becomes more and more confused.

The numerical values of the Reynolds number regimes presented in the table should not be considered inviolate. These ranges are estimations of the boundary values. Although the boundaries may have more precise values, no test has yet been performed that completely satisfies the ideal conditions assumed. In other words, no test exists in which the fluid flow is truly uniform and without turbulence, where the structure is completely fixed, smooth and infinitely long and where there are no boundary effects. Each one of these "real world" parameters affects the fluid-structure interaction and shifts the boundaries of the flow regime. This is reflected in the specified ranges. In addition, very few test programs have successfully entered the transcritical regime in a laboratory situation where these parameters can be suitably controlled.

Research has been performed which sought to identify the effects of independent nondimensional parameters with respect to the boundary values and subsequent fluid-structure interaction and restraint loads. An extensive list of important nondimensional parameters is presented in Table 7.2. Some of these parameters involve forced oscillation while others apply to free vibration of structures. Some parameters may be important for one set of structures but not significant for others. Three nondimensional parameters, namely, drag, lift and side force coefficients (included in Table 7.2) are not independent variables, but depend on other nondimensional parameters. The nondimensional parameters in Table 7.2 will be involved in the discussions in the following sections.

Chapter 7 Fluid Structure Interaction 253

Table 7.2: Summary of Nondimensional Variables

NONDIMENSIONAL VARIABLE	SYMBOL	DEFINITION	PARAMETERS
Reynolds Number	Re	$\dfrac{UD}{\nu}$	U = steady velocity ν = water viscosity D = structure dimension
Strouhal Number	St	$\dfrac{f_s D}{U}$	f_s = shedding frequency
Frequency Ratio	f^*	$\dfrac{f_c}{f_s}$	f_c = oscillation frequency
Shear Parameter	β	$\dfrac{dU}{dy}$	y = vertical coordinate
Aspect Ratio	R_A	$\dfrac{L}{D}$	L = cylinder length
Roughness Parameter	k_S	$\dfrac{K}{D}$	K = surface roughness
Reduced Velocity	V_r	$\dfrac{U}{f_n D}$	f_n = natural frequency of structure
Free Amplitude Ratio	R_y	$\dfrac{y_o}{D}$	y_0 = amplitude of free oscillation
Forced Amplitude Ratio	R_a	$\dfrac{a}{D}$	a = amplitude of forced oscillation
Elevation Ratio	R_E	$\dfrac{y}{L}$	
Turbulence Intensity	T_I	$\dfrac{U_{dev}}{U}$	U_{dev} = deviation of velocity from mean velocity
Drag Coefficient	C_D	$\dfrac{f_I}{1/2 \rho U^2 D}$	f_I = in-line force per unit length
Lift Coefficient	C_L	$\dfrac{f_L}{1/2 \rho U^2 D}$	f_L = transverse force per unit length
Pressure Coefficient	C_p	$\left[\dfrac{p - p_o}{1/2 \rho U^2}\right] + 1$	p_0 = stagnation pressure p = pressure around cylinder circumference

Table 7.2 (*Continued*)

NONDIMENSIONAL VARIABLE	SYMBOL	DEFINITION	PARAMETERS
Velocity Gradient	ε	$\dfrac{U_m}{U} - 1$	U_m = maximum velocity
Surface Effect Coefficient	C_s	$\dfrac{y_1}{D}$	y_1 = vertical distance from cylinder axis to free surface
Bottom Effect Coefficient	C_b	$\dfrac{y_2}{D}$	y_2 = vertical distance from cylinder axis to bottom boundary

The following sections discuss the fluid-structure interaction for both a vertical cylinder and a horizontal cylinder when placed in uniform fluid flow versus when placed in vertical shear flow.

7.2.1 Vertical Cylinder in Uniform Flow

Vertical cylinders, as members of an underwater support structure, experience loading from the flowing fluid. For an infinitely long, smooth, fixed vertical cylinder in uniform flow without turbulence, the fluid-structure interaction is defined by the Reynolds number [Hoerner (1965)]. Thus, for a constant viscosity fluid, the fluid-structure interaction remains the same, and the value of the steady drag (in-line) coefficient remains the same if the velocity is doubled while the cylinder diameter is decreased by a factor of two.

If the undisturbed flow is in the subcritical range, the vortices will continue to be well defined and regular, and the shedding frequency will be characterized by the Strouhal number (Eq. 2.19). In the subcritical regime, the Strouhal number is about 0.2 (Fig. 2.8). Within the transcritical regime, the Strouhal number is approximately 0.3. In the supercritical transition region, regular vortex shedding ceases, and the Strouhal number becomes ill-defined [Humphries and Walker (1987)]. The cylinder roughness and fluid turbulence tend to reduce the boundary value on either side of the

Chapter 7 Fluid Structure Interaction

supercritical region. The decrease is more pronounced on the upper boundary, and the supercritical region tends to become narrower.

End effects caused by the cylinder's finite length introduce the aspect of three dimensionality to the flow. The ends of the cylinder affect the pattern of vortex shedding and the values of drag and lift coefficients. Vortices are not shed uniformly along the cylinder length (L), and the values of the drag and lift coefficients are dependent on the location along the cylinder. For aspect ratios up to about $L/D = 20$, the vortex shedding pattern near the center of the cylinder is affected by the ends. This is illustrated in Fig. 7.1 [Mair and Stansby (1975)], where the data was obtained in shear flow. The base pressure along the x-axis is shown as a function of the normalized elevation for various values of aspect ratio, L/D. Note that the pressure is different near the middle region for small values of L/D.

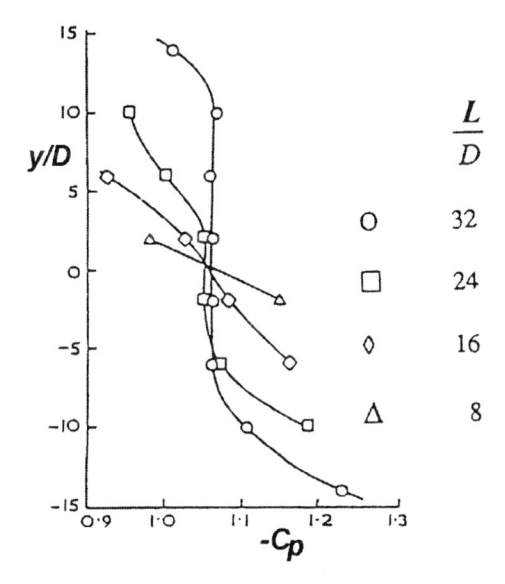

Fig. 7.1: Base Pressure C_p on Vertical Circular Cylinder vs. Spanwise Distance [Mair and Stansby (1975)]

The cylinder may have some freedom of movement because of its inherent flexibility or the flexibility of the mounting apparatus. Due to ease of control, many investigators have employed a forced oscillation test to duplicate this flexibility. In this case, reduced velocity, V_r (Eq. 2.18), which is proportional to the flow velocity and inversely proportional to the cylinder diameter and its structural natural frequency (Table 7.2), also becomes important. The amplitude of reduced velocity characterizes vortex-induced vibrations. Humphries and Walker (1987) separated the range of the reduced velocity into three different instability regions, as summarized in Table 7.3. The gaps in the ranges describe transition regions.

Within these three stability regions, the Strouhal number corresponding to the stationary cylinder is no longer valid. The shedding of vortices from the cylinder takes place at the structural vibrational frequency. This phenomenon where the vortex shedding frequency coincides with the modal frequency is commonly known as lock-in. The boundaries of these regions of instability are affected by the stability parameter, end effects, boundary effects and turbulence. As the amplitude of the cylinder motion increases, the range of lock-in increases, and lock-in can occur at one-half or one-third of the structural frequency [Stansby (1976)].

Table 7.3: Reduced Velocity Range for Vortex Induced Vibration of a Cylinder [Humphries and Walker (1987)]

Region	Reduced Velocity	Vortex Shedding	Type of Vibration
I	1.7 - 2.3	Symmetrical shedding	In-line oscillation only
II	2.8 - 3.2	Alternate shedding of vortices from two sides	Predominantly in-line vibration; some transverse vibration
III	4.5 - 8.0	Alternate shedding of vortices from two sides	Predominantly transverse vibration. In-line vibration at twice the transverse vibration frequency, resulting in a figure eight motion.

Chapter 7 Fluid Structure Interaction 257

7.2.2 Vertical Cylinder in Shear Flow

Shear current is a common phenomenon in nature. In fact, in deep oceans, flow is seldom uniform with depth. Positive shear where velocity is greatest near the surface, is most prevalent. However, shear can also be negative, especially in shallow water. When a vertical cylinder is in a shear flow, a three-dimensional flow regime will occur, with components of flow along the length of the cylinder. For a positive shear parameter, the variation of stagnation pressure on the upstream face of the cylinder promotes a vertical flow down the face of the cylinder. Conversely, the varying pressure in the wake of the cylinder promotes a vertical flow up the rear of the cylinder [Masch and Moore (1960)]. This effect is schematically shown in Fig. 7.2. The flow is vertically sheared from the upstream side of the cylinder to the downstream side, where a wake field exists. There is a downward flow on the upstream side with a corresponding upward flow on the downstream side.

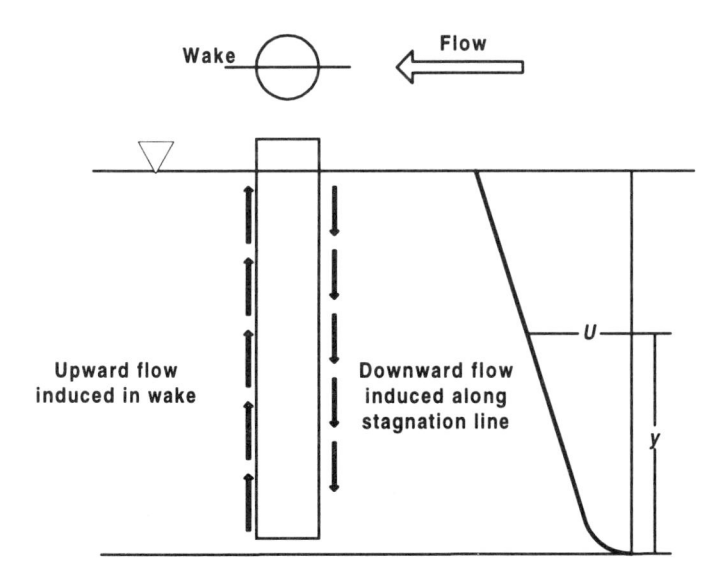

Fig. 7.2: Change of Stagnation Pressure on Vertical Cylinder in Shear Flow

258   *Theory and Practice of Vibration and Hydrodynamics*

These three-dimensional flow cells tend to reduce the expected dynamic pressure coefficient at the top of the cylinder, and increase the coefficient at its base. The drag coefficient has been found to change with the strength of shear, as established from experiment by Masch and Moore (1960). This variation of the local drag coefficient with shear parameter is demonstrated in Fig. 7.3. The drag coefficients are shown for four different shear parameters ranging from 0.058 to 0.117. The shear parameter is defined in Table 7.2 and normalized by U/D. The test data are fitted with mean lines. Note that the drag coefficient varies with cylinder depth as well as shear parameter.

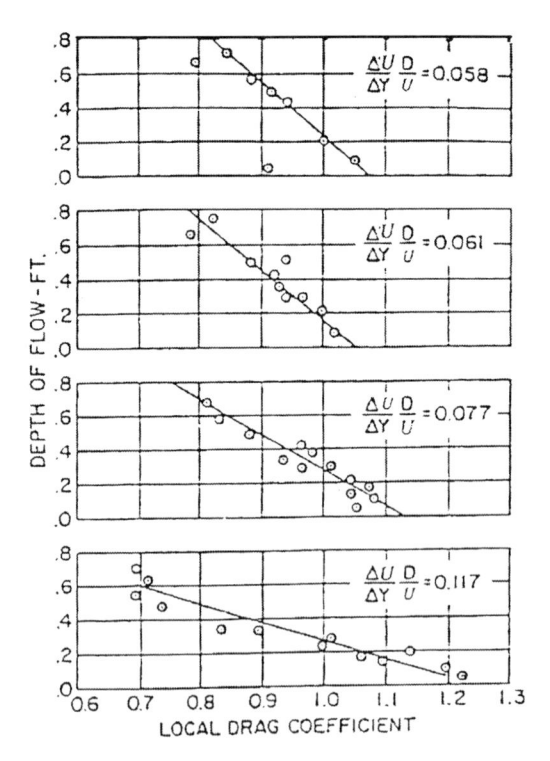

Fig. 7.3: Variation of Local Drag Coefficient with Shear Parameter [Masch and Moore (1960)]

Chapter 7 Fluid Structure Interaction

As already noted, a vertical cylinder in uniform flow will shed vortices at the same frequency, as defined by the Strouhal number, over the entire length of the cylinder, in the absence of end and boundary effects. In a shear flow, it may be expected that the vertical cylinder will shed vortices at a Strouhal frequency that is changing continuously with the velocity and a continuous spectrum of frequencies would arise. However, that is not the case. In a shear flow on a vertical cylinder, vortex cells are formed over which the shedding frequency does not vary continuously. The difference between the frequencies of uniform vs. shear flow is shown in Fig. 7.4. The number of cells and the extent of each cell are dependent on the aspect ratio, end and boundary effects, surface roughness and turbulence. There is a general trend toward decreasing cell length with increasing shear, and an increase in cell length with cylinder roughness [Griffin (1985)].

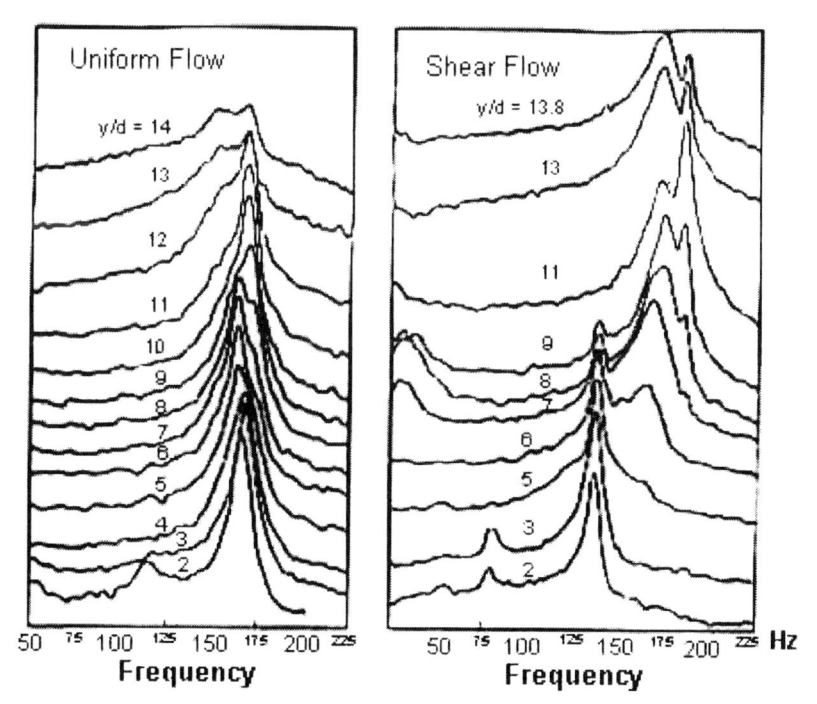

Fig. 7.4: Frequency Content of Wake Behind a Vertical Cylinder Due to Uniform Flow vs. Shear Flow [Maull and Young (1974)]

In a test with a vertical cylinder with a relatively small aspect ratio (L/D = 8) in shear flow [Mair and Stansby (1975)], only two stable end cells were formed. The bottom boundary layer carried vorticity which augments the shear, while the top boundary (when fixed) carried vorticity in the boundary layer of opposite sign. For a longer cylinder (L/D = 16), there was a region in mid-span extending about 8 diameters, which had cells distinct from the end cells, even though the cell structure was still dependent on the end conditions. Typically, the longest cells were on the order of $4D$ to $6D$.

At the intersection of the cells, there is a region of instability in which the shedding frequency tends to alternate between cells. If horizontal plates are attached to the two ends of the cylinder, these regions are stabilized and the cell structure becomes more distinct. In order to stabilize the cells, various investigators tried end plate configurations in which the end plate circumference was an integer multiple of the cylinder diameter. The optimum integer value appeared to be a function of the flow velocity. These plates, however, simply stabilize the cells; they do not completely remove the end effects.

In order to display the cellular structure of the vortices in shear flow, the Strouhal number is defined in terms of the local velocity or the mean velocity. If the local velocity is used in the definition, there will be a continuous change in the Strouhal number across the cell in order to maintain a constant frequency, and then a jump in value will occur at the discontinuous cell change. If a mean stream velocity, U, is used for the cell, the Strouhal number will remain a constant over the cell length and then jump to a new constant value at the next cellular region [Maull and Young (1974)]. In either case, the Strouhal number can not be considered as an independent variable in shear flow on a vertical cylinder, and can only be used as an indicator of the frequency range of vortex shedding that may exist (Fig. 7.4).

In the same way, the shear parameter, β is somewhat suspect as an independent variable. Apparently, no test data exists which demonstrates identical cellular flow patterns for setups with twice the diameter while half the shear. In fact, inspection of the Strouhal number would indicate that the cellular pattern would change in these setups.

7.2.3 Horizontal Cylinder in Uniform Flow

Subsea pipelines may be treated as horizontal cylinders, which encounter current in riverbeds and coastal regions. Horizontal cylinders are also found in subsurface and near surface regions as members of floating breakwaters, offshore structures, pontoons, etc. These structures are affected by current as well as waves. The stability of pipelines and their foundation design depends on the loading imposed on them by the fluid flow past them. This section describes the effect of uniform current flow on a horizontal cylinder placed across the flow.

The uniform velocity field U past a horizontal circular cylinder will generate waves near the upper boundary, i.e., free surface. This wave resistance is dependent upon the Froude number; $Fr = U^2/gy_1$ where y_1 is the depth of immersion measured from the cylinder axis to the free surface, and g is the gravitational acceleration. When y_1 is large compared to the cylinder radius, the maximum wave resistance occurs for $Fr = 1$ [Bishop and Hassan (1964)]. The wave resistance drag is negligibly small when $Fr < 0.375$ [Hoerner (1965)]. For towed ship shapes, the residual drag (i.e., form plus wave drag) is not a function of the Froude number for $Fr < 0.5$. Near the bottom boundary (or fixed top boundary without free surface), the resistance to flow causes a shear flow to develop. In this case, the vortex shedding pattern is affected by the proximity of the boundary.

Similar to a vertical cylinder, the boundary effects and end effects introduce three dimensionality in the flow past a horizontal cylinder of finite length. In a uniform flow in the absence of a boundary, the end effects are considered to be identical between a horizontal and a vertical cylinder.

In experiments with cylinders in a two-dimensional flow, the walls of the boundaries affect the pressure in the base region (180 degrees from the stagnation point). Use of end plates eliminates the boundary layer effects in the central part of the span of the cylinder, making the flow two-dimensional. However, even when the base pressure is uniform for most of the span, the flow may be far from two-dimensional [Stansby (1974)] for length-to-diameter (L/D) ratio of up to about 20. End plates of suitable design may make the flow nearly two-dimensional. The most optimum value of x/D (x = length of leading edge of base plate or its diameter) was found to be 2.5 [Stansby (1974)]. The effect of this plate on the base pressure along the length of the cylinder is shown in Fig. 7.5 from experiments. The Reynolds number was 4×10^4 for this case. For the smaller cylinder without end plates, the base pressure at the center of the span approaches the pressure of the 2-D flow (with end plates). However, the variation of the base pressure is considerably high, as the distance from the center increases.

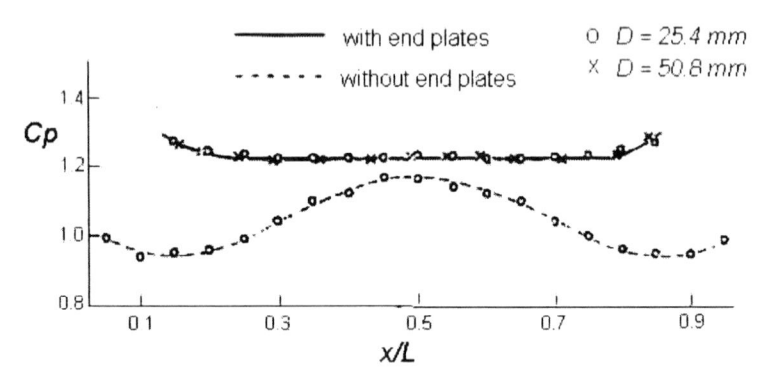

Fig. 7.5: Base Pressure Coefficient Along Cylinder [Stansby (1974)]

The major difference in the boundary effects between a horizontal and a vertical cylinder is the appearance of a lift force. As the horizontal cylinder approaches either boundary, the flow can no longer be symmetric, and the lift force becomes a function of the cylinder diameter and the distance to the boundary wall. A horizontal cylinder in uniform flow near a plane boundary can be treated as a pair of horizontal cylinders in the absence of a boundary. Data on the drag coefficient and Strouhal number as a function of the separation distance between the two cylinders [Hoerner (1965)] suggest that the proximity effect is negligible for gap/diameter (y_2/D) exceeding 4.0.

Pipelines in uniform flow may be subject to flow induced vibrations. Full-scale laboratory tests were performed at the Delft Hydraulic flume on flows around pipelines near the sea floor. Some of the pipes had vibration spoilers (to be discussed in Chapter 8) attached. In the experiment, the height of the pipe above the bottom and the height of the attached spoilers were varied. The conditions under which the alternating transverse forces on the pipe were suppressed, were determined for various steady flows. It was found that the spoilers on pipes increased the local scour when compared to a plain pipe, and increased the possibility of self-burial of the pipe in the ocean floor.

7.2.4 Horizontal Cylinder in Shear Flow

When a cylinder is mounted horizontally in shear flow, the relationship between the diameter D and the shear dU/dy takes on a physical significance in terms of velocity variation across the face of the structure. In this case, the mean velocity and the local velocity more nearly coincide. In the absence of boundary or end effects, the vortex cell structure exhibited by the vertical cylinder should disappear and the Strouhal number should regain its significance. However, the ratio of turbulence to velocity variation across a horizontal cylinder is much larger than for a vertical cylinder. It is important to recognize the importance of turbulence when comparing the results of shear flow to uniform flow. Ideally, for a one-to-one correspondence, both flows would have similar turbulence intensities.

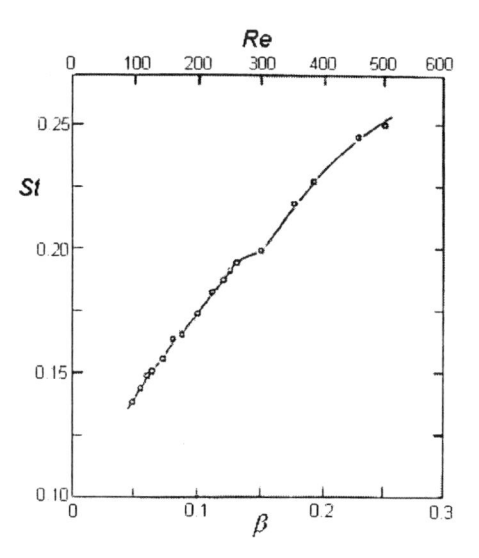

Fig. 7.6: Strouhal Number vs. Shear Parameter for Horizontal Cylinder [Kiya, et al. (1980)]

A test on a horizontal cylinder in shear flow was carried out by Kiya, et al. (1980). The wake characteristics behind the cylinder were measured as a function of the shear parameter and the Reynolds number. The aspect ratio in their tests varied from 2.0 to 12.5. In uniform flow, the transition into vortex shedding occurred at a slightly higher Reynolds number ($Re = 52$) than the generally accepted value of 40. This increase in the transition level was attributed to an (unmeasured) increased level of turbulence. The shear parameter ranged from 0.05 to 0.25. As the shear parameter increased, the Strouhal number, and hence the vortex shedding frequency, was found to increase as shown in Fig. 7.6. Similarly, the transition Reynolds number, at which vortex shedding appears, increased with the increase in the shear parameter, which is illustrated in Fig. 7.7.

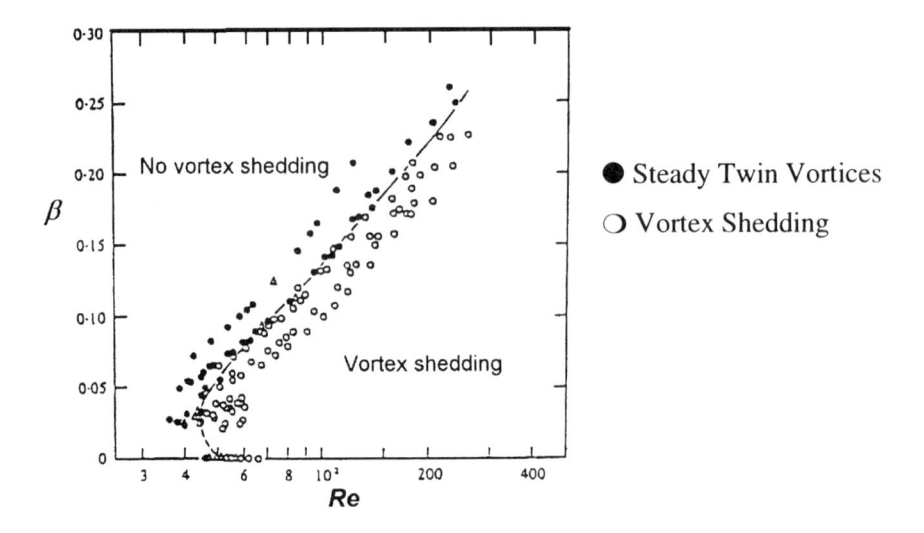

Fig. 7.7: Vortex Shedding Regions vs. Reynolds Number and Shear Parameter [Kiya, et al. (1980)]

In a test with a fixed horizontal cylinder [Chakrabarti, et al. (1991)], the loads on the cylinder in both uniform and shear current flow were measured

Chapter 7 Fluid Structure Interaction

to determine the effect of flow shear on the loads. The model was a smooth cylinder with an L/D ratio of 20, and was fixed at either end to the restraining wall. It was placed at elevations (y_1/d) of 0.5, 0.33 and 0.73 of the water depth. At the cylinder mid-span was a small instrumented section capable of measuring the local load on the cylinder in two orthogonal directions. A pluck test demonstrated that the model's first mode of vibration was more than double the computed Strouhal frequency of about 1 Hz for this test setup.

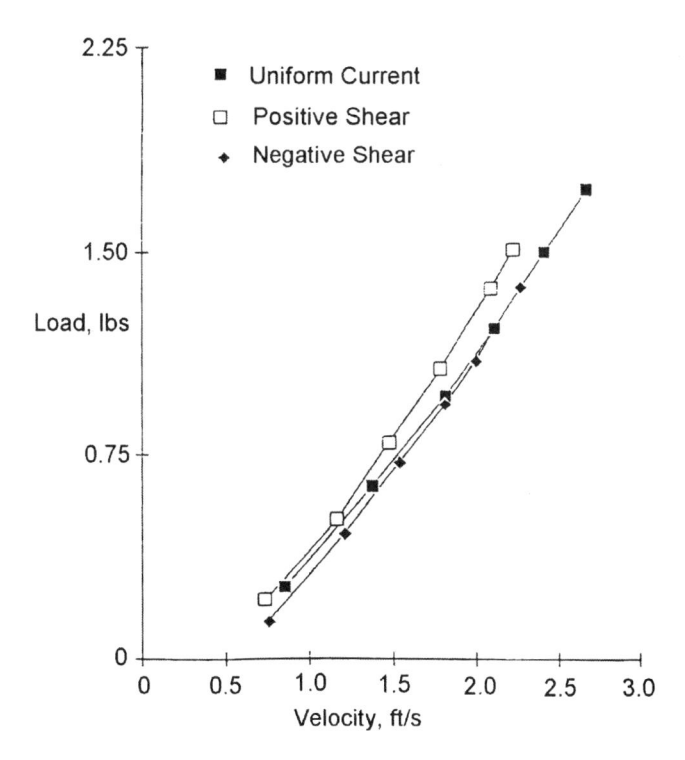

Fig. 7.8: In-Line Load on Horizontal Cylinder at an Elevation $y_1/d = 0.5$

The effect of shear was evident in the measured data. When the cylinder was at mid-depth ($y_1/d = 0.5$), the load at a given velocity from the positive shear was larger than the load created by the uniform flow or negative shear. The last two flows created approximately the same load on the cylinder except at low velocities where the negative shear loads were higher. The steady in-line load on the horizontal cylinder is presented as a function of velocity and shear pattern for the mid-depth elevation in Fig. 7.8. The same flow pattern was inverted to produce negative shear. As the velocity became larger, the positive shear produced higher loads compared to the loads from the uniform flow. In general, loads were relatively independent of elevation in positive and negative shear and depended only on velocity. In uniform current, however, the in-line load on the cylinder increased as the cylinder approached the free surface.

As the cylinder approached the free surface, the vertical load in both positive and negative shears increased dramatically, when compared with uniform flow. This increase is attributed to the three-dimensional effect from flow at the free surface. The vertical loads on the cylinder submerged at the free surface are shown in Fig. 7.9.

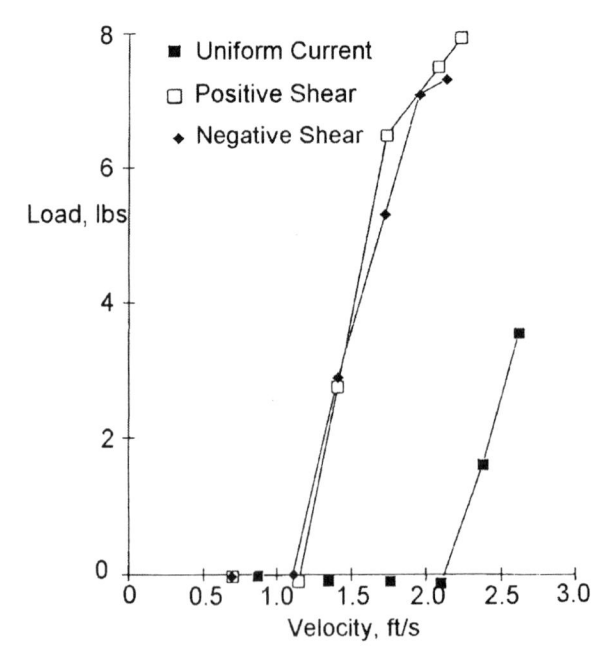

Fig. 7.9: Vertical Load vs. Shear Condition at the Free Surface

The vertical loads from the lift effect on the cylinder were oscillatory. Frequency domain analysis showed that at lower velocities, the vortex shedding was highly defined in terms of a single frequency. However, as the velocity increased, the flow became more turbulent. This is illustrated for negative shear at the mid-depth in Fig. 7.10. The load spectra are arranged at increasing values of the mean velocity. The spectral peaks represent the shedding frequency. The frequency of the periodic velocity fluctuations in the wake of the cylinder is represented by the Strouhal number. At higher velocities, the definition of shedding frequency became more undefined with multiple peaks in the spectra. The shear strength and Strouhal numbers are shown in each case. At the higher velocities, Strouhal numbers are undefined.

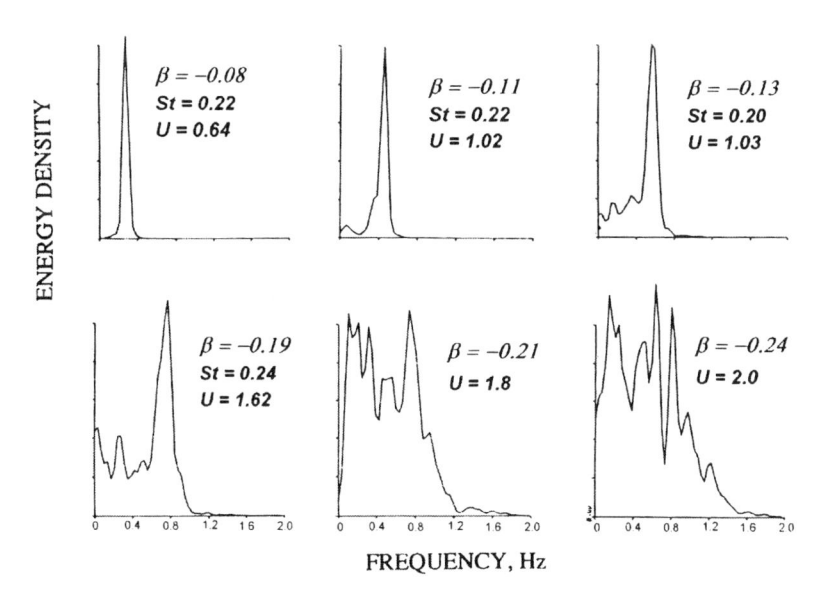

Fig. 7.10: Vertical Load Spectrum on a Horizontal Cylinder at Mid-Depth

The dynamic vertical loads on the cylinder at mid-depth in positive and negative shears contain greater energy at the higher frequencies and overall contain 30 to 40 percent more energy than the load in uniform current. Near the surface ($y = 0.73d$), there is a marked reduction (about 50 percent) in the significant value of the energy in comparison to the data at mid-depth. In this case, the surface tends to suppress eddy shedding and wake formation. Also, the non-zero steady component of the vertical load becomes large as the horizontal cylinder approaches the free surface. Strouhal numbers were calculated for the cylinder at mid-depth and velocities below 1.8 ft/sec (0.55 m/s) based on the peak spectral frequencies, such as shown in Fig. 7.10. These values are shown in Table 7.4 for different flow fields.

Kiya, et al. (1980) presented values of Strouhal number in the Reynolds number range, $35 < Re < 1500$, in which the Strouhal number was found to increase with shear parameter. The values presented in Table 7.4 for shear parameters in the range of 0.08 to 0.20 compare well with the results of Kiya, et al. A slight tendency for the Strouhal number to increase from uniform to shear flow is also indicated. Highly defined vortex shedding is apparent at mid-depth at a lower Reynolds number. The Strouhal number in this case tends to increase with the magnitude of the shear parameter. At higher Reynolds number and near the surface, the flow is more turbulent and the Strouhal number is less well defined.

Table 7.4: Strouhal Numbers for Various Shear Parameters

Current Shear	Velocity (ft/sec.)	β	Re, x10^4	Peak Freq. (Hz)	St No.
Uniform	0.82	0	4.10	0.35	0.21
	1.41	0	7.05	0.60	0.21
	1.80	0	9.00	0.74	0.21
Positive	0.64	0.09	3.20	0.28	0.22
	1.02	0.13	5.10	0.46	0.22
	1.57	0.15	7.85	0.70	0.22
	1.66	0.20	8.30	0.67	0.20
Negative	0.64	-0.08	3.20	0.28	0.22
	1.03	-0.11	5.15	0.46	0.22
	1.38	-0.13	6.90	0.56	0.20
	1.62	-0.19	8.10	0.77	0.24

Chapter 7 Fluid Structure Interaction

Comparison of the drag coefficient and Strouhal number from various investigations including numerical and experimental studies [see Siqueira, et al. (1999) for references] for a single cylinder at a Reynolds number of 200 is presented in Table 7.5. The difference in the values is attributable to different numerical and experimental techniques. However, in general, the consistency in the data is apparent with a mean Strouhal number of about 0.2 and an average drag coefficient of about 1.30. The cause of the deviations may be that when point vortices reach the large cells located far away from the cylinder, the lift and drag force experience very large oscillations.

Table 7.5: Single Cylinder C_D and Strouhal Number at $Re = 200$ [Siqueira, et al. (1999)]

Method	St No.	C_D
Experimental	0.17-0.19	-
Experimental	0.196	-
Numerical	0.195	1.32
Experimental	-	1.30
Numerical	0.196	1.30
Numerical	0.19	1.25
Numerical	0.195	1.30
Numerical	0.19	1.30

7.2.5 Blockage Factor in Steady Flow

A practical structure is often composed of relatively closely spaced small members connected in various orientations. These members may be circular or other geometric shapes. Whenever steady flow encounters a structure, the structure is not totally transparent to the flow, no matter how small its individual members are. The presence of the structure causes distortion in the fluid field around the structure. Since the global load is computed from the loads on the individual members of a structure, it is important to account for the distortion created in the flow field due to the presence of the structure itself. For example, if the structure is dense, then the steady flow actually slows down as it travels through the structure. In a practical design, this effect of blockage is accounted for with an overall factor, without describing the actual flow field. The actual velocity field within the structure is very complicated and does not appear in these calculations.

The term "blockage" is used with reference to the entire structure. A current blockage factor is introduced to account for the presence of the structure in the current flow field. This enables one to compute the true global load on the structure. The blockage factor is applied to the undisturbed current (i.e., steady flow) value in order to obtain an equivalent current velocity that accounts for the blockage by the structure. The reduced current value due to this factor does not represent the actual current field within the structure. The blockage factor for steady current past a dense structure consisting of many circular members (e.g., an offshore jacket type structure) may be estimated from

$$C_{BF} = \left[1 + \frac{\sum_i (C_D A)_i}{4\overline{A}} \right]^{-1} \qquad (7.1)$$

where the summed drag force is computed from each member in the dense structure, including the horizontal members in the flow, and the area $\overline{A}$ is the projected structure area normal to the flow. If the geometry of the structure changes significantly with water depth, then the blockage factor may be computed at several levels for the total force computation.

In practice, one often encounters a vertical array of cylinders in a steady flow, which interfere with the flow field (such as, a group of conductors or risers). The shielding from a vertical array of cylinders is given by the following simple empirical formula provided by the American Petroleum Institute [API (1992)], based on the relative spacing of the cylinders with respect to the diameter:

$$C_{BF} = \begin{cases} 0.25 S/D & \text{for } 0 < S/D < 4.0 \\ 1.0 & \text{for } S/D = 4.0 \end{cases} \qquad (7.2)$$

Table 7.6: Current Blockage Factor for Cylinder Group

No. of Cylinders	Current Heading	Blockage Factor
3	All	0.90
4	End-on	0.80
	Diagonal	0.85
	Broadside	0.80
6	End-on	0.75
	Diagonal	0.85
	Broadside	0.80
8	End-on	0.70
	Diagonal	0.85
	Broadside	0.80

The spacing S is the center to center distance of the cylinders of diameter D, which includes any marine growth on the surface of the cylinders.

According to the API guideline, the current blockage factor for a structure having up to 8 cylindrical members is determined from Table 7.6 based on the orientation of the structure relative to current.

7.3 Structures in Waves

Unlike steady flow past a cylinder, waves past a cylinder produce an oscillating in-line force on the cylinder. In addition, waves introduce a changing free surface, which introduces additional complexity of flow past a submerged structure. Large structures placed in the wave field will alter the form of the incident waves over a large area in its vicinity. If the dimension of the structure is large compared to the wavelength, the flow essentially remains attached to the body of the structure, and the flow may be described well by the potential flow. There are several numerical procedures (e.g., fluid finite element method to be introduced later) that can be used to describe the potential function generated in the vicinity of the structure, knowing the incident wave potential. One such method is known as the boundary element method (BEM) and will be briefly described here.

7.3.1 Linear Diffraction Theory

For structures in waves, the boundary element method is now a well-established numerical technique for the analysis of many engineering problems, in particular, linear and second-order problems. In potential theory, the method may be defined with a boundary value problem in terms of the Laplace partial differential equation and the appropriate boundary conditions (for example, see Sections 2.2 and 2.3). The basis of the method is to transform the original partial differential equation (PDE) or system of PDE's into an equivalent integral equation (or system of equations) by means of the corresponding Green's theorem, and the use of a fundamental solution. In this way, the resulting integral equation is numerically solved, instead of the original PDE. For some problems the use of this approach directly leads to surface integrals only, for which only the surface elements are necessary.

The boundary value problem is briefly derived in the following way. The analysis is based on potential theory and, therefore, the flow around the structure is assumed attached, and separation is neglected. It is assumed that the fluid is incompressible, flow is irrotational, wave amplitude is small and the structure is rigid. The consequence is that the fluid flow in the neighborhood of the submerged structure may be described by a velocity potential, Φ, which under linear theory is written as

$$\Phi = \phi \exp(-i\omega t) \tag{7.3}$$

in which ϕ is the spatial part of the total velocity potential, t is time and ω is the circular wave frequency corresponding to the incident wave. Note that it is convenient to assume Φ to be complex and work with the spatial component ϕ, recognizing that the time dependence appears as a factor exp $(-i\omega t)$, as shown in Eq. 7.3. The total potential is written as a sum:

$$\phi = \phi_i + \phi_s \tag{7.4}$$

where (known) potential ϕ_i is its incident wave component and (unknown) potential ϕ_s is its scattered wave component. The velocity potential due to the scattered waves, Φ_s, is represented by a continuous distribution of wave sources over the mean wetted structure surface. Due to the linear assumption, Φ_s can be described as a superposition of numerous wave sources, once the point sources of unit strength are known at points in the fluid domain. Under these assumptions, the differential equation is described by the Laplace's equation. The boundary conditions include the boundaries at the ocean floor, the free surface, and the submerged surface of the structure, and the radiation condition as the flow field approaches infinity. The bottom boundary condition assumes a flat bottom and states that the vertical component of the water particle velocity is zero at the bottom. The free surface boundary condition equates the vertical component of the water particle velocity to the rate of change of the free surface profile with time. Under linear theory, this condition is satisfied at the still water level (SWL). The far field flow condition states that the flow at infinity is only an outgoing flow and is satisfied by the Sommerfeld radiation condition.

The body surface boundary conditions have two forms, depending on whether the structure is fixed or is allowed to move with waves. In the first case, the normal component of the water particle velocity to the submerged surface of the structure is zero,

$$\frac{\partial \phi}{\partial n} = 0 \tag{7.5}$$

on the assumption that there can not be any flow of fluid into or out of the surface of the structure. The quantity n is the direction normal at the structure surface and defines the outward normal to a panel on the surface. For a moving structure, it becomes

$$\frac{\partial \phi_k}{\partial n} = u_{nk}; k = 1,2,..6 \tag{7.6}$$

where ϕ_k is the k-th radiation potential, n_k is the component of normal in the k-th direction, and u_{nk} is the normal velocity component of the fluid particle at the structure surface at a given degree of freedom k ($k=1,2,..6$).

A boundary integral technique based on the Green function, which represents both a far field wave potential and a near field potential is employed. According to this method, the structure surface is divided into small flat elements such that the geometry of the structure is approximately represented by these elements. These elements are referred to as lower order elements because of the approximate representation of the 3-D geometry and the method is referred to as lower order Boundary Element Method (BEM).

The problem is solved numerically by setting up a matrix equation in terms of the centers of the panels. For the purpose of evaluating the Green function, the field point and the source points are chosen in pairs as the coordinates of the panel center. Complex algebra is chosen for convenience and an NxN complex matrix is formed to describe the equation where N is the total number of panels. The solution for the source strengths at the center of each panel is obtained by a complex matrix inversion. For a large value of N required for accuracy, this computation is time consuming.

In addition, the 3-dimensional Green function is needed for each pair of panel centers. Certain symmetry exists so that the evaluation of the Green function is done only for one-half of the possible pairs including the diagonal points (where source and field points are the same). Several formulations for the 3-dimensional Green function are available. Most time-consuming, but most accurate is the integral representation of the Green function. An approximate series representation works quite well and with sufficient accuracy, as long as the source point and the field point are not too close to each other. It is found that for the values of $kr > 0.02$, (k = wave number, r = horizontal distance between the field point and the source point) the series form of G is satisfactory. The computation of the series G is considerably faster than the integral G. However, the form has a singularity problem as r approaches zero. The expressions for these two forms are given in Chakrabarti (1987).

While these two forms of the Green function are valid for finite water depth, considerable simplification may be achieved if the water depth is assumed infinite for all practical purposes. A depth parameter value of $kd > \pi$ or higher may be safely assumed to be deepwater. In this case, analytical expressions with known mathematical functions may be used in expressing G so that its computation is most efficient. For various formulations of infinite G, the reader may refer to the works of Newman (1984) and Noblesse (1994).

For all practical purposes for a large floating structure, the linear diffraction problem is assumed to be solved once the scattered potential, ϕ_s, is known. The details of the numerical method may be found in Chakrabarti (1987) and are not repeated here. Once the total potential is known from Eq. 7.4, the pressure in the fluid field is obtained from

Chapter 7 Fluid Structure Interaction

$$p = -\rho \frac{\partial \Phi}{\partial t} = i\rho\omega\phi \exp(-i\omega t) \tag{7.7}$$

in which ρ is the mass density of water. Knowing the pressure distribution at the center of each flat panel on the submerged surface of the structure, the forces in 6 degrees of freedom are computed by quadrature. Mathematically, the external force is obtained from the integral:

$$F_k = i\rho\omega \iint_S (\phi_i + \phi_s) n_k dS \tag{7.8}$$

where $k = 1, 2, \cdots 6$ and $S =$ submerged surface area of the structure.

The radiation problem is similarly computed in which the radiation condition on the structure surface is imposed, instead of the diffraction condition. In the absence of the incident wave, the structure is oscillated in a simple harmonic motion in a particular degree of freedom (e.g., surge, sway, heave, roll, yaw, and pitch) one at a time and the corresponding radiation potential is computed using the appropriate body surface boundary condition. The structure surface boundary condition is applied by equating the normal structure velocity to the normal velocity of the attached fluid at a given point on the structure (Eq. 7.6). The potential that is related to this problem is the radiated potential, which is obtained within the same formulation of the source matrix in which the source now represents the radiated source strength. This gives rise to six radiation potentials, ϕ_{rk}, for each of the 6 degrees of freedom. The total potential in waves including the scattered and radiation potential may be written as a sum of these potentials:

$$\phi = \phi_i + \sum_{k=1}^{6} \phi_{rk} + \phi_s \tag{7.9}$$

Here ϕ_{rk} is the potential due to forced oscillation of unit amplitude of the structure in still water for each degree of freedom. The radiated potential provides the added mass and damping coefficients of the structure in 6 degrees of freedom.

After hydrodynamic loads and coefficients are ascertained by the above method, the linearized equation of motion needs to be solved. The motion of the structure in a particular direction is obtained by setting up a 6x6 coupled equations of motion including the hydrodynamic loading and the hydrodynamic coefficients. The equations take the form of Eq. 3.183 for the six degrees of rigid body motion. An example of this numerical problem was included in Fig. 3.28 as the linear pitch.

7.3.2 Two-Body Interaction Problem

The linear diffraction/radiation theory described in the preceding section applies to a single rigid structure. Consider now the case of two large floating structures that move independently in the vicinity of one another in waves. The two-body motion problem is solved as a multi-degree of freedom system composed of rigid structures. In this case, the interaction problem is a little more complex. The incoming wave upon incident on one structure is scattered and becomes the incident wave for the second structure. This generates multiple scattering of waves between the two structures, which eventually die after a few cycles. Similarly for the radiated part, the motion of one structure creates radiated waves, which are incident on the second structure, altering the radiated potential field just as was the case with the scattered waves. In this case, the second vessel is considered fixed to first order, and its interaction with the first vessel is established. The scattered wave forces on the neighboring structures may be computed from the same numerical analysis of linear diffraction theory as above, since the structures are assumed fixed at their equilibrium positions for the first-order force computation. Therefore, the body-surface boundary condition to compute forces on the multiple modules, and hence, the BVP remain unaltered.

For the radiated potential, a modification is needed. A simple way to solve the above problem is to apply the principle of superposition applicable to the first-order theory. In computing the hydrodynamic coefficients associated with one structure, the radiation problem is solved to first order by introducing an additional boundary condition, which states that the second structure is fixed in its equilibrium position when the first structure is oscillated in a particular degree of freedom.

For two structures near each other, the total potential is similar to Eq. 7.9 and is written as:

$$\phi = \phi_o + \sum_{k=1}^{6} (\phi_{rk}^{(1)} + \phi_{rk}^{(2)}) + \phi_s \qquad (7.10)$$

Here $\phi_{rk}^{(1)}$ is the potential due to forced oscillation of unit amplitude of structure 1 in the k-th mode when structure 2 is held fixed; likewise for $\phi_{rk}^{(2)}$.

In order to compute the exciting forces on the fixed structures (or structures assumed fixed to first order), the diffraction condition that normal derivative of velocity potential equals zero, applies for both structures simultaneously, i.e., Eq. 7.5 holds. For the structure in motion, the radiation boundary condition becomes

$$\frac{\partial \phi_{rk}^{(i)}}{\partial n^{(j)}} = u_k^{(i)} n_k^{(j)} \delta_{ij} ; i, j = 1, 2 \qquad (7.11)$$

in which the potential for each degree of freedom is computed for $k = 1,2..6$, i and j take on the values 1 or 2 and $\delta_{ij}=1$ for $i = j$, while $\delta_{ij}=0$ for $i \neq j$. This means that when structure 1 is moving, structure 2 is fixed, and *vice versa*. The above, in principle, is not too different from the radiation boundary condition for a single structure problem, the only difference being the twofold increase in the number of terms involved.

The above boundary conditions give rise to a 6x6 added mass and damping coefficient matrix for one of the two structures due to its motion, and a cross 6x6 added mass and damping coefficient matrix on one structure due to the motion of the other structure. These cross terms in turn alter the size of the coefficient matrices that are to be handled by the numerical analysis. For example, for two structures, the order of the matrix is 12x12 instead of 6x6 for a single structure.

The equations of motion for the two floating structures are written in terms of 12 complex unknowns. In this case, the mass and stiffness terms of each structure are taken into account. In addition, the added mass and damping terms (from radiation forces) corresponding to each structure enter into the equation for the unknown amplitudes. The right hand side includes the exciting forces on the individual structure, including interaction forces from all modules:

$$\sum_{l=1}^{6} [-\omega^2 (m_{kl}^i + M_{kl}^{ij}) - i\omega N_{kl}^{ij} + C_{kl}^{ij}] x_l^j \exp(-i\varepsilon_l^j) = f_k^i \qquad (7.12)$$

in which the subscript k runs from 1 to 6 and superscripts i and j take on the values of 1 and 2. The quantity m_{kl}^i is the structure mass matrix and M and N are the added mass and damping matrix including cross terms. The matrix C includes hydrostatic as well as external stiffness terms. The solutions for 6 DOF motion x_k^i for the two floating structures are obtained from Eq. 7.12 by the inversion of the 12 x 12 matrix on the left-hand side. This equation is the generalization of the single structure motion equation and is valid for any number of structures in the system by generalizing the value of i.

7.3.3 Fluid-Structure Finite Element Method

In the fluid structure interaction problem, the structure is always finite in dimension and can be treated in finite element (as shown in sections 6.6.1) or discretized in surface element, such as in BEM. However, the fluid is often infinitely spaced. In these cases, the fluid may be represented numerically by finite elements. The fluid Finite Element Method (FEM) is a domain method in which the fluid domain is infinite. It is clear that some simplification is desired since an infinite number of finite elements are not practical in representing the fluid domain. In order to model such a domain numerically, a common practice is to truncate the fluid domain to a finite domain around the structure, and apply an appropriate truncation boundary condition at this boundary. This truncation boundary condition takes the form of the radiation boundary condition applied in the BEM.

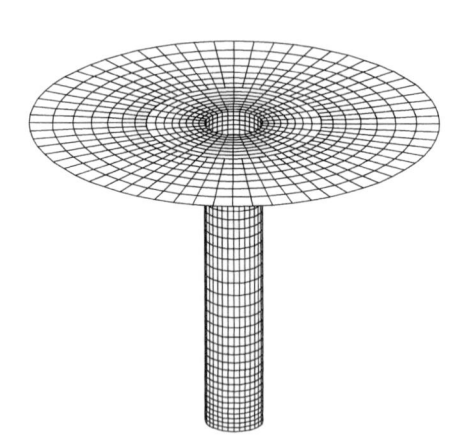

Fig. 7.11: Example Fluid-Structure Finite Element Mesh (Finer Mesh on the Structure Surface) [Kim, et al. (2001)]

The basic hydrodynamic equation and surface boundary conditions are similar to the BEM except for this far-field boundary. Unlike the BEM, FEM does not require a fundamental solution, such as the Green function. The truncation boundary divides the fluid domain into a finite inner domain and an infinite outer domain. The inner domain being a finite domain may be conveniently modeled by the fluid finite elements. Clearly the larger this domain is, the numerical error in the results is reduced while the computation time increases. One method in treating the outer domain is to apply a fictitious absorbing boundary condition at the junction, similar to an artificial beach. The truncation boundary considers only the outgoing waves, similar to the Sommerfeld radiation condition. There are a variety of numerical methods that are available to apply this boundary condition. The mathematics of boundary dampers is rather complex and beyond the scope of this book. For a review of these boundaries, the reader should consult the works of Givoli (1991). A finite element mesh describing the fluid and the body of a structure is illustrated in Fig. 7.11. The finite element approximation of the Laplace equation for the diffraction/ radiation problem with appropriate boundary conditions may be derived using the standard variational approach. The numerical method solves for the potential in the fluid field including the structure boundary. Once the potential function is known as a function of time on the body surface, the computation of forces is straightforward. Additionally, the added mass and damping coefficients and the motions of the structure are computed the same way as for the BEM. The results of the numerical routine are similar.

7.3.4 Stability of Motion

The preceding analysis is based on linear systems where the stiffness terms are linear. For nonlinear stiffness terms, the above simple solution for motion is not possible. This is shown in Chapter 3, in particular, in describing Duffing's solution. Such a system may introduce instability in the structure motion. This is illustrated further here with a Single Point Mooring (SPM) system. For the transfer of oil produced offshore, a shuttle tanker is commonly used. In this case, the shuttle tanker may be connected to an articulated tower by a single synthetic hawser line, which has highly nonlinear stiffness characteristics (Fig. 4.9). This type of system may be adequately described by the Duffing's equation in which the tension can be expressed as a cube of the oscillation angle in the form, $K\kappa^2(\psi_0-\psi)^3$ (see Section 3.4.5). The quantity ψ is the oscillation angle and ψ_0 is the maximum oscillation angle beyond which the line goes slack. Then, the restoring moment from the tower buoyancy and hawser stiffness is given by

$$M_R = \begin{cases} K\psi - K\kappa^2(\psi_0-\psi)^3 & \text{for } \psi < \psi_0 \\ K\psi & \text{for } \psi \geq \psi_0 \end{cases} \quad (7.13)$$

Assuming that the tower is inclined at a steady angle ψ_s due to steady environmental loads, the restoring force of the tower will be

$$K\psi_s - K\kappa^2(\psi_0-\psi_s)^3 = 0 \quad (7.14)$$

Assuming a small oscillation ψ of the tower around the steady angle ψ_s, an approximation is made retaining terms up to second order in ψ. Then the restoring moment of the tower may be written as

$$M_R = K\psi' + K\kappa^2[3(\psi_0-\psi_s)^2\psi' - 3(\psi_0-\psi_s)\psi'^2] \quad (7.15)$$

where ψ' is the dynamic angle, $\psi' = \psi - \psi_s$. The hydrodynamic force on the tower is computed at its instantaneous inclined position so that it depends on the pitch angle of the tower. The inertia force on the tower is written in terms of moment about its bottom articulated joint, and to first order it becomes:

$$M_w = M_1 \cos(\omega t + \varepsilon_1) + M_2 \psi \cos(\omega t + \varepsilon_2) \quad (7.16)$$

Introducing these quantities in the equation of pitch motion of the tower about its bottom joint one obtains:

$$I\ddot{\psi}' + C\dot{\psi}' + K\psi' + K\kappa^2[3(\psi_0-\psi_s)^2\psi' - 3(\psi_0-\psi_s)\psi'^2]$$
$$= M_1 \cos(\omega t + \varepsilon_1) + M_2 \psi \cos(\omega t + \varepsilon_2) \quad (7.17)$$

Further simplification is made at this point to linearize the equation by replacing the second order term ψ'^2 by $2\psi_L\psi'$ in which ψ_L is the linear pitch amplitude of the tower (which is unknown at this point). Upon dividing both sides of Eq. 7.17 by I and introducing the damping factor ζ and natural frequency ω_n, one gets

Chapter 7 Fluid Structure Interaction

$$\ddot{\psi}' + 2\zeta\omega_n\dot{\psi}' + \omega^2_n[1 + \beta\cos(\omega t + \varepsilon)]\psi' = \frac{M_1}{I}\cos(\omega t + \varepsilon_1) \tag{7.18}$$

in which

$$\omega^2_n = \frac{K[1 + 3\kappa^2(\psi_0 - \psi_s)^2]}{I} \tag{7.19}$$

and β is the parametric excitation obtained from the following relationship:

$$\omega^2_n \beta\cos(\omega t + \varepsilon) = [-6K\kappa^2(\psi_0 - \psi_s)\psi_L - M_2\cos(\omega t + \varepsilon_2)]/I \tag{7.20}$$

It is important to examine the stability of the solution for Eq. 7.17. Therefore, the right hand side of Eq. 7.18 is set to zero without loss of generality. Equation 7.17 becomes a classical Mathieu equation with the following substitutions:

$$2\tau \equiv \omega t + \varepsilon; \qquad \delta = (2\omega_n/\omega)^2 \tag{7.21}$$

$$\mu \equiv \zeta\sqrt{\delta} \tag{7.22}$$

and

$$\beta_1 = \beta\delta/2 \tag{7.23}$$

Using the above approximation, the equation of motion, Eq. 7.18 becomes

$$\ddot{\psi}' + 2\mu\dot{\psi}' + [\delta + 2\beta_1\cos 2\tau]\psi' = 0 \tag{7.24}$$

Note that in the above equation, τ is equivalent to the nondimensional frequency, δ is the normalized natural frequency and μ is the normalized damping. The existence of stable or unstable solutions depends on the relative values of δ and β_1. If $\beta_1 = 0$ and $\delta > 0$, the solution reduces to a stable simple harmonic motion. If $\beta_1 \neq 0$ but is small enough that $|2\beta_1| < \delta$, the coefficient of ψ fluctuates but is always positive. The solution is expected to be stable. Combinations of a β_1 lying in the range -0.5δ and 0.5δ where $\delta > 0$ is expected to lead to stable solutions. Outside this range, the solution is unstable. When damping is present, the relative size of these regions depends on the magnitude of damping.

The boundaries of the stability of this equation are examined surrounding the value of $\delta = 1$, which means that the stability region of the tower is considered for waves having twice the natural frequency of the tower. Using Eq. 7.24, the value of δ on the stability boundaries is given by

$$\delta \equiv 1 \pm (\beta^2_1 - 4\mu^2)^{1/2} \qquad (7.25)$$

Then, from Eqs. 7.22-23 and 7.25, one obtains

$$\beta = 2\{[(\delta-1)/\delta]^2 + (2\zeta/\sqrt{\delta})^2\}^{1/2} \qquad (7.26)$$

The stability boundaries from this equation are plotted as β vs. δ in Fig. 7.12 for different values of the damping factor, ζ. It is observed that the motions of the tower remain stable below the boundary. However, inside the boundary, the motion is unstable and unbounded at half the wave frequency. According to the Duffing solution shown in Fig. 3.20, the solution is expected to jump from the unstable region to a stable region or back and forth between two stable regions if this situation arises. It is also seen that the stable region increases with the increase of the damping in the system. In other words, if the damping in the system is large enough, the solution for the most part will be stable, even in the presence of Matheiu type instability. In practical applications, this type of instability is not that prevalent due to the presence of finite damping.

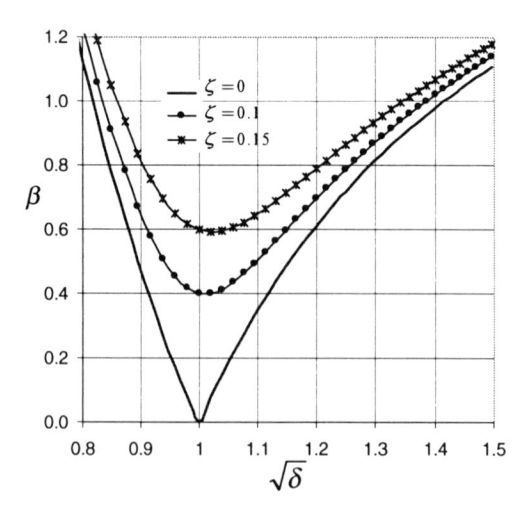

Fig. 7.12: Stability Regions of Matheiu Equation with Linear Damping

Chapter 7 Fluid Structure Interaction

An excellent example of this behavior is obtained from a wave basin model test of a single point offshore mooring system [Chantrel & Marol (1987)]. In this case, an articulated tower model was moored to a tanker model with a simulated single mooring line with nonlinear restoring force characteristics. The system was subjected to regular waves. The motion of the tower as well as the hawser tension was monitored. In the system, the hawser was allowed to go slack when the force on the system reversed in the oscillatory wave. Since the wave was regular, one would expect a regular response in the tower pitch and in the hawser load. However, it was found from measurements (Fig. 7.13) that the tower oscillation and the hawser load became large for no apparent reasons in the middle of the test run. After a few cycles of large oscillation (and corresponding large hawser load) the solutions returned close to their original values. This is a clear illustration that the solution moved from a stable to an unstable region (similar to the Duffing solution in Fig. 3.20) with a small change in the excitation, and returned back to the original curve with a small decrease in the excitation.

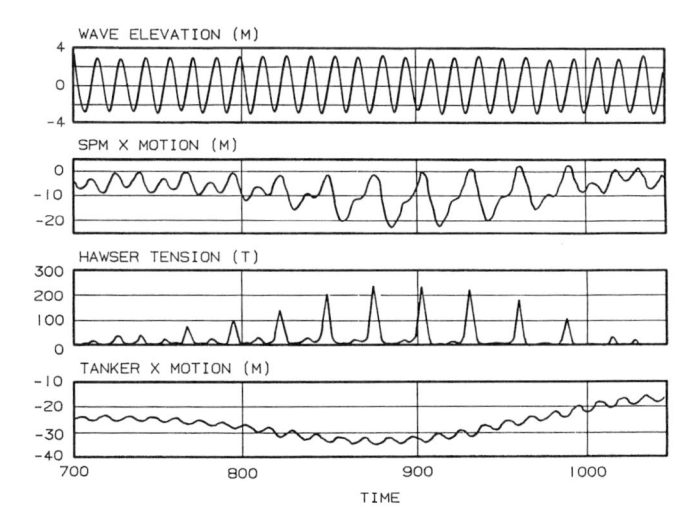

Fig. 7.13: Instability in the Motion of an SPM [Chantrel & Marol (1987)]

7.4 Structure Damping

The above illustration demonstrates how important the damping is in limiting excitation in a system. Structures are normally designed so that the natural frequencies of the various modes of motions are well beyond the dominant frequencies of the dynamic excitation, such as, from the environment. This is not always possible, particularly if the structure is flexible and is subject to higher modes of vibration. Even if this is achieved in a design, nonlinear effect may impose a dynamic motion whose frequency is different from the excitation frequency, and may coincide with one of the natural frequencies of the structure. In these cases, the motion is limited by the damping present in the system. This was illustrated in section 3.1 for a single degree of freedom system.

It is therefore important that the damping of a system is accurately evaluated. Damping in a system comes from primarily material and hydrodynamic sources. The material damping is present in the structural material and the mechanical connection. This damping often takes the form of Coulomb friction. The material damping is generally small.

There are several sources of hydrodynamic damping. For an oscillating circular cylinder in simple harmonic motion at small amplitudes in water, the Stokes damping force on a cylinder of unit length in laminar flow is given by the relation:

$$f = \sqrt{\frac{\pi}{\beta}} \rho D^2 \omega_n^2 X \qquad (7.27)$$

where $\beta = Re/KC$, $KC = 2\pi X/D$ = Keulegan-Carpenter number, and X = motion amplitude. This damping force is also given in terms of drag coefficient C_D:

$$C_D = \frac{3\pi^2}{2KC\sqrt{\pi\beta}} \qquad (7.28)$$

It has been experimentally demonstrated that, in still water, the C_D values may be accurately estimated by the Stokes solution (Eq. 7.28) for $X/D < 0.1$ and $\beta > 50000$, but the values of C_D are overestimated for $X/D > 0.1$ or $\beta < 25000$.

In steady and oscillatory flows, the total damping of a structure, which is subject to various environmental loads, comes from the following sources:

Chapter 7 Fluid Structure Interaction 285

$$C = C_R + C_{SW} + C_W + C_C \qquad (7.29)$$

where C_R = wave radiation damping, C_{SW} = still water viscous damping, C_W = wave drift damping, C_C = steady flow (or current) damping. In waves, the hydrodynamic damping appears in the form of linear radiation damping, linear viscous damping and nonlinear viscous damping. The radiation damping arises from the linear diffraction theory as described in section 7.2. The loads created by the linear damping are proportional to the velocity of the structure. The loads created by the nonlinear damping are proportional to the higher power (e.g., squared and cubic) of structural velocity, the exponent depending on the type of damping. These quantities have been discussed in Section 3.4.4.

7.4.1 Damping of a TLP in Heave

Tension Leg Platforms are floating structures held in place by a group of vertical tendons (tensioned members). They generally have small motion in the heave direction due to high tendon tension. However, the tendon loads vary considerably in oscillatory heave motion in waves. The damping of the cylindrical members in a TLP has been shown to be small. The contribution from the radiation and material damping for a TLP is generally extremely small. For example, the material damping for a TLP is of the order of 0.1 percent, whereas the radiation damping is about the same or slightly higher. Experimental results for the hydrodynamic damping ratio provide the following values in Table 7.7 for the members of a TLP model.

Table 7.7: Percent Damping Factor for a TLP in Model and Full Scale

	Model Scale	Model Scale
Members	Round Section	Square Section
Vertical Column	0.049	0.049
Horizontal Pontoon	0.176	0.278

7.4.2 Damping in Ship Roll

For a large floating structure in waves, damping is computed by the linear diffraction/ radiation theory. For most degrees of freedom of a large floating structure, this radiation damping is adequate for an accurate prediction of the rigid body motions at the wave frequencies. This is not particularly true for the roll motion of a long floating structure. The general equation of motion for roll was introduced in Chapter 3. For ships, barges and similar long offshore structures, the roll radiation damping is generally quite small compared to the total damping in the system. For these structures, the roll damping is highly nonlinear and depends on several factors beyond the radiated waves. Moreover, the dynamic amplification in roll may be large for such structures, since the roll natural period generally falls within the frequency range of a typical wave energy spectrum experienced by them. Therefore, it is of utmost importance that an accurate estimate of the roll damping is made for such structures.

The prediction of actual roll damping is a difficult analytical task. The fluid flow effect around the structure in roll is complex and no analytical theory exists that may be used to compute the total roll damping. Therefore, several nonlinear components of roll damping are determined empirically from the model and full-scale experiments. The forms of these components are described in this section [see Ikeda, et al (1978) for further details]. These empirical coefficients were derived from model tests of a limited number of ship/barge hull forms, including cylinders and flat plates. Thus, their application in the case of a special hull form may be inappropriate. For conventional hull shapes, they may be considered adequate. In these experiments, each source of damping was generally isolated and the interaction among the various sources was ignored. They are expressed approximately in terms of an equivalent linear damping coefficient or damping factor. For the application of the empirical formulas based on model tests to the full-scale structures, the scale effect is inherently assumed small.

In the following, the equivalent coefficient of roll damping components is expressed in terms of its various contributions. The total damping coefficient is obtained from

Chapter 7 Fluid Structure Interaction

$$B_{eq} = B_f + B_e + B_w + B_L + B_{BK} \qquad (7.30)$$

in which the component damping coefficients are as follows: B_f = hull skin friction damping, B_e = hull eddy shedding damping, B_w = free surface wave damping, B_L = lift force damping, and B_{BK} = bilge keel damping. The last quantity is present when the hull is equipped with a pair of bilge keels in the beam area. The expressions are derived for a ship or a barge, either stationary or moving in the seaway with a steady forward speed. For zero forward speed of the vessel, an additional subscript 0 is added to the terms in Eq. 7.30. This subdivision of roll damping may not be justifiable hydrodynamically since the hydrodynamic interaction among these components is unavoidable. However, it is a convenience, which allows computation of the individual components from limited experiments and provides practical results.

7.4.2.1 Skin friction

Frictional damping is caused by the skin friction stress on the hull surface of the ship form as the ship rolls. Thus it is conceivable that this damping will be influenced by waves. The presence of the bilge keel also alters the skin friction.

For zero speed ($U=0$), the friction damping in a laminar flow field in terms of an equivalent linear damping coefficient is written as follows:

$$B_{fo} = \frac{4}{3\pi} \rho S r_e^3 \psi_0 \omega C_f \qquad (7.31)$$

in which the friction coefficient C_f is given by

$$C_f = 1.328 \left[\frac{2\pi v}{3.22 \, r_e^2 \psi_0^2 \omega} \right]^{1/2} \qquad (7.32)$$

The effective bilge radius in the above formula is computed from

$$r_e = \frac{1}{\pi}[(0.887 + 0.145 C_B)\frac{S}{L} - 2OG] \qquad (7.33)$$

where B = beam, D = draft, L = lateral dimension of the ship, C_B = block coefficient of the ship, ψ_0 = amplitude of roll motion (in radians), and U =

forward speed (or steady current speed). The quantity OG is the vertical distance from the origin O (at still water level) to the roll axis, G, which is measured positive downward ($OG = D - KG$). The quantity S is the wetted surface area, which is calculated from

$$S = L(1.7D + C_B B) \tag{7.34}$$

Note that the skin friction coefficient is a function of the fluid viscosity or an equivalent Reynolds number defined as $Re = (r_e \psi_0)^2 \omega / \nu$. Therefore, the skin friction is higher in the model scale compared to that in the full-scale, and scaling of skin friction by Froude's law is clearly not applicable. The values of C_f are given as a function of Re in Fig. 7.14. This effect is qualitatively similar to the steady towing force coefficient.

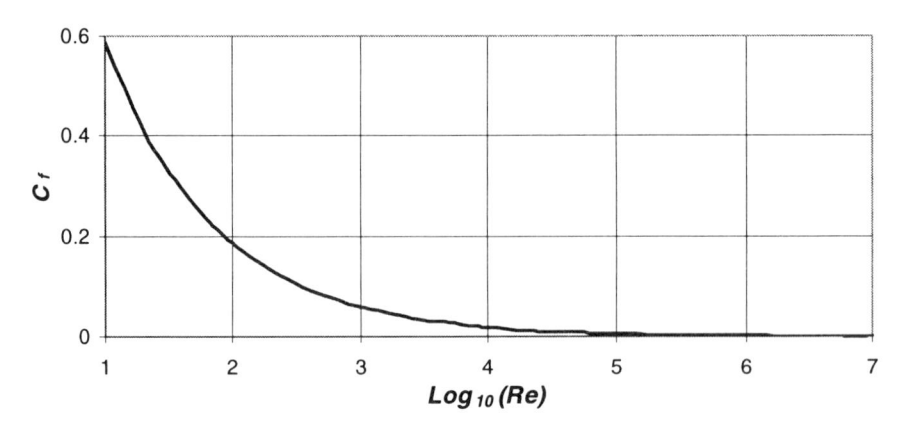

Fig. 7.14: Values of C_f as Function of Re

Adjusting this form to account for the turbulent flow, the formulation for the skin friction damping coefficient has been proposed as:

$$B_{fo} = 0.787 \, \rho S r_e^2 \sqrt{\omega \nu} \left\{ 1 + 0.00814 \left(\frac{r_e^2 \psi_0^2 \omega}{\nu} \right)^{0.386} \right\} \tag{7.35}$$

The first term in the above expression arises from the laminar flow past the ship, which is linear and independent of the roll amplitude. The second term is nonlinear and gives the modification due to turbulent flow. Note that the first term in Eq. 7.35 can be shown to be the same as the expression in Eq. 7.31 once Eq. 7.32 is substituted in it.

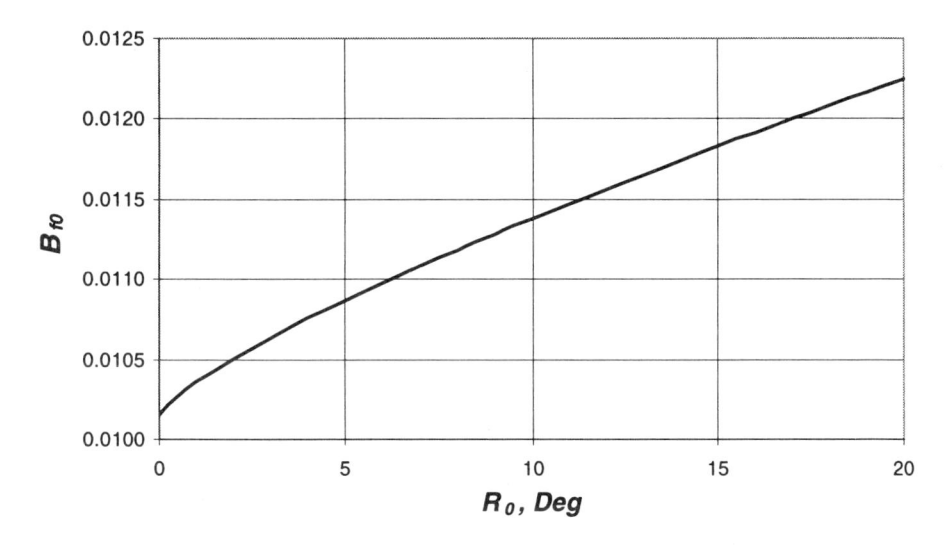

Fig. 7.15: Skin Friction Damping Coefficients in Laminar and Turbulent flow

The effect of turbulence on the skin friction damping coefficient is shown in Fig. 7.15 for different roll angles. The roll angle is varied from 0 to 20 degrees. The skin friction is seen to increase slowly but steadily with the roll angle.

In the presence of a non-zero forward speed U, a simple modification to the above formula is

$$B_f = B_{fo}(1 + 4.1\frac{U}{\omega L}) \qquad (7.36)$$

where the constant 4.1 is an experimentally determined value for an elongated spheroid. This empirical formula shows that the skin friction increases slightly in forward speed.

7.4.2.2 Eddy making damping

Viscous eddy (or vortex) damping arises from the pressure variation due to separation at the sharp corners and the associated vortices. This component of damping for a bare hull is found to be a square function of the roll frequency and roll amplitude. The damping is large near the bow and stern for slender ships, while it has a greater contribution near the midship section for a barge-like structure. The eddy-making drag is computed from a formula similar to the velocity-squared drag force based on a drag coefficient:

$$F = 1/2 C_e \rho S (r\omega\psi_0)^2 \qquad (7.37)$$

where r is the radial distance from the CG of the ship to the corner where eddies are shed (local radius). The drag coefficient C_e is obtained from the formula provided for a U shaped or a V shaped hull:

$$C_e = C_1 C_2 \exp(-\beta r_e / D) \qquad (7.38)$$

in which KG = distance from the keel to the CG, α = angle (deg) between the hull surface at the water line and the vertical, and β = exponential parameter. The quantity r_e = effective bilge radius defined in this case as follows:

Table 7.8: Values of Coefficient C_1 vs. B/KG

B/KG	C_1
0	0.50
0.25	0.61
0.50	0.62
1.0	0.61
1.5	0.53
2.0	0.40
2.5	0.35
3.0	0.32
3.5	0.29
4.0	0.26

$$r_e = \begin{cases} 0.5B[4.12 - 3.69(KG/B) + 0.823(KG/B)^2] & \text{for } KG/B < 2.1 \\ 0 & \text{for } KG/B \geq 2.1 \end{cases} \quad (7.39)$$

The value of C_1 depends on the function B/KG and is given in Table 7.8. The values of C_2 are functions of α and r_e/D and are given in Table 7.9. The intermediate values may be interpolated from this table.

Table 7.9: Values of Coefficient C_2

α(deg) ↓ r_e/D →	0.0	0.0571	0.1142	0.1713
0	1.0	1.0	1.0	1.0
5	0.86	0.75	0.74	0.70
10	0.77	0.67	0.72	0.72
20	0.68	0.75	0.89	1.20
30	0.65	0.92	1.34	1.94

The exponential parameter β is calculated by the formula:

$$\beta = 14.1 - 46.7\psi_0 + 61.7\psi_0^2 \quad (7.40)$$

For a rectangular section, e.g., a barge, $C_2 = 1.0$ and $\beta = 0$. Then, $C_e = C_1$ and the quantity r in Eq. 7.37 for the drag force becomes the distance from the roll axis to the corner. For a triangular cross-sectioned ship, the drag coefficient is determined from

$$C_e = 0.438 - 0.449\,(B/KG) + 0.236\,(B/KG)^2 \quad (7.41)$$

Alternately, the formula for the eddy-making damping per unit ship length was derived empirically [Ikeda, et al. (1978)]:

$$B_{eo} = \frac{4}{3\pi}\rho D^4 \psi_0 \omega C_p C_R \quad (7.42)$$

where

$$C_p = 0.5[0.87\exp(-\gamma) - 4\exp(-0.187\gamma) + 3] \quad (7.43)$$

in which the velocity increment ratio is expressed as

$$\gamma = \frac{\sqrt{\pi} f_3}{2[D - OG]\sqrt{H_0'\sigma'}} \left[r_{max} + \frac{2H_1}{H_2} \sqrt{A_1^2 + B_1^2} \right] \quad (7.44)$$

where

$$H_1 = \frac{B}{2(1 + a_1 + a_3)} \quad (7.45)$$

$$H_2 = 1 + a_1^2 + 9a_3^2 + 2a_1(1 - 3a_3)\cos 2\varepsilon - 6a_3 \cos 4\varepsilon \quad (7.46)$$

$$A_1 = -2a_3 \cos 5\varepsilon + a_1(1 - a_3)\cos 3\varepsilon - \{(6 - 3a_1)a_3^2 + a_1(a_1 - 3)a_3 + a_1^2\}\cos\varepsilon \quad (7.47)$$

$$B_1 = -2a_3 \sin 5\varepsilon + a_1(1 - a_3)\sin 3\varepsilon - \{(6 + 3a_1)a_3^2 + a_1(a_1 + 3)a_2 + a_1^2\}\sin\varepsilon \quad (7.48)$$

and C_R is computed at incremental ship stations

$$C_R = \left[\frac{r_{max}}{D}\right]^2 \left[1 - f_1 \frac{R_b}{D}\right] \left[1 - \frac{OG}{D} - f_1 \frac{R_b}{D}\right] + f_2 \left[H_0 - f_1 \frac{R_b}{D}\right]^2 \quad (7.49)$$

The quantities R_b = bilge radius, OG = distance (positive downward) from O to G, and H_0 = half the beam-draft ratio at different stations of the ship (a variable depending on its shape):

$$H_0 = \frac{B}{2D} \quad (7.50)$$

and

$$H_0' = \frac{H_0 D}{D - OG} \quad (7.51)$$

The other quantities are defined as

$$\sigma' = \frac{\sigma D - OG}{D - OG} \quad (7.52)$$

and

$$R_b = \begin{cases} 2d\sqrt{\dfrac{H_0(\sigma-1)}{\pi-4}} & for \ \ R/D<1, R/D<H_0 \\ D & for \ \ H_0 \geq 1, R/D > 1 \\ B/2 & for \ \ H_0 < 1, R/D > H_0 \end{cases} \quad (7.53)$$

where σ = area coefficient at a cross section along the hull (σ = area/(BD)). The functions, f_1, f_2, and f_3 are

$$f_1 = 0.5[1 + \tanh\{20(\sigma - 0.7)\}] \quad (7.54)$$

$$f_2 = 0.5[1 - \cos\pi\sigma] - 1.5[1 - \exp\{-5(1-\sigma)\}]\sin^2\pi\sigma \quad (7.55)$$

$$f_3 = 1 + 4\exp\{-1.65 \times 10^5 (1-\sigma)^2\} \quad (7.56)$$

$$r_{max} = M[\{(1+a_1)\sin\varepsilon - a_3 \sin 3\varepsilon\}^2 + \{(1-a_1)\cos\varepsilon + a_3\cos 3\varepsilon\}^2]^{1/2} \quad (7.57)$$

$$\varepsilon = \frac{1}{2}\cos^{-1}\frac{a_1(1+a_3)}{4a_3} \quad (7.58)$$

The constants a_1, a_2, and a_3 are defined as the extinction coefficients from fitting the extinction curve in roll with a three degree polynomial in the roll angle.

For a three-dimension ship form, the above sectional coefficients are integrated over the length of the ship. Thus, the section damping coefficient considers the section geometry of the ship along its length.

The above formulation was modified by Ikeda to take into account the rectangular cross section of a barge and the sharp corners present at the base. The result is based on experiments on the two-dimensional models of rectangular cross sections having different breadth to draft ratios. The tests included free decay and forced roll test runs. This formula is much simpler and depends only on the quantity H_0, but is independent of the bilge radius (on the assumption that the corner is sharp). The eddy-making damping coefficient is computed from

$$B_{eo} = \frac{2}{\pi}\rho LD^4 (H_0^2 + 1 - OG/D)[H_0^2 + (1 - OG/D)^2]\psi_0\omega \quad (7.59)$$

This formula is applicable to the rectangular shaped barge. Note that both formulations give the roll damping coefficient as a linear function of the roll amplitude and frequency, and are more applicable for smaller roll angles (about 5 degrees). At higher roll angles, the formulas seem to overestimate the experimental damping.

In the presence of steady current or forward speed U, the eddy making damping decreases and the following empirical formula is proposed:

$$B_e = B_{eo}\left[\frac{(0.04\omega L/U)^2}{1+(0.04\omega L/U)^2}\right] \quad (7.60)$$

Equation 7.60 shows that the eddy-making damping decreases rapidly with the forward speed and becomes negligible at a large value of $\omega L/U$.

7.4.2.3 Lift damping

Lift damping in roll occurs in the form of a lift moment created on the ship structure as the ship moves with a forward speed. A simple expression of this complex phenomenon is provided [Ikeda, et al. (1978)] in terms of an equivalent linear damping as:

$$B_L = \frac{0.15}{2}\rho ULD^3 k_N\left[1 - 2.8\frac{OG}{D} + 4.667\left(\frac{OG}{D}\right)^2\right] \quad (7.61)$$

The slope constant of the lift coefficient,

$$k_N = 2\pi\frac{D}{L} + \kappa\left(4.1\frac{B}{L} - 0.045\right) \quad (7.62)$$

where

$$\kappa = \begin{cases} 0 & \text{for } \sigma \leq 0.92 \\ 0.1 & \text{for } 0.92 < \sigma \leq 0.97 \\ 0.3 & \text{for } 0.97 < \sigma \leq 0.99 \end{cases} \quad (7.63)$$

and where σ (<1.0) is the mid-ship area coefficient (area/BD). Since the lift damping is a function of U and independent of the frequency ω, it is zero for a ship at zero forward speed. At high forward speeds of a ship, the contribution to the total damping due to lift is significant.

7.4.2.4 Wave damping

Wave damping is caused by the free surface waves created by the ship, and is thus a function of the wave parameters. It is computed quite accurately from the linear diffraction/ radiation theory discussed earlier, and is generally known as the radiation damping, B_{w0}.

In the presence of ship forward speed, this damping is modified. A simple analytical expression for this damping may be derived for a flat plate. The wave damping from ship forward speed or current is obtained by the following formula:

$$B_w = B_{w0} \frac{1}{2}\{[(A_2+1)+(A_2-1)\tanh 20(\tau-0.3)]+(2A_1-A_2-1)\exp(-150(\tau-0.25)^2)\} \tag{7.64}$$

where

$$A_1 = 1 + \xi_d^{-1.2}\exp(-2\xi_d) \tag{7.65}$$

$$A_2 = 0.5 + \xi_d^{-1.0}\exp(-2\xi_d) \tag{7.66}$$

$$\xi_d = \omega^2 d/g \tag{7.67}$$

$$\tau = U\omega/g \tag{7.68}$$

Note that this damping coefficient is maximum at $\tau = 1/4$ and asymptotically approaches a constant value at large values of τ. This equation may be applied to structures which are undergoing a slow drift motion (considered an equivalent steady speed) in waves.

7.4.2.5 Bilge-keel damping

The bilge keel is a simple, effective and efficient way to increase the overall damping and, thus, stabilize excessive roll motion of a ship or a barge. When the bilge keel is present, it produces additional damping due to normal force on the keel plus pressure variation on the hull surface caused by the presence of the bilge keel. It includes the damping of the bilge keels themselves and the interaction effects among the bilge keels, hull and waves. These contributions vary with the amplitude of roll and the wave frequency.

Neglecting the wave effect, the bilge keel damping is written in terms of the contribution from the normal and hull pressure as

$$B_{BK} = B_{BKN} + B_{BKH} \tag{7.69}$$

The normal force component per unit length is written as

$$B_{BKN} = \frac{8}{3\pi} \rho r_{cb}{}^3 b_{BK} \omega \psi_o f^2 C_D \tag{7.70}$$

in terms of an equivalent drag coefficient

$$C_D = 22.5 \frac{b_{BK}}{\pi r_{cb} \psi_o f} + 2.4 \tag{7.71}$$

where b_{BK} = breadth of the bilge keel and f is a correction factor to take into account the increase in the flow speed at the bilge keel,

$$f = 1 + 0.3 \exp\{-160(1-\sigma)\} \tag{7.72}$$

On the other hand, the pressure component of damping per unit length due to hull surface is obtained from the pressure measurement on a 2-D hull surface, which is caused by the presence of the bilge keels.

$$B_{BKH} = \frac{4}{3\pi} \rho r_{cb}{}^2 D^2 \omega \psi_o f^2 \left\{ -(-22.5 \frac{b_{BK}}{\pi r f \psi_o} - 1.2) A_2 + 1.2 B_2 \right\} \tag{7.73}$$

where

$$A_2 = (m_3 + m_4) m_8 - m_7^2 \tag{7.74}$$

$$B_2 = \frac{m_4^3}{3(H_0 - 0.215 m_1)} + \frac{(1-m_1)^2 (2m_3 - m_2)}{6(1 - 0.215 m_1)} + (m_3 m_5 + m_4 m_6) m_1 \tag{7.75}$$

$$m_1 = R_b / D; \ m_2 = OG / D; \ m_3 = 1 - m_1 - m_2; \ m_4 = H_0 - m_1 \tag{7.76}$$

Chapter 7 Fluid Structure Interaction

$$m_5 = \frac{0.414H_0 + 0.0651m_1^2 - (0.382H_0 + 0.0106)m_1}{(H_0 - 0.215m_1)(1 - 0.215m_1)} \qquad (7.77)$$

$$m_6 = \frac{0.414H_0 + 0.0651m_1^2 - (0.382 + 0.0106H_0)m_1}{(H_0 - 0.215m_1)(1 - 0.215m_1)} \qquad (7.78)$$

$$m_7 = \begin{cases} S_0/D - 0.25\pi m_1 & \text{for } S_0 > 0.25\pi R_b \\ 0, & \text{otherwise} \end{cases} \qquad (7.79)$$

$$m_8 = \begin{cases} m_7 + 0.414m_1 & \text{for } S_0 > 0.25\pi R_b \\ m_7 + \sqrt{2}\{1 - \cos(S_0/R_b)\}m_1, & \text{otherwise} \end{cases} \qquad (7.80)$$

and where

$$S_0 = 0.3\pi fr_{cb}\psi_o + 1.95b_{BK} \qquad (7.81)$$

The bilge-circle radius R_b and the mean distance r_{cb} from the roll axis to the bilge keel are given by

$$R_b = \begin{cases} 2D\sqrt{\dfrac{H_0(\sigma - 1)}{\pi - 4}} & \text{for } R_b/D < 1, \ R_b/D < H_0 \\ D & \text{for } H_0 \geq 1, \ R_b/D > 1 \\ B/2 & \text{for } H_0 \leq 1, \ R_b/D > H_0 \end{cases} \qquad (7.82)$$

and

$$r_{cb} = D[\{H_0 - 0.293R_b/D\}^2 + \{1 - OG/D - 0.293R_b/D\}^2]^{1/2} \qquad (7.83)$$

7.4.2.6 Damping results

The expressions presented in the preceding section provide five contributions to the total damping in roll. For the eddy making component, the expressions are given by different sets of equations, depending on whether the vessel is a barge or a ship shape. In all cases, the effect of current or forward speed has been shown explicitly with additional formulations. When bilge keels are present, roll damping depends on the shape and position of the bilge keel on different ship shapes having a small to large block coefficient. The calculations consider the detailed cross-section of the vessel. The roll predictions from these calculations are considered to be accurate and the method is usually robust. The empirical formulas have been derived from many experiments with flat plates, cylinders, ship and barge models. However, as with any such empirical formulations, the formulas are based on experimental data on particular ship shapes and generalization to all shapes may pose some problems.

An example of the computed damping coefficients for a container ship is shown in Fig. 7.15. The difference between the presence and absence of a bilge keel is clear in the figure. The bilge keel almost doubles the total damping in roll for the stationary (zero forward speed) ship. The container ship data represents the geometry of a particular container ship model of dimensions 1.75m x 0.254m x 0.095m draft. The ship model had a block coefficient of 0.571 and was described by the cross sections at 21 stations numbered from 0 to 10. The KG was assumed to be the same as the draft. The bilge keel width was taken as 0.0045m and its length spanned the stations from 3.75 to 6.25. The wave radiation-damping coefficient was set at .0009562 obtained from the diffraction/ radiation theory. The roll mass coefficient was chosen as 2.0 and the roll amplitude was assumed to be 10 deg. The moment of inertia was computed using the simple formula for a rectangular box.

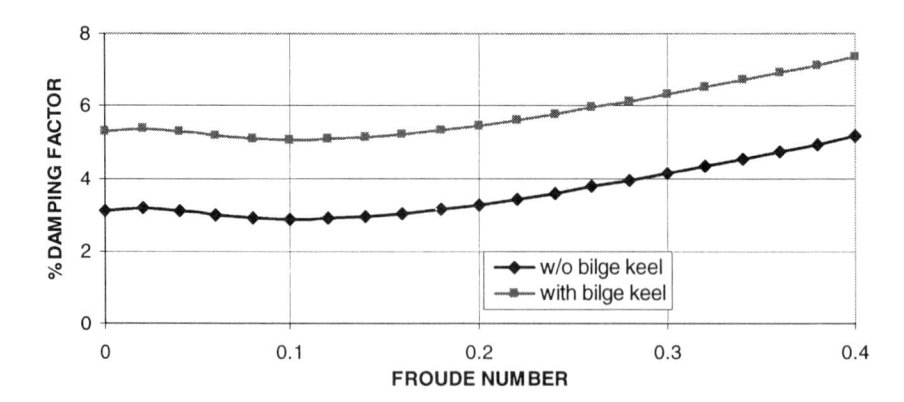

Fig. 7.16: Total Damping Coefficient in Roll for a Container Ship With and Without Bilge Keel

Chapter 7 Fluid Structure Interaction

The data represents an equivalent damping factor, normalized by $2\omega_n I$, where I is the roll moment of inertia and ω_n is the roll natural frequency. The total damping shown in Fig. 7.16 is contributed by five damping terms discussed above. The individual contribution of the five components to this damping coefficient in roll versus the Froude number is illustrated for the container ship in Fig. 7.17. The wave damping component is comparatively small. It is found that the forward speed increases the lift component significantly. The most important contribution at zero forward speed is the eddy-making damping coefficient. The presence of the bilge keel provides a steady increase in the damping coefficient.

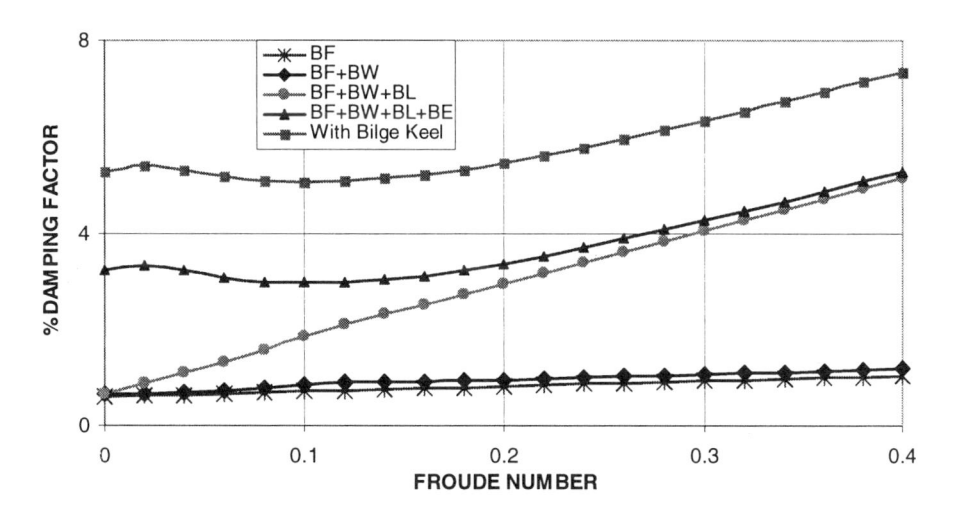

Fig. 7.17: Components of Roll Damping for a Container Ship

For another example, consider a more rectangular shaped derrick barge with length, width and draft of 400ft x 120ft x 19ft (122m x 36.5m x 5.8m), and a block coefficient of 0.923. This type of barge has a high fore and aft rake or slant height. Near the section with rakes, the breadth remains unchanged, with the only difference being the depth of the section, which reduced to an end value of 3.25ft (1m) within two stations over a total of 21 stations. The bilge radius is 1.97ft (0.6m) running from station 3 to station

14. This configuration results in a high cross sectional area coefficient (σ) of 0.999. The roll amplitude is chosen to be 10 deg.

Without any modifications to the end geometry, the damping factor for the barge geometry due to eddy-making at zero forward speed turns out to be very large at 59% of critical (see Table 7.10 below). Two factors contribute to the large eddy making damping: the cross sectional area coefficient and the fore & aft rakes.

If the area coefficient is changed to 0.8, the damping factor due to eddy making reduces to 18%. If the rakes are further reduced (from an end depth of 3.25ft or 1m to 15.1ft or 4.6m at fore and 14.5ft or 4.4m at aft), the damping factor becomes 9.6%. For the case with 0.99 area coefficient and the reduced rakes, the damping factor is 22.5%.

The effect of the area coefficient on the eddy making damping is obvious. The higher the area coefficient, the sharper is the corner, which in turn results in more fluid separation and eddy making. However, a 59% damping factor is obviously too large, and is erroneous. Besides, the large effect of fore and aft rakes on the damping coefficient is impractical. This effect is attributed mainly to the three-dimensionality and end effects of the barge. This is illustrated by an example of the distribution of the damping coefficient along the ship length. The variation of the damping coefficient on the stations of the barge geometry is shown in Fig. 7.18. Note that the end effect is significant in this example.

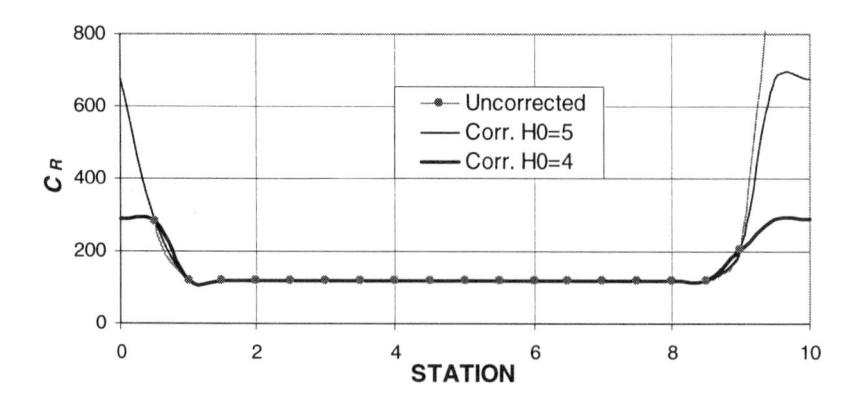

Fig. 7.18: Distribution of Eddy-Making Damping Coefficient Along a 3-D Barge

Chapter 7 Fluid Structure Interaction 301

The empirical formulas for the eddy-making damping were derived from experiments on models of 2-dimensional shapes in which care was taken to provide end plates to the model to minimize the end effects. The 3-dimensional structures of finite length are expected to suffer from significant end effects. This is particularly true for barges with rakes. Thus, the eddy-making damping calculation with the above formulation is modified for the outer stations of a three-dimensional barge to reduce this large end effect. In this modification, the unrealistic increase of the sectional damping is avoided by reducing the cross-sectional coefficients at the two end stations by 30%. This reduction factor is strictly empirical in nature in order to reduce the two-dimensional effect of the rakes for a rectangular barge. The cases shown in Fig. 7.18 are for the value of $H_0 = 4$ and 5. This modification produces more reasonable results for a rectangular barge, and has much smaller end effects for a three-dimensional ship shape.

The original formula for the eddy component of damping does not appear suitable for certain rectangular barges with high area coefficients, and the revised correction formula is proposed. Table 7.10 summarizes the results before and after the modification.

Table 7.10: Comparison of Eddy-Making Damping for a Derrick Barge

Vessel	Rake Fore & Aft	Area coeff. σ	Damping Factor Original formula	Modified w/ end factor	New formula & end factor
Barge	High	0.999	0.59	0.38	0.20
Barge	High	0.8	0.18	0.14	0.20
Barge	Reduced	0.8	0.096	0.049	0.16
Barge	Reduced	0.99	0.225	0.11	0.16
Ship	Actual	varies <0.97	0.0528	0.025	-

The revised formulation is used to estimate the roll damping for the above mentioned derrick barge with an actual loading condition at the draft of 16.9 ft (5.2m). The revised formula for the rectangular barge is used in the calculation. The total damping factor with and without a bilge keel is shown in Fig. 7.19. Note that at zero forward speed (i.e., zero Froude number), the damping factor without the bilge keel is about 14%. This value increases to about 18% for the derrick barge when a bilge keel of the section dimension stated earlier is included.

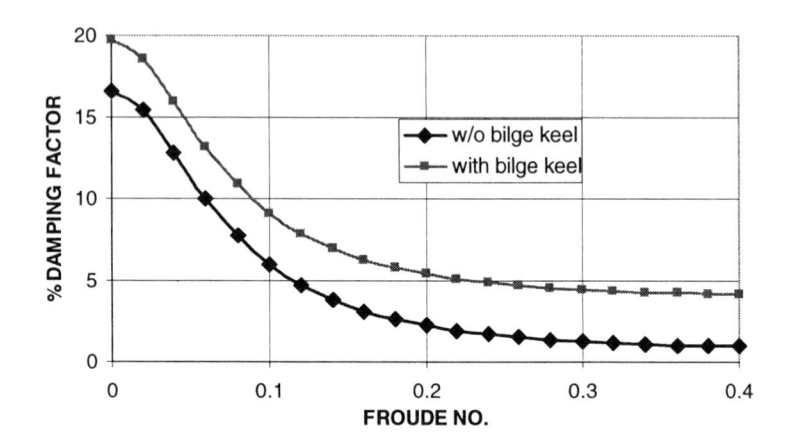

Fig. 7.19: Total Roll Damping Factor for a Derrick Barge

7.4.3 Roll Suppression

There are several methods that can be used to reduce the amount of roll oscillation. One of the obvious solutions is to move the roll natural period from the wave energy frequency regime to a value outside of this range. Since this is not an easy task for a ship shaped structure, there are other techniques employed to minimize the effect of natural period of oscillation of a long ship shaped structure in roll. One method already described is the introduction of the bilge keel. The size of the bilge keel may be increased to introduce additional damping near the roll natural period.

A special purpose flume tank is often built and placed on the middle of the ship deck. The flume tank is carefully designed with many baffles in the longitudinal and transverse directions, and is partially filled with water. As the ship rolls, the water in the flume tank oscillates with it, introducing anti-roll moment. The baffles eliminate any standing wave generation in the flume tank, which would otherwise hamper the roll suppression effect. The water level in the flume tank may be varied with the predominant wave frequency to tune the anti-roll effect, allowing a broader damping effect over a band of wave frequencies.

Sometimes, it had been necessary to add compartmented sponsons to either side of the beam of an existing ship so that the displacement of the ship increased, allowing placement of additional payload on the deck of the converted ship. One example of this is the conversion of a drill ship to a production platform. This modification increases the inertia as well as the cut-water plane of the ship. The consequence often is to move the natural period in roll to a lower value within the area of high wave energy. One common method of introducing additional damping in this case in roll is to open up part of the sponson at the bottom, allowing air to vent at the top.

Another variation of this technique is the introduction of an anti-roll system consisting of an air chamber on both the port and the starboard sides of the ship. The air chambers on either side contain low-pressure air and are cross-connected. The air pressure acting on the underside of the chamber top is maintained as a constant, independent of the waterline at the side of the ship and the roll motion of the ship. Therefore, the chambers, if placed outside of the ship, do not introduce any additional moment. If the chambers are submerged, the wave pressure on the top of the chambers produces an anti-roll moment. The submerged chambers enhance damping by producing additional radiated waves. Additionally, it has been found that the natural period for the system is moved up and away from the highest energy area of the waves. The above elements of the modification have been found to be quite efficient in reducing roll. Experiments have shown significant reduction in the roll motion in this system compared to other systems. In model tests with drill ships including these modifications, the damping has been found to increase five folds.

7.5 Tensioned Riser Analysis

Many slender structures placed in fluid flow experience large deformation. The flexibility of these structures affects the fluid structure interaction problem, and the deflections of these structures can not be neglected in analyzing the effect of fluid flow around the structure. Examples of these structures include deep water jacket structures, flexible risers, mooring lines, tethers, and hanging pipelines. The behavior of these structures during installation and operation is highly nonlinear including large deflections and finite rotations. For structures that are flexible, the stiffness of the structure must be included in the equation of motion. This is illustrated with the dynamic analysis of a conventional marine riser. Even though a specific example has been chosen, the derivation of the equation of motion is applicable to a variety of submerged structures of this kind.

A riser is a unique common element to many floating offshore structures. Risers connect the floating drilling/ production facility with subsea wells and are critical to safe field operations. For deepwater operation, design of risers is one of the biggest challenges. Several alternatives to the conventional risers are being investigated currently, such as flexible pipes and steel catenary risers, which will require a similar treatment. A conventional riser is a long slender vertical cylindrical pipe placed at or near the sea surface and extending to the ocean floor (see Fig. 7.20). Numerically, marine risers are characterized by a line topology, i.e. all of the nodes follow each other in a single sequence from the top to the bottom. Production risers often have buoyancy and additional attached masses at irregularly spaced intervals along its length. Furthermore, marine risers are quite flexible, and the axial tension which, in many cases, provides the main contribution to the total stiffness, varies along the riser.

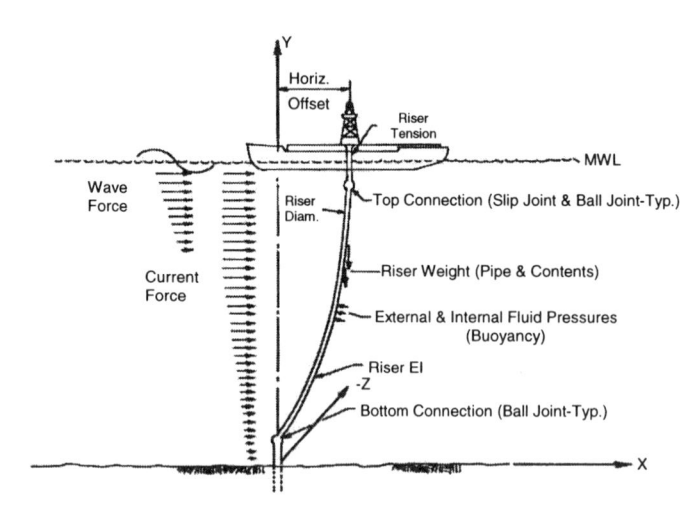

Fig. 7.20: Schematic of a Vertical Tensioned Riser

Chapter 7 Fluid Structure Interaction

The outer geometry of the riser is not uniform because of various elements attached to it. Here, however, for simplicity in the derivation, it is considered to be a uniform pipe. Since a small segment of the pipe will be chosen for the balance of the forces, generalization to a non-uniform pipe is rather straightforward. For the purposes of the derivation of the equation of motion for a riser and its different components, it is assumed that the riser represents a bent tubular member in one plane and only one plane of motion is considered. Similar equations may be applied to the orthogonal plane, and the two motions may then be combined with the coupling between them coming from the external forces.

Consider the free body diagram of a bent tubular structure shown in Fig. 7.21. The vertical equilibrium equation is obtained by equating the internal and external forces. The total external forces on the segment s of length ds in the horizontal and vertical directions are written as F_{xs} and F_{ys}. The weight of the segment is F_w acting at the midpoint s of the segment. The internal forces on the segment acting at its ends are the shear, axial forces and bending moment, F_s, F_A and M_B respectively.

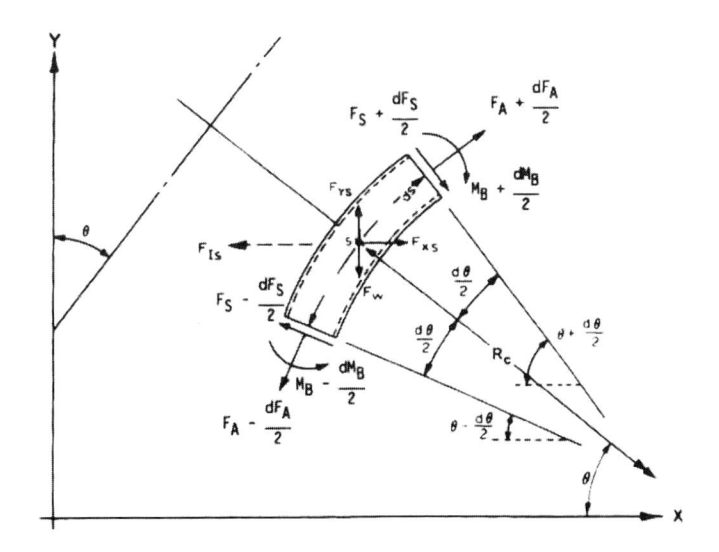

Fig. 7.21: Free Body Diagram of a Bent Tubular Segment

The length of the segment is assumed small, so that the trigonometric function of $d\theta$ may be approximated by their first series term. Upon introducing this approximation and dividing by ds, the three equilibrium equations in the horizontal, vertical and angular directions may be expressed as

$$f_1 \cos\theta - f_2 \sin\theta - f_w + f_{ys} = 0 \qquad (7.84)$$

$$f_1 \sin\theta + f_2 \cos\theta + f_{xs} - m\ddot{x} = 0 \qquad (7.85)$$

$$\frac{dM_B}{ds} + F_s = 0 \qquad (7.86)$$

in which

$$f_1 = \frac{dF_A}{ds} - F_s \frac{d\theta}{ds} \qquad (7.87)$$

$$f_2 = \frac{dF_s}{ds} + F_A \frac{d\theta}{ds} \qquad (7.88)$$

Here, m is the mass per unit length of the segment, including added mass effect, and $\ddot{x}$ represents the acceleration of the segment at point s, and f_w, f_{xs} and f_{ys} are the weight and force intensities.

In the subsequent derivation, the pipe deflection is assumed small so that the small deflection beam theory is applicable. Also, the products of differentials are neglected as negligible second order effects. Then, $\cos\theta$ and $\sin\theta$ may be replaced by dx/ds and dy/ds. Furthermore, with the small angle assumption, the following relationship is obtained between the bending moment and curvature:

$$M_B = EI \frac{d\theta}{ds} = EI \frac{d^2x}{ds^2} \qquad (7.89)$$

Applying the above, the vertical force, horizontal force, and moment equilibrium equations, respectively, become:

$$\frac{dF_A}{dy} - \frac{d}{dy}(F_s \frac{dx}{dy}) - f_w + f_{ys} = 0 \qquad (7.90)$$

Chapter 7 Fluid Structure Interaction

$$\frac{d}{dy}(F_A \frac{dx}{dy}) + \frac{dF_s}{dy} + f_{xs} - m\ddot{x} = 0 \qquad (7.91)$$

$$\frac{d}{dy}(EI \frac{d^2x}{dy^2}) + F_s = 0 \qquad (7.92)$$

In order to develop the horizontal equation of motion of the marine riser, the above three equations are combined into one. Then the horizontal equilibrium equation reduces to

$$[\frac{d^2}{ds^2}(EI \frac{d\theta}{ds}) - F_a \frac{d\theta}{ds}]\sec\theta - (f_w - f_{ys})\tan\theta + m\ddot{x} = f_{xs} \qquad (7.93)$$

The weight intensity, f_w is written in terms of the internal and external cross sections, A_i and A_o, of the tube segment. The hydrostatic load components are derived due to linear internal and external fluid pressures acting on the tube segment. The hydrostatic load due to internal fluid opposes the hydrostatic load from the external fluid. The statically equivalent pressure load becomes

$$f_{xp} = (A_o \overline{P_o} - A_i \overline{P_i}) \frac{d^2x}{dy^2} - (A_o \overline{P_o} - A_i \overline{P_i})g \frac{dx}{dy} \qquad (7.94)$$

Introducing these expressions into the horizontal equation of motion and noting the simplifying assumptions made above, the riser equation becomes

$$\frac{d^2}{dy^2}(EI \frac{d^2x}{dy^2}) - (F_A + A_o \overline{P_o} - A_i \overline{P_i}) \frac{d^2x}{dy^2}$$
$$- [\rho_s g(A_o - A_i) - f_{ys} - g(A_o \rho_o - A_i \rho_i)] \frac{dx}{dy} + m\ddot{x} = f_{xs} \qquad (7.95)$$

The expression inside the parenthesis in the second term of Eq. 7.95 is often described as effective tension $F_e(y)$. Introducing this substitution, the equation of motion is explicitly written as

$$\frac{d^2}{dy^2}(EI(y) \frac{d^2x}{dy^2}) - \frac{d}{dy}[F_e(y) \frac{dx}{dy}] + m(y)\ddot{x} = f_{xs}(x, y, t) \qquad (7.96)$$

The first term represents the resistance of the riser due to its flexural rigidity. The second term is the loading from the axial force and the internal and external fluid pressure. The third term is the riser's inertial resistance and the fourth term on the right hand side is the applied horizontal force intensity.

This horizontal equation may be solved for both the static and dynamic analysis of the riser. For the static analysis, the riser inertia is absent and the external loading is due to the current load. In this case, the equation becomes

$$\frac{d^2}{dy^2}(EI(y)\frac{d^2x}{dy^2}) - \frac{d}{dy}[F_e(y)\frac{dx}{dy}] + m(y)\ddot{x} = \frac{1}{2}\rho C_D(y)D(y)|U(y)|U(y)$$

(7.97)

where $U(y)$ is the current velocity as a function of the vertical coordinate y, and $C_D(y)$ is the corresponding drag coefficient for the riser. Additional constraints are needed to solve this equation, which are specified at the top and bottom joints as end restraints. The restraints could to fixed, pinned, free or a specified top offset from the vessel displacement. In order to solve the dynamic problem due to oscillatory excitation, the right hand side of Eq. 7.96 should represent the dynamic load, e.g., from wave and vessel motion. The two solutions may be combined into one, when the static and dynamic external loads on the right hand side are combined. While the analytical solution is limited in scope, either finite difference or finite element methods are used to solve for the deflected riser mode shapes and structural properties under static or dynamic loads. Because of its versatility, the finite element method (FEM) becomes an obvious candidate for the numerical tool. Indeed, most of the general-purpose riser analysis packages are based on the FEM, and the reader is referred to the vast literature that exists on the FEM [see, for example, Meirovich (1997), Moe, et al. (2000)] for details of these analyses.

7.6 Flexible Line Analysis

A floating structure is generally held in place with a (6 to 8 line) spread or taut mooring system. The spread mooring system takes the form of a catenary, while the taut mooring system has a large pre-tension. The

Chapter 7 Fluid Structure Interaction

mooring lines are made of a combination of cables and chains, or polyester lines and chains.

The characteristics of mooring lines are highly nonlinear based on their shapes under load, and their solutions are almost invariably made numerically. The numerical analysis for the mooring line itself is based on a lumped mass method. In this case, the mooring line is divided into a finite number of non-elastic lumped masses, similar to a vertical riser configuration, and these masses are connected with massless linear springs of given stiffness. The lumped masses introduce inertia and damping (both hydrodynamic and material) in the dynamic system. The damping has both linear and nonlinear components arising from the wave damping and damping based on vortex shedding. The geometry of the mooring line is updated at each time step based on the wave action on the mooring line and the vessel motion. The force components acting on each lumped mass are the tension, gravity force, and hydrodynamic drag and inertia force.

Typically, a large portion of the catenary mooring line remains near or on the ocean floor (Fig. 7.22). Thus, the interaction of the line with the foundation is an important consideration in the floating structure analysis. Accurate modeling of the mooring line on the seabed requires knowledge of the soil parameter. Generally, the foundation is considered elastic. The interaction of the mooring line with the seabed introduces nonlinear damping in the system. The force in the mooring line on the sea bottom is altered due to viscosity, shear and density of the seabed media.

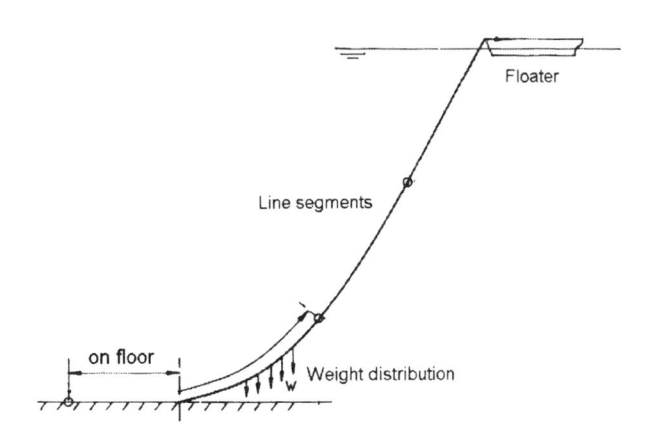

Fig. 7.22: Typical Spread Mooring Line Configuration

An idealized lumped mass system for a mooring line partially on the seabed is shown in Fig. 7.23 [Inoue & Sunderan (1994)]. The mooring line is divided into N segments, and the mass of each segment is lumped at the intersection point. This provides N-1 inelastic lumped masses. The line between the masses is considered to be a massless elastic element. The forces at each mass are equated to the inertia, and the boundaries between masses are matched to the mooring line displacement. The seabed is assumed to consist of a series of linear springs with dashpot, supporting the mooring line in touch with the bottom. The touchdown point is a variable during the oscillating excitation, which is included in the analysis. On the assumption that the flexural rigidity is negligible and the transverse displacement is small, the Hamilton's principle provides the equation for the line displacement.

In a simple mooring line analysis in which the top end of the line is subjected to a harmonic heave motion, Inoue and Sunderan (1994) studied the effect of various seabed conditions. The dynamic tension is reduced in magnitude when the soil-line interaction is included. For the example chosen, the amount of reduction was between 15 to 30 percent, depending on the imposed frequency.

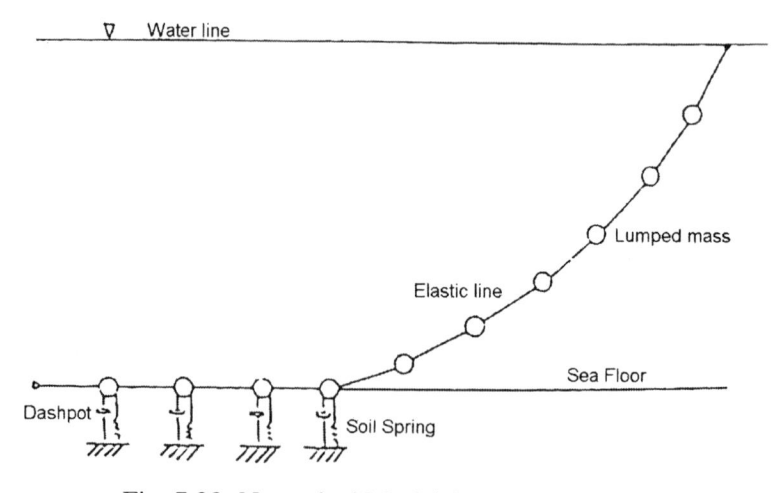

Fig. 7.23: Numerical Model for a Mooring Line

In soft soils, such as clay or silt, the lack of interaction effect may lead to overprediction of the mooring line forces. The energy absorption characteristics of the anchor and the embedded portion of the mooring system should be properly understood. In sandy or gravelly soil, the assumption is reasonably valid since the line will mostly remain on the sea-floor. In reality, the anchor and part of the mooring chain on the sea-floor are embedded in the soft soil. The embedded portion of the chain will provide significant energy absorption [Wung, et al. (1995)] through hysteretic, viscous and radiation damping effects under cyclic loading. Wung, et al, modeled the interaction as a nonlinear plastic spring and dashpot in parallel, connected to a second dashpot, the values of which were established through model tests. The soil flexibility, nonlinear soil behavior and limit in the anchor capacity that were neglected in their analysis may, however, be important. The proximity of the line to the bottom, but not touching the bottom, will also have some influence on the dynamics of the line.

The transverse and in-line vibration of the mooring line due to vortex shedding from the line is generally evaluated independently. Inclusion of these effects may be important for the coupled motions of certain system and will make the floating body dynamics very complex.

7.7 Coupled Dynamic Analysis

So far, the discussion has been limited to the dynamics of structure systems in which one component has been considered independent of the other. For floating structure analysis, the mooring/ riser system is simply treated as an external nonlinear stiffness term. The effect of moorings and risers are applied as position-dependent forces. For the flexible elements, such as risers, the motions of the floating structure under the environmental loads are added to the attachment point as an externally defined oscillation. This includes any low frequency oscillations as well. Thus the two systems are designed based on two stand-alone design tools. These two independent steps are illustrated in Fig. 7.24 as step 1 and step 2 respectively. For many structural systems, this method of uncoupling provides acceptable results for the responses of the elements of a floating system. However, in structural systems where the relative size and weight of various elements, such as the floating structure and the connectors are comparable, the influence of one on the other increases. In these instances, the simultaneous coupling at an instant of time may not be ignored. The method of superposition of individual responses described earlier may not be adequate, and the instantaneous responses of the coupled system may have to be solved simultaneously.

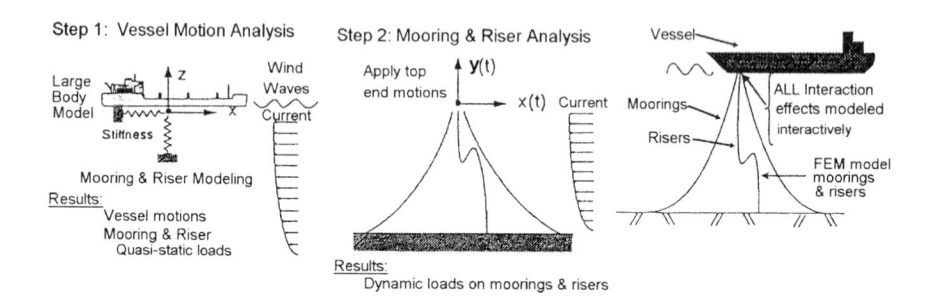

Fig. 7.24: Illustration of Coupled vs. Uncoupled Response Analysis [after Ormberg, et al. (1997)]

The analytical model in this regard must solve the problem as a coupled system where the dynamics of the floating structure are combined with the dynamics of the connectors, mooring lines and risers. This is shown on the right side of Fig. 7.24. In coupled analysis, the mooring lines and risers are included in the numerical model along with the floating structure. Since the deflection of the riser and lines have influence on the damping and inertia of the system, and since they change continuously during oscillation, a time domain finite difference scheme appears appropriate for the solution. The lines and riser may be modeled as finite elements within this formulation. Various nonlinearities present in these components may be considered in the analysis, including the VIV effect on the riser and lines. The contact of the lines with the seabed may be modeled by nonlinear springs and damping elements. Numerical analysis has shown that the simultaneous dynamics of lines and risers may significantly affect the motion of the floating structure. This is particularly true for present-day deep-water structures. In systems where the mooring lines become an integral part of the system (rather than a stiffness term only), instability in motion of the floating structure may be experienced.

7.8 Exercises

Exercise 1

Compute the roll damping components of a rectangular barge of dimensions 100m long, 20m wide and 10m draft. Assume that the barge is equipped with bilge keels having a width of 0.5m.

Exercise 2

Assume a tube segment for a riser, similar to the one in Fig. 7.21 and make the same assumptions regarding the small deflection angle. The segment is subjected to the hydrostatic internal and external fluid pressure. Show the detailed derivation for the expression of the horizontal and vertical forces on the segment (given in Eqs. 7.90-91).

Exercise 3

Describe a finite difference scheme that can be used in solving the riser horizontal equation for displacement. State the assumptions made in your numerical scheme.

Exercises 4

Formulate a numerical scheme that can be used in solving the coupled motion of a floating structure connected to two mooring lines one fore and one aft in a head sea environment. State the assumptions made in your numerical scheme. Assume that the only motion of interest is oscillatory surge motion. Apply a Morison type force on the floating structure as well as the lines.

7.9 References

1. American Petroleum Institute, "Recommended Practice for Planning, Designing and Constructing Fixed Offshore Platforms", 20th Edition, API-RP2A, 1992.

2. Bishop, R.E.D. and Hassan, Y., "The Lift and Drag Forces on a Circular Cylinder Oscillating in a Flowing Fluid", Proceedings of the Royal Society (London), Vol. 277, pp. 51-75, 1964.

3. Chakrabarti, S.K., Hydrodynamics of Offshore Structures, WIT Press, Southampton, UK, 1987.

4. Chakrabarti, S.K., Nonlinear Methods in Offshore Engineering, Elsevier, The Netherlands, 1990.

5. Chakrabarti, S.K., Cotter, D.C., and Palo, P., "Shear Current Forces on a Submerged Cylinder", Proceedings of 10th International Offshore Mechanics and Arctic Engineering, Vol. 1-A, 1991, pp. 147-151.

6. Chantrel, J., and Marol, P., "Subharmonic Response of Articulated Loading Platform", Proceedings of the Sixth Conference on Offshore Mechanics and Arctic Engineering, Houston, TX, 1987, pp.35-43.

7. Givoli, D., "Nonreflecting Boundary Conditions", Journal of Computational Physics, Vol. 94, 1991, pp. 1-29.

8. Griffin, O.M., "Vortex Shedding from Bluff Bodies in a Shear Flow: a Review", Journal of Fluids Engineering, Vol. 107, ASME, 1985, pp. 298-306.

9. Hoerner, S.F., Fluid-Dynamic Drag, published by the author, 1965.

10. Humphries, J.A. and Walker, D.H., "Vortex Excited Response of Large Scale Cylinders in Shear Flow", Proceedings of Sixth International Offshore Mechanics and Arctic Engineering Conference, ASME, Vol. 2, Houston, TX, 1987, pp. 139-143.

11. Ikeda, Y., Himeno, Y., and Tanaka, N., "A Prediction Method for Ship Roll Damping", Report No. 00405, Dept. of Naval Arch., Univ. of Osaka Prefecture, Osaka, Japan, 1978.

12. Inoue, Y. and Sunderan, S., "Dynamics of the Interaction of Mooring Line with the Sea Bed'" Proceedings of the Fourth International Offshore and Polar Engineering Conference, Osaka, Japan, April, 1994, pp. 317-323.

13. Kim, M.H., Arcandra, Kim, Y.B., "Variability Of Spar Motion Analysis Against Design Methodologies/Parameters", International Conference on Offshore Mechanics And Arctic Engineering, OMAE/OFT 1064, ASME, 2001.

14. Kiya, M., Tamura, H. and Arie, M., "Vortex-Shedding from a Circular Cylinder in Moderate-Reynolds Number Shear-Flow", Journal of Fluid Mechanics, Vol., 101, 1980, pp. 721-736.

15. LeMehaute, B., *An Introduction to Hydrodynamics and Water Waves*, Springer Verlag, New York, 1976.

16. Mair, W.A. and Stansby, P.K., "Vortex Wakes of Bluff Cylinders in a Shear Flow", SIAM Journal of Applied Mathematics, Vol. 28, 1975, pp. 519-540.

17. Masch, F.D. and Moore, W.L., "Drag Forces in Velocity Gradient Flow", Proceedings of the ASCE, Journal of the Hydraulics Division, Vol. 86, No. HY 7, 1960, pp. 1-11.

18. Maull, D.J. and Young, R.A., "Vortex Shedding from a Bluff Body in a Shear Flow", Flow-Induced Structural Vibrations, E. Naudascher, ed., 1974, pp. 717-729.

19. Meirovich, L. *Principles and Techniques of Vibrations*, Prentice Hall International, Upper Saddle River, NJ, 1997

20. Moe, G., Cheng, Y., Vandiver, J. K., "Riser Analysis by Means of Some Finite Element Approaches", Proceedings of ETCE/OMAE2000 Joint Conference, ASME, 2000.

21. Newman, J.N., "Algorithms for the Free Surface Green Function", Journal of Engineering Mathematics, Vol. 19, 1985, pp. 57-67.

22. Noblesse, F., "The Green Function in the Theory of Radiation and Diffraction of Regular Water Waves by a Body", Journal of Engineering Mathematics Vol. 16, 1982, pp. 137-169.

23. Ormberg, H., Fylling, I.V., Larsen, K., and Sodahl, N., "Coupled Analysis of Vessel Motions and Mooring and Riser System Dynamics", 16th International Conference on Offshore Mechanics and Arctic Engineering, ASME, 1997.

24. Siqueira, C. L. R., Meneghini, J. R., Saltara F., Ferrari Jr., J. A., "Numerical Simulation of Flow Interference Between Two Circular Cylinders in Tandem and Side By Side Arrangements", 18th International Conference on Offshore Mechanics and Arctic Engineering, ASME, July, 1999.

25. Stansby, P.K., "The Locking-On of Vortex Shedding Due to the Cross-Stream Vibration of Circular Cylinders in Uniform and Shear Flows", Journal of Fluid Mechanics, Vol. 74, 1976, pp. 641-667.

26. Wung, C.C., et al., "Effect of Soil on Mooring System Dynamics", Proceedings on Offshore Technology Conference, Houston, TX, May 1995, pp. 301-307.

Chapter 8

Fluid Induced Vibration

8.1 Introduction

This chapter is intended to give an overview of the fundamental aspects of fluid induced vibration. This vibration, commonly referred to as Vortex Induced Vibration (VIV), is created as a result of the vortex shedding from the surface of the structure due to fluid flow past the structure. The formation of vortices past a cylindrical structure in a steady flow has been introduced in Chapter 2. This chapter will cover a relatively broad range of subjects associated with the VIV of structures. The fundamental aspects of VIV on bluff bodies will be described. In particular, the flow interference with circular cylinders and cylinder arrays will be discussed. A few empirical models that attempt to describe the complicated flow field will be given. The numerical modeling of VIV will be formulated even though the details of the computation fluid dynamics (CFD) will be left out. The VIV of cables, pipelines and marine risers will be discussed. This has a wide range of practical applications including the recent deepwater development of offshore exploration and production platforms by the oil industry. Research continues in the area of suppressing the VIV problem in structures subject to fluid flow. Several techniques of mitigating the VIV will be discussed.

 The engineering applications of fluid-structure interaction frequently involve long cylindrical structures, as in power transmission lines, heat exchanger tubes, and pipelines. In the offshore application, one of the areas of flow-structure vibration coupling is the vortex-induced vibration of long slender structures such as pipelines, risers, tendons, spar platforms, etc. When exposed to fluid flow, such aerodynamically bluff bodies experience separation of flow as it passes the body surface and the formation of a trailing vortex wake behind the body. The VIV is caused by the shed vortex behind a structure in fluid flow. This vortex shedding results in alternating pressure field on the surface of the bluff body and oscillating lift and drag forces. These phenomena have been observed in various engineering fields, such as the famous Tacoma Narrows Bridge that failed by the wind-induced VIV. Vortex shedding at a frequency near a torsional frequency of the bridge was thought to be partially responsible for the bridge failure. Observed amplitudes of a torsional mode were as large as 45 deg.

The flow field generated by flow separation around a body is a very complex fluid dynamics problem. The vortex shedding mechanism represents one of the most challenging problems in fluid dynamics. There are numerous fascinating fluid mechanical effects associated with this flow-structure interaction [Bearman (1984)]. Much progress has been made toward the understanding and prediction of flow around bluff bodies and the associated interaction of the body with the fluid flow.

In practical applications, the influence of the vortex-induced unsteady fluid forces must be considered. This is particularly so because when the frequency of vortex shedding approaches a structural frequency, synchronization takes place and the structure undergoes severe vortex-induced vibration. It is associated with the phenomenon known as lock-in, where the vortex shedding frequency is "captured" by, or coincides with the structure vibration frequency. It is well known that large cross-flow response can occur when the vortex shedding frequency f_s is close to the structural natural frequency f_n. Response of a cylinder of diameter D occurs in a steady flow of velocity U for a range of reduced velocity V_r ($V_r = U/f_nD$). The response depends on the mass and damping of the cylinder. This phenomenon is well documented in wind engineering studies where slender structures with low structural damping have been observed to respond transverse to the wind direction. It is also the reason that many drilling risers have failed in high currents. Dynamic response due to this phenomenon has been the subject of intensive research effort over the years, and yet true prediction of this vibration remains elusive.

Vortex shedding from a circular cylinder produces alternating (transverse) forces on the cylinder. If the cylinder is free to move, then these forces will produce cylinder vibration. If the cylinder vibrates at, or near, one of its structural natural frequencies, then the vibration is initially limited only by the cylinder damping. However, once the amplitude reaches about 1 to 1.5 times the cylinder diameter, the boundary layers are altered enough by the cylinder motion that the vibration becomes self-limiting. If the cylinder does not vibrate at or near one of its structural natural frequencies, then the vibration is simply a forced vibration without much amplification, and is typically limited to about 0.1 to 0.2 times the cylinder diameter. This is mainly because the lift forces are not sufficient to produce larger motions than the above limits.

8.2 Flow-Induced Vortices

Vortices are formed at rigid boundaries and transported in the fluid as they decay and lose their strength by diffusion within the fluid interior. The wake is defined as a low-pressure region near the boundary of a submerged body. Experimental observation shows that the wake of a bluff body is comprised of an alternating vortex formation in the interior of a homogeneous fluid. Under normal conditions, vorticity is neither created nor destroyed. Vorticity is produced only at the boundaries resulting from tangential acceleration at the boundary, and from tangential pressure gradients acting along the boundary.

Numerous theoretical, experimental and numerical efforts have been carried out to understand the near and far wake fields of a bluff body, in particular a body of cylindrical shape. This shape has many practical applications. Moreover, it is a simple shape to understand the phenomenon of vortex formation, including separation, transition, shear layer evolution, wake instabilities, and fluid-structure interactions.

8.2.1 Formation of Vortices

When the fluid encounters a solid boundary, such as an immersed cylinder, a real fluid cannot have a motion relative to the boundary of the cylinder. When an adverse pressure gradient is created from the flow at the boundary, the flow separates the boundary. This causes the inception of a vortex.

The flow at the surface of the cylinder becomes slower due to friction between the cylinder and fluid when compared to the free stream flow as the flow moves around the cylinder. The outer flow eventually cannot stay attached to the cylinder and separates from behind it. This creates a low pressure at the cylinder beyond the separation point. The flow near the cylinder moves slower than the flow further away, forming what is known as a boundary layer. At low Reynolds numbers, the vortex pattern in the wake is steady. As the Reynolds number increases, the vortex on one side of the cylinder grows stronger compared to that on the other side due to even a small imperfection of the two flows due to surface imperfection, out-of-roundness or roughness. If the bluff body roundness and surface smoothness can be made within a small tolerance, then this alternate shedding may be reduced or eliminated. This effect will be discussed subsequently. The stronger vortex continues to build rapidly until it cannot stay attached to the body and separates. This is known as shedding of vortices. After this side sheds one vortex, the other side takes over and its vortex starts to become stronger and eventually sheds. Thus, the alternating shedding process continues. Due to complex flow around the bluff body, this process is not very predictable and the vortex pattern behind the bluff body is irregular. The shed vortices are swept away by the ambient flow in its field, and the strengths of the vortices reduce by their loss of energy with time, eventually dissipating in the flow.

Gerrard (1966) provided an excellent explanation of this phenomenon of generation of vortices and stabilizing effect. According to him, the mutual interaction between the two separating shear layers on the opposite sides is a key factor in the formation of a vortex, which grows by gaining circulation. When the growing vortex is strong enough, it draws the opposing shear layer across the near wake. The fluid particles of the opposite shear layer can be either entrained into the growing vortex reducing its strength, or find their way into the other shear layer of opposite sign, or fed back into the near wake region. In sufficient concentration, this vorticity cuts off further supply of circulation to the growing vortex, which is then shed and moves downstream. The amount of fluid that follows controls the shedding frequency, the strength of vortices shed and the base pressure. The entrainment of opposite vorticity into the shear layer is an important stabilizing mechanism. For example, the separating shear layer becomes stronger when the base pressure falls. The growing vortex becomes stronger drawing more of the opposite shear layer across the wake and more of it is entrained into the shear layer. Therefore, the reduction of the shear layer circulation reduces the strength of the growing vortex, thus acting as a stabilizing factor.

8.2.2 Reynolds and Strouhal Dependence

It has been known that for stationary cylinders, the Strouhal number varies with Reynolds number and is particularly variable in the critical Reynolds number range of 1×10^5 to 6×10^5. Data on Strouhal number for stationary cylinders as a function of Reynolds number and cylinder roughness were introduced in Fig. 2.8. In the critical Reynolds number regime of about 1×10^5 to 6×10^5, the Strouhal number varies dramatically. For a smooth cylinder and around 2×10^5, the Strouhal number is about 0.2. It remains constant up to a Reynolds number of approximately 2×10^6, then rises to about 0.3 beyond a corresponding Reynolds number of 3×10^6. With increasing roughness (K/D = 0.00075, where K is the average "peak to trough" height of the surface protrusions), the Strouhal number rises from 0.2 at about 1×10^5, with values reaching as high as 0.5, then dropping down to about 0.26 at a Reynolds number of 5×10^6. There is considerable scatter in Strouhal number, with values between 0.3 and 0.5 in the Reynolds number range of 1×10^5 to 5×10^6. At a roughness of K/D = 0.003, the Strouhal number rises from 0.22 at around 1×10^5 to 0.25 at 1.5×10^5. After that it is essentially constant at slightly less than 0.25 all the way up to 5×10^6. This data [Achenbach & Heineche (1981)] shows that roughness is critical in determining the Strouhal number in the range of Reynolds numbers from 1×10^5 to about 3×10^6. It implies that greater roughness of rigid fixed cylinder translates to lower Strouhal numbers and therefore lower vortex shedding frequencies.

8.2.3 Vortex Formation in Steady Uniform Flow

It has been shown that the wake of a cylindrical body subjected to a uniform steady flow is comprised of an alternating vortex sheet. The character of the vortices that are shed immediately behind the body and in the wake further downstream depends on the Reynolds number, degree of roughness of the body surface and the intensity of turbulence in the ambient flow.

The separation point in a steady flow may remain fixed, such as at the edge of a plate or a fin, or it may move, such as on the surface of a circular cylinder. The variation of the drag coefficient for a fixed separation point is considerably less than that for a moving separation point. The problem, however, is still quite complex to understand and formulate. For a moving separation point, the shedding vortex is a function of the point where the flow separates at an instant of time.

The resonant flow-induced vibration of a stationary structure occurs when the vortex shedding frequency due to the flow of fluid past the structure approaches one of the natural frequencies of the structure. This "lock-in" of the structure captures the vortex shedding frequency by the vibration frequency of the structure over a range of flow speeds. The lock-in effect causes the wake and the structure to oscillate in unison. This causes a marked increase of structural oscillation.

Transverse VIV leads to substantial increases in the drag coefficient. In the case of a long slender member, there are potentially many modes of vibration that can be excited. In addition to transverse oscillations, in-line oscillations of smaller amplitude can also be excited. However, they take place at roughly half the flow speed of transverse oscillations and hence are more likely to occur.

Dynamic response of a structure can occur in other ranges of V_r as well, associated with sub and super harmonics of the vortex shedding frequency, some aspects of which will be discussed below. As shown in Chapter 7, the wake behind a cylinder is different along its length, i.e., the wake pattern is generally three dimensional. With lock-in, the wake becomes almost perfectly correlated spanwise (i.e., lengthwise); that is, it becomes effectively two-dimensional.

Experiments were conducted by Allen and Henning (2001) to determine the effect of surface roughness on vortex-induced vibration and drag of flexible cylinders at critical and supercritical Reynolds numbers. The flow was generated by the cylinder specimen moving with a rotating arm in a circular tank and thus had a linear shear. Four levels of surface roughness K/D were tested for their effect on VIV as shown in Fig. 8.1. Cylinders experienced vibrations of up to the 3rd mode (in transverse bending). Very smooth cylinders experienced low drag and virtually no VIV, as the boundary layers became turbulent in the critical Reynolds number range.

As the surface roughness increased, the displacement also increased. However, once a cylinder was sufficiently rough, further increase in the surface roughness had little effect on the displacement. This is illustrated in Fig. 8.1 from the above experiment. The drag coefficient appeared to be strongly coupled with displacement, with the smoothest cylinder having the lowest drag.

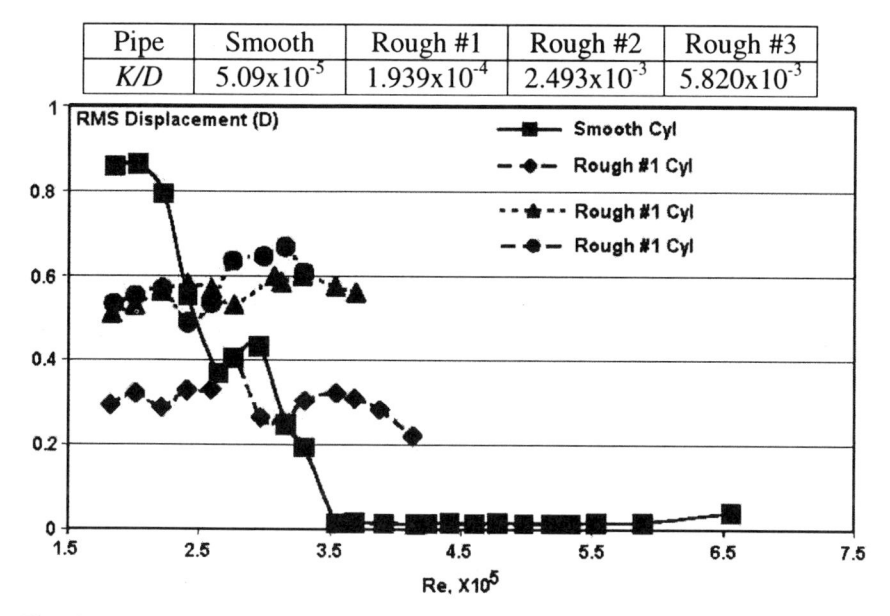

Pipe	Smooth	Rough #1	Rough #2	Rough #3
K/D	5.09×10^{-5}	1.939×10^{-4}	2.493×10^{-3}	5.820×10^{-3}

Fig. 8.1: Displacement of Test Cylinder at Different Roughness [Allen and Henning (2001)]

8.2.4 Vortex Formation in Shear Flow

When the flow is not uniform, there is an added effect of vorticity in the approaching flow. If the ambient flow has a shear with depth (whose strength is given by the shear parameter β), then different modes are likely to be excited at different depths. This can generate a traveling wave along the length of the structure, which may result in difficulty in identifying lock-in boundaries due to the interference between modes [Bearman (2000)]. For a vertical cylinder in shear flow, the ratio of length-to-diameter ratio, L/D, is an important parameter, in addition to the shear (strength) parameter. In a shear flow, a changing cellular pattern of vortex shedding exists along the span of a stationary vertical cylinder. Over each cell the vortex shedding frequency is constant. The Strouhal number and base pressure coefficient are also found to be constant over the cell.

324   *Theory and Practice of Vibration and Hydrodynamics*

The experimental results are generally limited by the small cylinder aspect ratio (L/D <20). In a highly turbulent shear flow ($\beta = 0.18$), the critical Reynolds number Re_{cr} has been observed to be reduced by a factor of 10. In low and moderate-turbulence shear flow, the vortex shedding pattern is cellular. The cell length has been found to decrease with increasing shear. For a long flexible cable ($L/D = 100$), a discernible cell structure existed at moderate subcritical Reynolds number for $\beta = 0.005$ [Peltzer (1982)]. The base pressure coefficient at the midspan of a cylinder in a highly turbulent shear flow for a shear parameter $\beta = 0.18$ and a turbulence level (U_{dev}/U) of 5 percent, is shown in Fig. 8.2 for both uniform and shear flow [Davies (1976)]. It is clear that the critical Reynolds number for the onset of the vortex shedding from a vertical circular cylinder is reduced in the shear flow by a factor of ten from the uniform smooth flow value of $Re = 2 \times 10^5$ to 3×10^5.

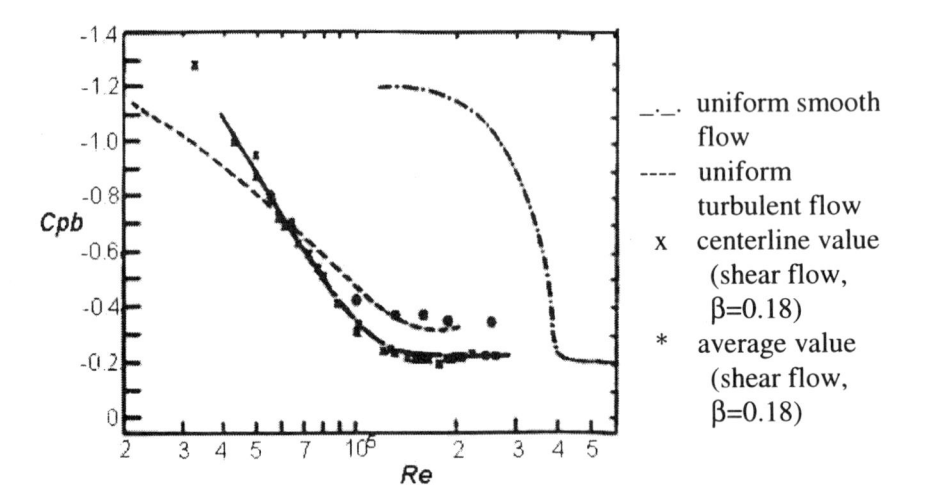

Fig. 8.2: Midspan Base Pressure Coefficient for a Circular Cylinder in a Shear Flow [Davies (1976)]

Chapter 8 Fluid Induced Vibration 325

The cell structure at the subcritical Reynolds number is well-defined with strong vortex shedding and associated predominant wake frequency. As the Reynolds number increases, the shedding patterns become irregular, disappearing into a turbulent background at $Re = 10^5$. The distribution of the base pressure coefficient C_{pb} measured on a vertical cylinder in a linear shear flow is shown in Fig. 8.3 for $Re = 2 \times 10^4$ and $\beta = 0.015$ [Peltzer and Rooney (1980)]. In this case, the length ratio for the cylinder was $L/D = 48$, and roughness parameter $K/D = 0.001$. In general, the vortex shedding patterns were free of constant frequency cells. Note that this result typically shows the end effect adjacent to the end plates of the cylinder.

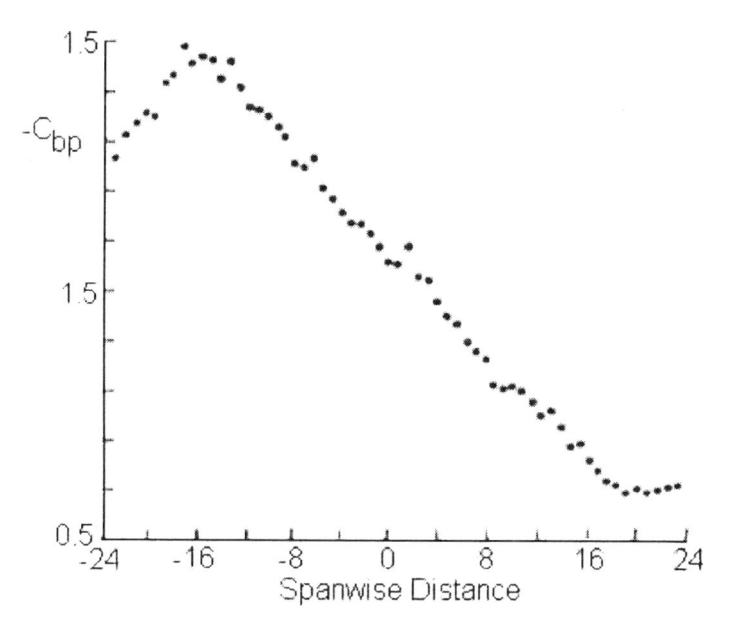

Fig. 8.3: Spanwise Variation of C_{pb} on a Rough Circular Cylinder in a Shear Flow [Peltzer and Rooney (1980)]

8.2.5 Vortex Formation in Oscillatory Flow

For a time dependent flow, further complications arise in establishing the true picture of the vortex formation. If the flow is time dependent, such as waves rather than steady uniform flow, the vorticity formation also depends on the past history of the flow.

In an oscillatory flow, the vortices are shed alternately on both sides during each half-cycle. Through flow visualization during an oscillatory wave past a vertical cylinder, the vortex pattern shown in Fig. 8.4 was discovered [Iwagaki, et al. (1976)]. The figure gives the vortex pattern observed at the surface of the water as a function of the surface Keulegan Carpenter number (KC). At low KC number (< 3) the flow is almost laminar and no wake region or flow separation exists. Thus one should expect symmetric pressure distribution about the direction of flow, and little transverse force generated. For surface KC number between 3 and 8, the vortices appear to form, but still appear symmetrically attached to the cylinder surface. During flow reversal, they are quickly thrown out of the newly formed vortices indicated by the arrow. As the surface KC number increases, the separation point moves on the cylinder surface and vortices are shed alternately from the two sides of the cylinder, causing an asymmetric pressure distribution. Thus in these cases, the cylinder will experience an inline force and a transverse force. Moreover, since the flow reverses every half cycle for a harmonic flow, the vortex pattern in Fig. 8.4 will reverse every half cycle as well. In this cycle, the earlier vortices tend to sweep back onto the cylinder causing the flow field to be asymmetric and time dependent. The inline force is mostly regular superimposed with minor irregularity. The lift force time history has an irregular form.

It has been found from experimental measurements that the frequency of the lift force is a direct function of the number of vortices shed. In particular, the frequency is (N-1) times the fundamental (oscillatory) frequency, where N is the number of vortices shed. Since the number of vortices goes up with the increase in KC number, the lift force frequency increases with the KC number as well. Also, since the number of vortices present in the flow field for a given KC number varies with time, it is expected that the irregular lift force will contain multiple frequencies. These frequencies, however, are multiples of the fundamental frequency. This has been verified with measured data on pressures and forces on a cylinder in waves. An example of the predominant frequencies in oscillatory flow on a cylinder as functions of KC number and Reynolds number was shown in Fig. 2.15.

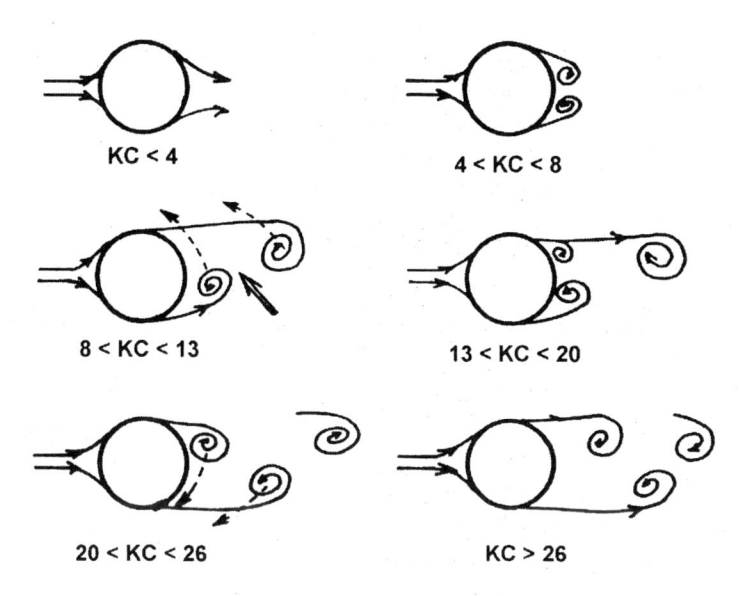

Fig. 8.4: Vortex Pattern Past a Vertical Cylinder in Waves

8.3 Theory of Vortex-Induced Vibration

The vibration caused by the vortices from the fluid flow past a structure depends on several factors. The important hydrodynamic quantities that influence VIV are

- Shedding frequencies and their interactions,
- Added mass (or mass ratio) and damping,
- Reynolds number,
- Lift coefficient, and
- Correlation of force components.

When the frequency of vortex shedding is close to the natural frequency of the structure, it may jump to the structural frequency and become locked to this frequency over a finite range of reduced velocity. The range of this lock-in depends on several parameters, including principally the fluid inertia force and damping force. For very low damping, the oscillation occurs not only in the transverse direction, but also in-line with the flow. The in-line oscillation amplitude, however, is much reduced compared to the transverse one. In-line response appears at twice the vortex shedding frequency, giving a "figure-eight" type of response. The oscillation amplitudes recorded in an experiment for a flexible cylinder under tow are shown in Fig. 8.5. The Y-axis represents normalized transverse amplitude (A/D) while the X-axis is the normalized inline amplitude (A/D) and reduced velocity (V_r). The reduced velocity varied from 2.5 to 8.3. The results show that the in-line oscillation is small compared to the transverse oscillation. The oscillations increase with the reduced velocity to a maximum near $V_r = 7.2$ to 7.4, before decreasing again, and for most part, the motion follows a figure-eight configuration.

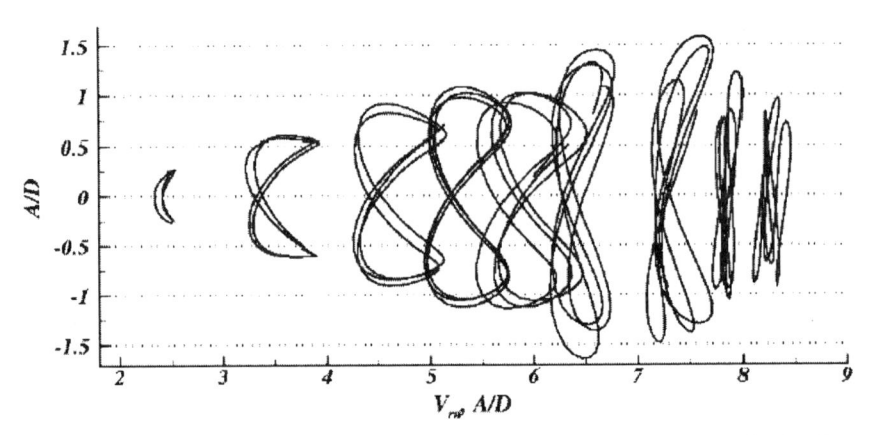

Fig. 8.5: In-Line vs. Transverse Displacement of a Flexible Cylinder for Various Reduced Velocities [Davis, et al. (2000)]

In addition to damping, the maximum amplitude of response is influenced by the magnitude of the mass ratio. The mass ratio is related to the ratio of the structure density to the fluid density surrounding it. A decreasing value of this ratio has been found to lead to increasing response. For a fixed structure the vortex shedding frequency is proportional to the fluid velocity, in agreement with the Strouhal number relationship. Therefore, for a flexible body, if the fluid behaves in a linear fashion, there would be a simple resonant behavior at a given critical flow speed. In practice it is observed that the fluid interacts strongly with the structural motion and the vortex shedding frequency is captured by the body frequency over a range of flow speed. As the vortex shedding frequency fs, as defined by the Strouhal number relationship, approaches and eventually synchronizes with the natural frequency of the body, lock-in resonance phenomenon develops and the vibration amplitude and drag force increase sharply. Within this frequency synchronization region, a sudden transition between two primary vortex wake patterns is observed. The extent of this range depends on the values of mass ratio and damping. One of the most significant differences between flexible structures in air and in water is that in the latter case, the mass ratio is much smaller, leading to a stronger fluid/structure interaction [Bearman (2000)]. This results in larger VIV amplitude in fluid, and significant amplitudes of oscillation over a broader range of flow speed. Additionally, the phase angle by which the transverse force leads the displacement varies significantly through the lock-in range. This force is the sum of the vortex shedding force and the inertia (or added mass) force proportional to the acceleration of the body. In most studies these two forces are combined and referred to as the transverse force. Lock-in has been observed whenever vortex shedding occurs, from 2-D laminar conditions to post-critical conditions ($Re \geq 10^6$).

At low Reynolds number (e.g. $Re = 100$), numerical computations for steady flow past cylinders are easier and computations have verified experimental results. Maximum cross-response amplitude of half the diameter (½ D) has been found for cylinders having small mass and damping. For sub-critical Reynolds number (in the range of 10^3 to 10^5), maximum amplitudes of one diameter have been measured experimentally and for post-critical field conditions, amplitudes of up to two diameters have been recorded for piles responding in cantilever mode (CIRIA Report UR8).

The theory developed for VIV falls into two main categories: empirical and numerical. Because of the complex flow field, as already mentioned, an analytical solution even in a very simple case is not possible. Numerical solution for vortex-induced vibration has been developed for low values of Reynolds number. Empirical models are generally based on experimental observations.

8.3.1 Empirical VIV Models

VIV analysis methods are based on computation fluid dynamics as well as empirical models. The CFD models are more elegant and generally based on the complete Navier-Stokes equation. These models are both two-dimensional and three-dimensional. The 2-D models are simpler and allow higher Reynolds number flow. The 3-D models are more complex and are necessarily limited to lower Reynolds number. Both models have insurmountable difficulties in describing the complex flow field experienced by the prototype Reynolds number in the range of 10^6 to 10^7.

Empirical models, on the other hand, depend on the empirical data and may account for higher Reynolds number. Empirical models for prediction of vortex-induced vibrations have been developed by various investigators since the middle seventies, and have traditionally been based on measured data from small scale oscillation tests with short cylinder sections. These models are simple and make no attempt to describe the flow field. The empirical methods are based on the assumption that VIV appears at discrete frequencies. They have been solved in the frequency as well as time domain. Mode superposition and frequency response methods are most common in the frequency domain approach [Larsen (2000)]. The benefit of the time domain is not only that the hydrodynamic load and damping parameters are easier to apply, but that a nonlinear finite element code may be used. This type of analysis may include, for example, the sea floor interaction as found in catenary risers. Time domain analyses determine the time history of the force on the structure due to these flows.

A few of these research efforts may be found in the references cited. The simplest models apply to uniform current and uniform cross sections, and are based on the assumption that the response occurs at an eigenfrequency and have the shape of the associated eigenmode. These earlier models have been improved substantially through continued research. Halse (2000) showed reasonably good comparisons between predictions and full-scale measurements. However, the results from these models still show large differences among them. An overview of the VIV related to empirical models may be found in ISSC (2000).

These models are generally used to analyze the long hanging flexible members of submerged structures, such as, risers. A free-span pipeline is another slender structure that is prone to significant damage potential from

VIV. The analysis of pipeline VIV in shear flow is somewhat simpler than that of the riser VIV, since the shear current on a pipeline is not a major concern. On the other hand, the variable gap between pipe and sea floor adds complexity to the problem. This problem, however, is considered more mature than the riser problem, and practical design guidelines are available [Mørk, et. al (1998)].

8.3.2 Transverse Oscillator Model

For a horizontal cylinder in a steady flow, the vortices are shed in an orderly fashion and depend on the Strouhal number. The inline drag force has a steady component and an oscillatory component that has a frequency that is twice the vortex shedding frequency. The spanwise correlation of shedding along the length of the structure is generally limited. For example, for a horizontal circular cylinder, this length is about 3 to 4 times the diameter of the cylinder. However, if the structure responds to the shedding frequency by oscillation, the correlation length is much longer.

It is well known (Chapter 2) that the hydrodynamic coefficients for an oscillatory flow around a fixed cylinder is a function of the Keulegan-Carpenter number and the Reynolds number (or β parameter). They have additional dependence on the nondimensional mass, the damping and the reduced velocity based on maximum flow velocity.

Consider a circular cylinder free to oscillate in a transverse direction in a steady flow or a harmonically oscillated flow. Assume that the cylinder is fixed in the inline direction, but free to respond in the transverse direction. The time history of the transverse force on the cylinder may be constructed if the lift coefficients are known. Since the transverse force is irregular, this would require the knowledge of the components of the lift coefficients. If the Fourier components of the lift coefficient are known, then the lift force time history may be written as

$$f_y = \frac{1}{2}\rho A U^2 \sum_{m=1}^{M} C_L(m) \cos(m\omega t + \varepsilon_m) \tag{8.1}$$

in which M is the number of lift force components chosen to represent the time history, ω is the fundamental frequency, $C_L(m)$ are the Fourier components of the lift coefficients and ε_m are the corresponding phase angles. The time history may be computed for a steady flow or an oscillatory flow. For the steady flow, U is the mean flow velocity. For the oscillatory flow, U is replaced by the maximum flow velocity u_0. The phase angles are chosen randomly for each component. In general, the lift coefficients are functions of KC and Re (Chapter 2). Of course, the difficulty in computing this time history is the knowledge of the values of C_L. The experimental values of five components of C_L vs. KC for a vertical cylinder in waves are given in Fig. 8.6 along with the fitted lines. The dependence on the Reynolds number is not known here for a small range of Re. The component lift coefficients are found to peak in succession with the increase in the KC number. This is typical for these coefficients. Additional data on rms C_L included in Chapter 2 is more extensive and shows similar trend. They also show that lift coefficients are weakly dependent on the Reynolds number. While the component lift coefficients in Fig. 8.6 are not directly applicable in practice because of the small range of KC, they show the type of variation in the C_L values expected.

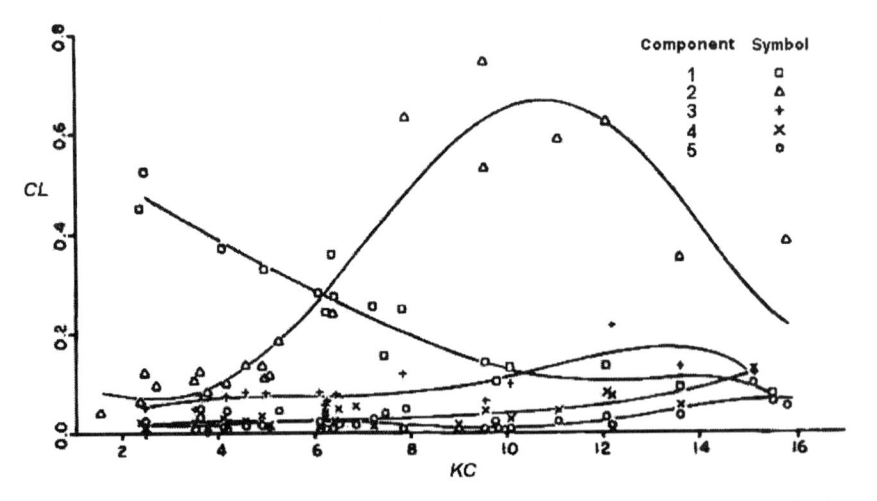

Fig. 8.6: Five Components of Lift Force Coefficients for a Vertical Cylinder in Waves

When these values are known in a particular case, then the transverse force time history may be computed from Eq. 8.1. The transverse oscillation of the cylinder may then be determined if the cylinder is considered to be free to respond in the transverse direction as a one-degree oscillator model. The equation of motion due to the lift force in this case is described by

$$m\ddot{y}(t) + c\dot{y}(t) + ky(t) = f_y(t) \tag{8.2}$$

This is a simple damped single degree of freedom equation for the transverse vibration of the rigid cylinder. The transverse motion may be determined from this equation if the right hand side is substituted from Eq. 8.1. The solution of this equation is straightforward by a simple numerical scheme. However, it is limited by the availability of practical values of lift coefficients.

A more simple empirical model may be developed to describe the transverse force on a circular cylinder in a mean and fluctuating fluid flow that depends on a single value of the lift coefficient. In this case the lift force on the cylinder [Vaicaitis (1976)] is computed from the empirical form

$$f_y = \frac{1}{2}\rho C_L A[U + u(t)]^2 f(y,t) \tag{8.3}$$

where U is the current velocity, u is the oscillatory velocity, and C_L is the empirical lift coefficient.

The quantity $f(y,t)$ is the function representing the vortex shedding process past the cylinder due to steady and oscillating flow. The function $f(y,t)$ is assumed to be an oscillatory function based on shedding frequency:

$$f(y,t) = \cos(\omega_L t) \tag{8.4}$$

When the vortex shedding frequency approaches the natural frequencies of the vibrating structure, synchronization occurs and the vortex shedding frequency is locked to the structural frequency. The vortex shedding process is a function of the shedding frequency ω_s. For steady, uniform flow the shedding frequency is related to the Strouhal number based on the steady velocity. Extending this concept to the oscillatory flow, the shedding frequency is defined as

$$\omega_L = \frac{2\pi St}{D}|U + u(t)| \qquad (8.5)$$

Therefore, the shedding frequency is a random process in the space-time domain. The lift force computed by the above formulation may then be used in the analysis of the structural response. The time length for the vortex shedding function is taken as about one period of the dominant wave in a random sea. For longer time simulation, the process is repeated starting at time $t = 0$ for each dominant period. The total time history of the cross-flow force is obtained by connecting these segments. It is recognized that this is an oversimplification of the complex flow field, but permits a simple calculation of the cylinder response.

A similar quasi-steady analytical model was presented by Bearman (1984). The model is based on the steady flow on a cylinder and describes the lift force at a high Keulegan Carpenter number. For steady flow, $u(t)$ in Eq. 8.3 is set to zero and ω_L is the oscillation frequency based on the Strouhal number St for the steady flow. The lift coefficient and St are functions of Reynolds number. This model may also be used in an oscillatory flow. In oscillatory flow, the flow velocity is replaced by $u(t)$. Then generalizing the expression in Eq. 8.3 for oscillatory flow, the lift force profile for each half cycle is given by

$$f_y = \frac{1}{2}\rho C_L A u^2(t) \cos[\int_0^t \frac{2\pi St}{D}|u(t)|dt] \qquad (8.6)$$

in which the absolute value of $u(t)$ in the integrand is the amplitude of the instantaneous velocity irrespective of the sign. The lift force frequency in this model is assumed to vary continuously according to Eq. 8.5 for $U = 0$. If the motion is assumed to be periodic with a frequency ω, then the expression for the lift force for each half cycle becomes

$$f_y = \frac{1}{2}\rho C_L A u_0^2 \sin^2(\omega t + \varepsilon_1) \cos[St(KC)(1 - \cos\omega t) + \varepsilon_2] \qquad (8.7)$$

If St, C_L and KC are assumed constant for each half cycle having a period T, then the lift force profile may be created, which will be frequency and amplitude modulated. An example of this empirical model will be given later in the exercise section.

8.3.3 Analytic VIV Model

The above formulation provides simple expressions for the lift force on a cylindrical structure, but is not rigorous and fails to give an insight into the actual flow field. For the understanding of the flow effect on the structure it is important that the actual flow field be described. This is only possible with a more accurate analytical model that takes into account the mechanism of the vortex shedding. In Chapter 7, the potential theory for the fluid structure interaction problem was described. In this case the fluid flow could be described by a potential function, since the flow remained attached to the structure as it flows past it. The flow was assumed to be inviscid so that the potential theory describes the flow well. The numerical technique adopted to define the problem was a boundary element method (BEM). The same problem may be solved by a finite element approach in which the structure boundary and the fluid field are both described by finite elements (FEM). These methods are applicable to large structures where the flow separation is minimal and gives rise to a steady state condition. However, when the flow is allowed to separate upon flowing over a body surface creating a wake, as is expected in a viscous flow, the flow becomes unsteady in that at each time step, the wake field takes on a different shape.

The basic equations that are solved in two dimensions are given here. These are the flow continuity equation (mass conservation) and the momentum equation. The governing equations are written in the vorticity and stream function form. For incompressible, homogeneous fluids, the vorticity vector $\vec{\omega}(=\nabla\times\vec{u})$ is generated at the boundary of the fluid regions. For a two-dimensional flow, the vorticity transport equation is derived from the momentum equation:

$$\frac{\partial \omega}{\partial t} + \vec{u}.\nabla\omega = 0 \qquad (8.8)$$

Unlike potential flow equation, Eq. 2.9, the differential equation becomes the nonhomogeneous Poisson's equation. In terms of the stream function ψ, it is written as

$$\nabla^2\psi = -\omega \qquad (8.9)$$

Once ψ is known, then the velocity components (u, v) are written as

$$u = \frac{\partial \psi}{\partial y} \quad \text{and} \quad v = -\frac{\partial \psi}{\partial x} \tag{8.10}$$

The nondimensional form of the Navier-Stokes equation in two dimensions is written in Cartesian coordinate system (XY) in terms of the particle velocities u and v. The mass conservation or the continuity equation has the form

$$\frac{\partial u}{\partial x} + \frac{\partial v}{\partial y} = 0 \tag{8.11}$$

The momentum equations in the X and Y directions are written as

$$\frac{\partial u}{\partial t} + u\frac{\partial u}{\partial x} + v\frac{\partial u}{\partial y} = -\frac{\partial p}{\partial x} + \frac{1}{Re}\left(\frac{\partial^2 u}{\partial x^2} + \frac{\partial^2 u}{\partial y^2}\right) \tag{8.12}$$

$$\frac{\partial v}{\partial t} + u\frac{\partial v}{\partial x} + v\frac{\partial v}{\partial y} = -\frac{\partial p}{\partial y} + \frac{1}{Re}\left(\frac{\partial^2 v}{\partial x^2} + \frac{\partial^2 v}{\partial y^2}\right) \tag{8.13}$$

where u and v are Cartesian velocities normalized (with respect to the characteristic length D of the body and the free stream velocity U_∞), p is the normalized pressure, and Re is the free-stream Reynolds number:

$$Re = \frac{U_\infty D}{\nu} \tag{8.14}$$

One of the computationally simpler numerical methods to solve this problem is the discrete vortex method using the vortex-in-cell formulation [Meneghini and Bearman (1995)]. In this numerical method, the continuous distribution of vorticity is represented by a finite number of discrete vortices. The vorticities are generated from the tangential acceleration at the rigid boundaries, and its subsequent decay is caused by diffusion. Vorticities are shed from the separation points, which must be known apriori. Additionally, they may be shed from the entire body surface. The difficulty arises from knowing the separation point (or line) and the rate at which they are shed. The time requirement for computation increases rapidly with the number of vortex elements in the fluid field, and is one of the major weaknesses of the computational methods. The detailed numerical formulation is quite complex and beyond the scope of this book. For those details, the reader may refer to the works of Sarpkaya (1989), Meneghini & Bearman (1995), Price (1995) and others referenced at the end of this chapter.

In the discrete vortex method, the vortices are discretely placed in cells dividing the flow field behind the structure. The convection of vorticity and their diffusion are described to define the boundary layer development. The boundary conditions at the cylinder surface are imposed as zero normal and zero tangential (no slip) velocity, relative to the motion of the cylinder. At each numerical step, vortices are assigned to fluid mesh points using a vortex-in-cell approach. In its simplest form, the vorticity of the n-th vortex in a given cell is allocated to the four surrounding mesh points according to an area-weighting scheme. New vorticity is introduced at the cylinder wall to satisfy the no-slip condition. The diffusion of all the vorticity at a point in time in the flow field is calculated using a finite difference scheme. Poisson's equation is solved. The new flow velocity at each mesh point is computed, which defines the flow field at that instant of time. The computation domain is chosen large enough to allow no vortices to cross the outer boundary during the course of calculations.

A contour plot of the near wake vorticity at a Reynolds number of 200 is shown in Fig. 8.7. The fluctuating lift and drag coefficients are shown for this case against the nondimensional time Ut/D in Fig. 8.8 [Meneghini and Bearman (1995)]. Note that the lift coefficient has a large fluctuation about a mean value of zero, while the fluctuation in the drag coefficient is small about a mean value of 1.0.

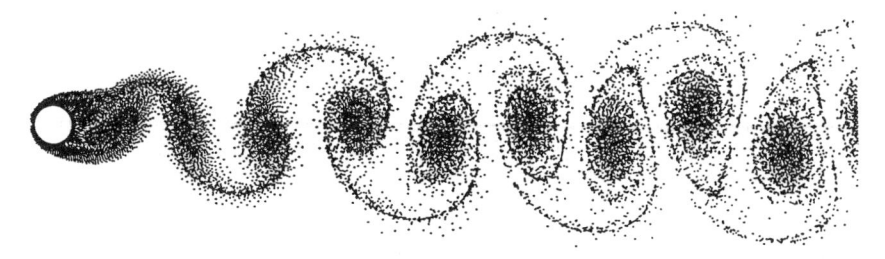

Fig. 8.7: Wake Structure for Flow Past a Cylinder for $Re = 200$ [Meneghini and Bearman (1995)]

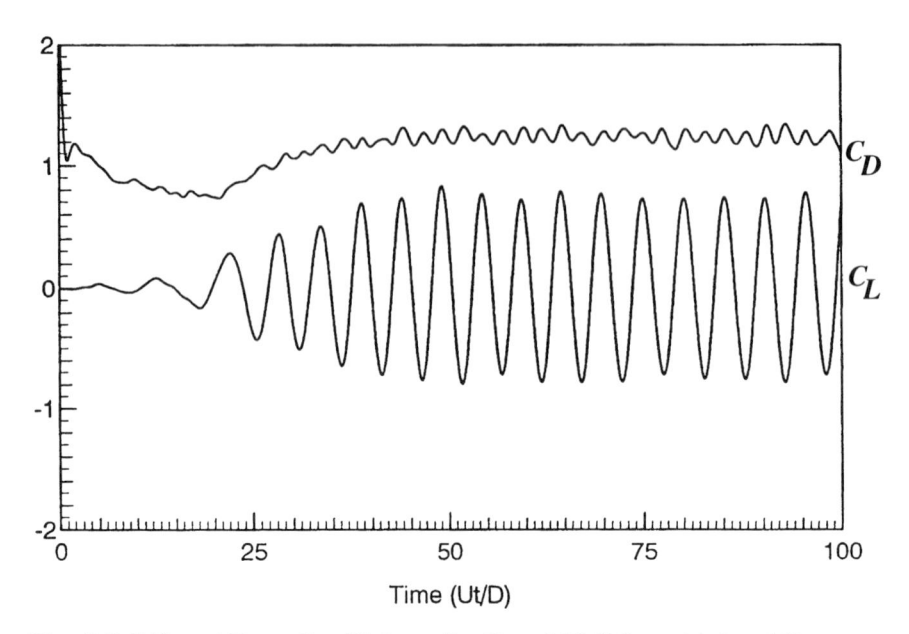

Fig. 8.8: Lift and Drag Coefficients for $Re = 200$ [Meneghini and Bearman (1995)]

In a numerical analysis, a freely supported cylinder in a planar harmonic flow [Bearman, et al. (1995)] was allowed to oscillate in the transverse direction. In this case, the unsteady two-dimensional Navier-Stokes equation was solved in conjunction with the transverse equation of motion. The discrete vortex method provided the transverse force. The equation of motion of the cylinder was represented as a simple spring-mass system with a damper (Eq. 8.2). The transverse fluid force was input at each time step from the discrete vortex method. Once the cylinder velocity and acceleration were known, the cross flow velocity was adjusted to match the cylinder velocity at the given time step.

The lock-in boundaries for a cylinder in steady flow in which the cylinder was fixed inline, but forced to oscillate transversely were derived numerically [Bearman, et al. (1995)]. The numerical results are shown in Fig. 8.9 as mean lines in terms of the nondimensional cylinder amplitude of oscillation (A/D) versus the nondimensional frequency (forcing frequency/vortex shedding frequency, ω/ω_s). The mean fitted curves show

Chapter 8 Fluid Induced Vibration

the lock-in boundaries for two separate A/D regions. Lock-in appears in the inner (upper) region of the boundaries. For lower amplitudes (A/D < 0.6) the region is found large, while it is smaller for A/D > 0.6. For A/D < 0.6, the lock-in is more extensive for $\omega/\omega_s < 1.0$ than for $\omega/\omega_s > 1.0$.

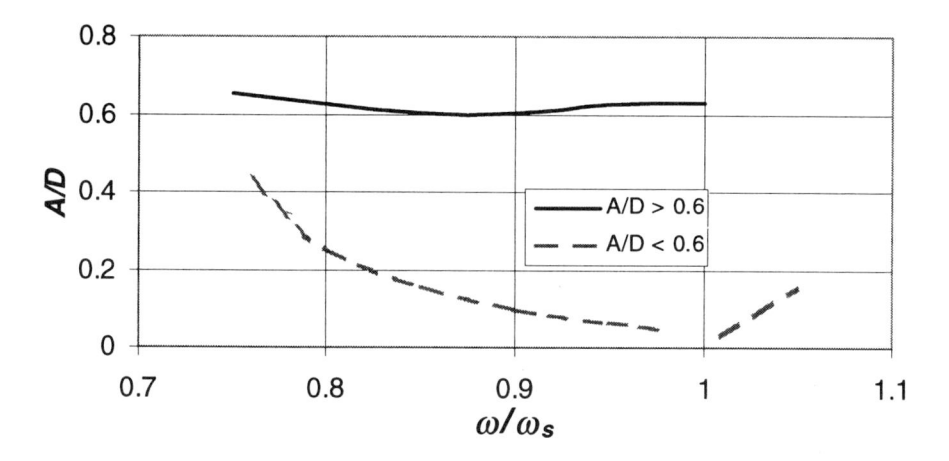

Fig. 8.9: Lock-in Boundary for Different Normalized Amplitudes [after Bearman, et al. (1995)]

For the transverse response in oscillatory flow, comparison was made with experimental results [Bearman, et al. (1995)]. The experiment was conducted in a U-tube in which a cylinder was mounted across a harmonically oscillated flow. The cylinder was fixed in the inline direction but allowed to oscillate in the transverse direction. The nondimensional mass of the cylinder was 2.63 and the damping factor in air was 0.0048. The ratio of the cylinder oscillation frequency in still water to the U-tube frequency was 1.81. The value of the Stokes viscous parameter β was 750. The rms values of the cylinder response in the transverse direction, y_{rms}, was numerically computed. The comparison of the normalized transverse response vs. *KC* number is shown in Fig. 8.10. The comparison shows that the numerical results captured the complex fluid structure interaction well.

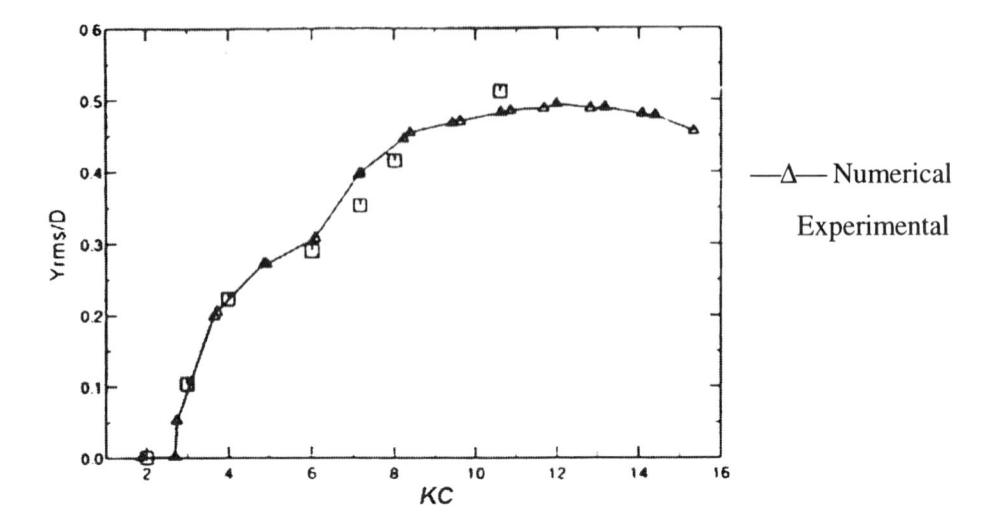

Fig. 8.10: Normalized Transverse Response vs. *KC* number, [Bearman, et al. (1995)]

In a numerical approach developed by Schulz and Kallinderis (1998), Eqs. 8.11-13 were solved by a numerical integration procedure known as Euler forward marching scheme. An elastically mounted cylinder was chosen for the analysis. The two-dimensional oscillation of the cylinder due to this excitation was determined. The purpose of the analysis was to derive the drag force coefficient C_D and transverse force coefficient C_L due to the flow around the cylinder. Note that once these coefficients are known, the forces on the cylinder may be constructed. For example, the transverse force is given by an equation similar to Eq. 8.6 for relative oscillatory flow past the cylinder. Then the transverse motion of the cylinder may be computed using Eq. 8.2.

Schulz & Kallinderis (1998) provided results on the motion of an elastically mounted cylinder applying the above method. They used three different Reynolds number ranges: a low range of $90 \leq Re \leq 140$, an intermediate range of $6.83 \times 10^3 \leq Re \leq 1.85 \times 10^4$, and a high range of $2.25 \times 10^5 \leq Re \leq 4.75 \times 10^5$ near the critical regime. All three ranges showed that

Chapter 8 Fluid Induced Vibration 341

the natural period of the structure dominates the vortex shedding frequency and each exhibits the lock-in phenomenon associated with VIV. Some of these results are of interest here.

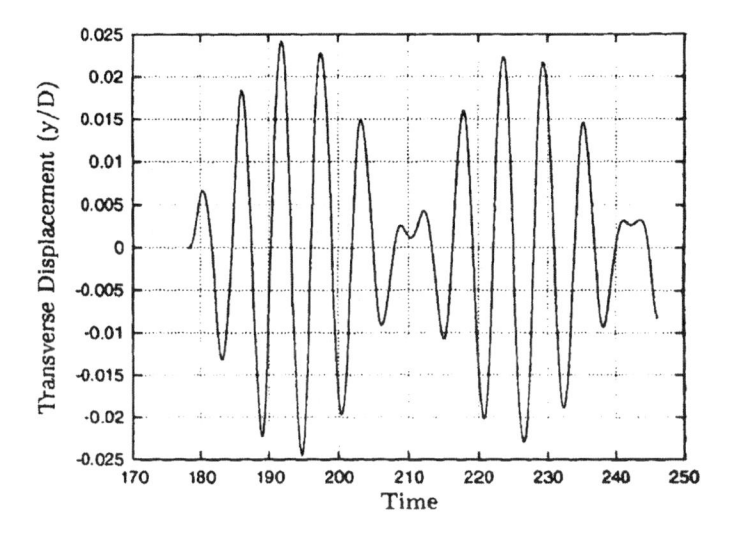

Fig. 8.11: Transverse Oscillation of an Elastically Mounted Cylinder Outside Lock-in [Schulz & Kallinderis (1998)]

In the low Reynolds number range, examples of cylinder displacement for the values of reduced velocity inside and outside the lock-in regime are presented in Figs. 8.11 and 8.12. The time history of the transverse cylinder oscillation outside the lock-in regime for $V_r = 5.02$ (and $Re = 90$) is shown in Fig. 8.11. The displacement y is normalized with the cylinder diameter D. In this case, the transverse displacement has a beating effect around the vortex shedding frequency and its magnitude is small. The same cylinder displacement is shown within the lock-in regime in Fig. 8.12 where $V_r = 6.13$ (and $Re = 110$). Here, the transverse displacement is quite large, increasing monotonically until it reaches a steady peak amplitude of over 40% of the cylinder diameter compared to only 2.5% in the previous case. Notably for the latter case, if the vortex shedding frequency between the fixed vs. oscillating cylinder is compared, it is found that the frequency for the fixed cylinder is 0.190 Hz. However, for the oscillating cylinder, the frequency moves to a value of 0.162 Hz, which happens to be the natural frequency of the structure. Thus, the vortex shedding frequency was altered from its natural state to match the natural structural frequency, indicating that a lock-in has taken place.

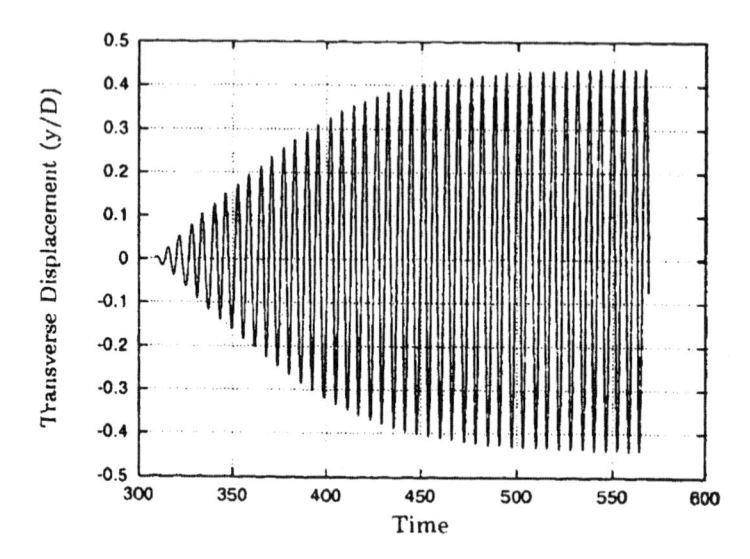

Fig. 8.12: Transverse Oscillation of an Elastically Mounted Cylinder at Lock-in [Schulz & Kallinderis (1998)]

The above illustrations are for the transverse oscillation of the elastically mounted cylinders. As mentioned earlier, the cylinder undergoes a two-dimensional motion when not constrained in the inline direction. An equation similar to Eq. 8.2 can be written for the in-line direction in terms of the normalized drag coefficient, C_D. This equation, when solved, will describe the inline motion of the free cylinder. The two-dimensional motion of the cylinder may be obtained by linear superposition of the cylinder motion in the inline and transverse directions. The steady state oscillation of the free cylinder is shown in Fig. 8.13. In this case, the reduced velocities in the inline and transverse directions $V_{rx} = V_{ry} = 6.13$, and the Reynolds number $Re = 110$. The initial displacement (Fig. 8.13a) starts out elliptical. However, the frequency of the transverse oscillation is twice that of the inline oscillation. Hence, the ellipse changes to a figure eight as the final shape, which is shown in Fig. 8.13b. This is a characteristic of a two-dimensional motion during lock-in.

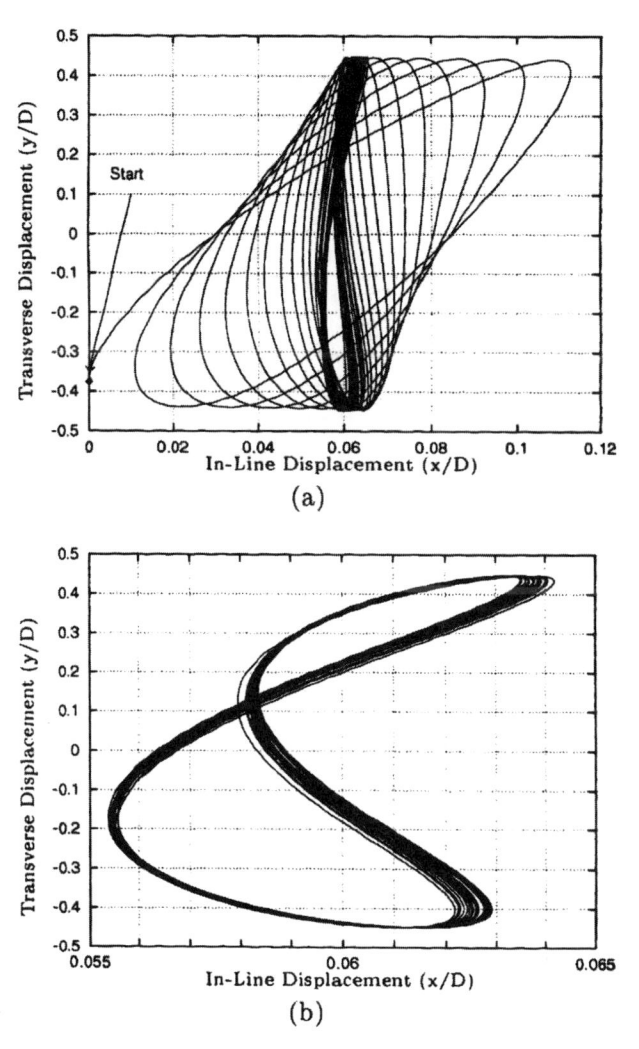

Fig. 8.13: Two-Dimensional Displacement History of a Freely Vibrating Elastically Mounted Cylinder [Schulz & Kallinderis (1998)]

8.4 Eigenfrequencies of Structures

Many long slender structures resting horizontally or vertically in a fluid may behave like an elastic structure, and its mode shapes and eigenfrequencies, at least in the lateral bending, may be an important hydrodynamic factor. These eigenfrequencies will produce interaction with the fluid, giving rise to added mass and damping forces. A simple approach considers a potential flow for the hydrodynamic interaction with the elastic structure oscillating at its eigenfrequency.

The elastic structure is assumed to be composed of several rigid elements connected by elastic connectors. Generalized modes, also known as generalized coordinates, can be used in various ways to represent the different rigid-body modes of multiple modules, which are connected with elastic springs. The structural deflections of these modules at the rigid body modes are determined. One or more of these physically different modes may be analyzed simultaneously with a logical framework. For example, consider the hydroelastic analysis of multiple modules connected by hinges. In this case, hinge deflection modes and bending modes may be combined in a single set of generalized mode shapes.

It is straightforward to express the conventional rigid-body modes of motion in terms of a set of generalized modes. If the modules are connected by structural constraints such as hinges, and the deflections occur only at these hinges, then the possible number of modes can be reduced. Appropriate modes for representing a structure having up to five hinged modules are shown in Fig. 8.14. In general, there will be $N+1$ generalized longitudinal modes used to represent the motions of N identical bodies connected in an array by $N-1$ simple hinges. These mode shapes are defined to be either symmetric or antisymmetric about the middle hinge. The first two modes correspond to global heave and pitch without hinge deflections. The remaining modes represent the hinge deflections with zero displacement at the ends. Each of these modes will have a structural natural frequency associated with it

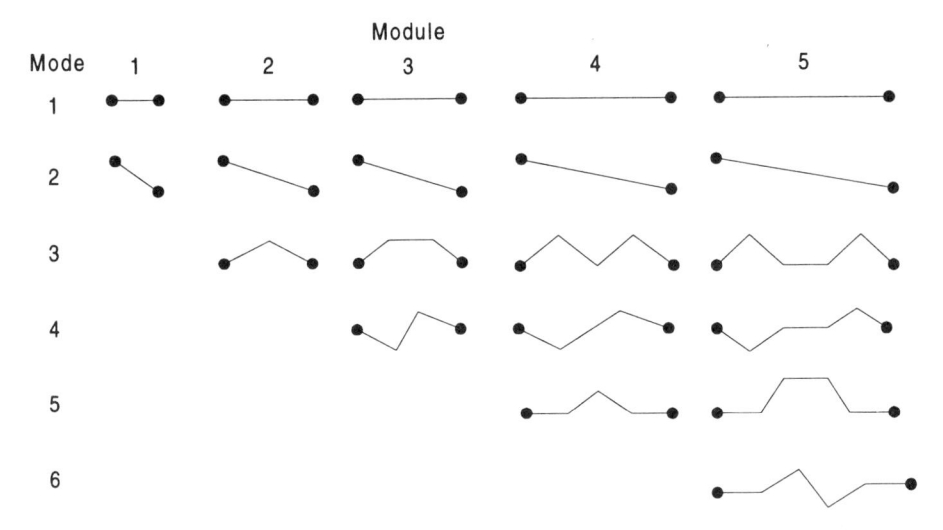

Fig. 8.14: Generalized Mode Shapes for Five Hinged Modules

For each one of these mode shapes, the added mass and damping coefficients for the elastic structure in water may be computed in much the same way as the added mass and damping coefficients for a rigid body oscillation for the six degrees of freedom. In this case, the boundary condition at the body surface will correspond to the motion of the structure in the particular mode of oscillation. The added mass and damping coefficients will correspond to the frequency of oscillation that coincides with the modal frequency of the structure at the selected generalized mode. These may then be introduced in the equations of motion for the structural frequency corresponding to the mode shape.

Thus, the approximation inherent in the above analysis is that the elastic structure is composed of N rigid modules attached to one another by flexible connectors whose stiffness in 6 DOF are linear and known. The structural components undergo mode shape oscillations at the prescribed dry modes specified by their frequencies. In general, as many mode shapes (up to $N+1$) as needed may be analyzed based on the number of rigid modules (N) present in the structure. An example of this technique will be presented for the rigid body modes in Chapter 9.

8.5 Vortex-Induced Vibration of Risers, Cables and Pipelines

The preceding simplification is applicable to long structures that are representative of a combination of a limited number of rigid body modules that are interconnected with hinges. A long flexible structure, such as a cable, riser or a pipeline between two supports, is difficult to represent by such a limited number of modules. These structures have an infinite number of eigenfrequencies and many of them may excite the structure due to VIV. Vortices are shed from a rigid fixed structure with a characteristic frequency. For flexible structures, the eigenmodes additionally influence the shedding frequency, leading to a possible lock-in at one of these eigenfrequencies. The prediction of amplitude of the cylinder response in this case is quite complex.

8.5.1 VIV of Risers

Top Tensioned Risers (TTRs) are long flexible circular cylinders used to link the seabed to a floating platform. In the offshore industry, risers are used to convey hydrocarbons from the seabed to the platform. These risers are subject to steady current with varying intensity and oscillatory wave flows. Hydroelastic interactions take place between the riser and the fluid. Computational Fluid Dynamics (CFD) is used to compute this interaction. The response of the risers is estimated in a quasi-three-dimensional fashion, where hydrodynamic forces are assessed through two-dimensional strips using the discrete vortex method. The riser structure is represented using a model based on the finite element technique applied to beam elements.

The flow field around risers is extremely complex, and risers have a large aspect ratio. This makes a complete three-dimensional simulation of the riser response at realistic prototype conditions infeasible. A structural model for the single riser may be based on the finite element technique applied to beam elements. A static model may be used to obtain the eigenvalue solution and to evaluate the eigenfrequencies and eigenmodes. In this case, the nodes are assumed to be free in the vertical, transverse and inline directions.

For the dynamic response, forces may be assessed through the Morison equation and a lift force formulation, respectively. A lumped mass model may be based on the static response of the riser. The hydrodynamic forces may also be assessed through the discrete vortex method. This method is based on a Lagrangian numerical technique for simulating two-dimensional, incompressible and viscous fluid flow based on the potential flow theory (see Section 8.3.3).

A vertical marine riser can be represented as a beam column under lateral loading and subjected to both internal and external hydrostatic pressure. The governing equation has already been derived in section 7.5. A finite element analysis method is usually adopted for solving this equation. Each beam element involves six degrees of freedom, two translations and one rotation at each end.

A weighted residual method, known as the Galerkin method [Zienkiewicz and Morgan (1983)], is used to obtain the basic equation for the riser. In order to find the static configuration of the riser, an incremental iterative procedure is adopted. This iterative procedure makes successive corrections to a solution until equilibrium under a total load vector is satisfied.

For the dynamic analysis [e.g., Yamamoto, et al. (2001)], the differential equation for motion for a system with many degrees of freedom can be written as a matrix equation. In the formulation of the beam element, mass matrix, a lumped mass or a consistent mass approach may be used. In the lumped mass approach, the entire mass is assumed to be concentrated at nodes, and only the translational degrees of freedom are defined. In this kind of system, the mass matrix has a diagonal form. Off-diagonal terms disappear since the acceleration of a nodal point mass only produces an inertia force at that point. The consistent mass approach, however, makes use of the finite-element technique and requires that the mass matrix be computed from the same shape functions that are used in deriving the stiffness matrix. Coupling due to off-diagonal terms exists, and rotational as well as translational degrees of freedom need to be considered. The consistent mass approach can lead to greater accuracy. The lumped mass formulation is easier to apply since fewer degrees of freedom are involved, leading to a simpler definition of element properties.

The application of the rigid cylinder data on vortex shedding (Section 8.2.2) to full-scale design has been difficult due to the lack of information on the effect of cylinder motion on the Strouhal number. A field experiment addressed this area. As part of the Norwegian Deepwater Project, VIV on the Helland-Hansen drilling riser was recorded for more than two months [Halse (2000) and Kaasen, et al. (2000)]. The Reynolds number range was from about 0.9×10^5 to 6×10^5, overlapping the critical Reynolds number range for stationary cylinders. The field results on the Helland-Hansen riser suggested that the Strouhal number-Reynolds number relationship for the vibrating riser was very similar to that of a stationary cylinder. The roughness for the riser (K/D) was around 0.1 percent and the apparent Strouhal number varied from 0.22 at a Reynolds number of about 1×10^5 to 0.33 at 3×10^5, and dropped back to 0.25 at around 6×10^5.

8.5.2 VIV of Cables

Cables suspended in fluid flow are subject to vortex induced vibration similar to risers. In steady flow, the VIV of cables is a function of Reynolds number, Strouhal number, reduced velocity and mass ratio. In uniform flow, the response of the cable is dominated by resonance at a single peak when the vortex shedding frequency coincides with the natural frequency of the cable. This relationship may be written as

$$\frac{USt}{D} = \frac{n}{2L}\sqrt{\frac{T}{m}} \qquad (8.15)$$

where T is the tension in the cable of length L, diameter D, and mass per unit length m. The quantity n denotes the mode (= 1,2,3...) of the eigenfrequency. The velocity of the uniform flow is U. The lift force acting on a unit length of the cable is obtained from Eqs. 8.3 - 8.4 after setting $u(t) = 0$ and ω_L is the vortex shedding frequency. The lift coefficient is a function of Reynolds number and has a typical value of 0.5 for $300 < Re < 2 \times 10^5$. The dependence of St on Re for a rigid fixed cylinder has been shown in Fig. 2.8.

In shear flow, the cable response is expected to occur at multiple frequencies because different parts of the cable experience different shedding frequencies based on the local Strouhal number. The range of excitation frequency may be estimated from the strength of shear velocity ΔU as

$$\Delta f_s = \frac{St(\Delta U)}{D} \qquad (8.16)$$

Damping plays an important role in computing the response of the cable to the force generated at the vortex shedding frequencies. The damping arises from the internal structural response as well as external hydrodynamic effect. In shear flow this damping varies due to the multi-modal response. The hydrodynamic damping is generally dominant in this case. The solution for the case of uniform flow is relatively straightforward since the response is predominantly in one resonant mode. However, in sheared flow, the non-resonant modes should be taken into account as well. These components contribute significantly to the total response, especially when the damping at resonance is large. A simple superposition of normal modes may be used to develop the solution in shear flow. However, it should include contributions from several non-resonant modes as well. It introduces several peaks in the response spectrum of the cable. Measurements in experiments with cylinders in shear flow have shown responses at such multiple peaks, with the larger peaks near the resonant frequency.

There are two different numerical computational methods applicable in these cases [Le Cunff, et al. (1999)]. In the first method, a modal approach may be used to determine the modes of oscillation, which are excited by the incoming flow. In the modal approach, the eigenmodes are obtained, taking into account the added mass of the structure. The excited modes are determined with the reduced velocity criterion. The lift and damping coefficient at each point are defined based on the amplitude of vibration at the previous time-step. The time response of the structure for each mode is obtained. The structure is then decomposed in several domains where one of the eigenfrequencies may be locked-in.

A second method consists of exciting the structure at the local shedding frequency, and computing the response as the sum of the resulting waves propagating from each point. The method is based on the idea that the structure is excited by forces at different frequencies, generating waves which propagate along the structure. The lift coefficient in phase with the velocity is assumed to decrease with the amplitude, becoming negative when the amplitude is greater than one diameter. This coefficient and the local Strouhal frequency are used to define the external forces.

In this method, the deflection equation for a cable is given by:

$$m\frac{\partial^2 y}{\partial t^2} + C\frac{\partial y}{\partial t} - \frac{\partial}{\partial x}(T\frac{\partial y}{\partial x}) = f_L \qquad (8.17)$$

in which m is the mass of the cable element, and T is the tension of the element of cable. The Strouhal number is obtained from Eq. 8.5 in which the frequency is considered a function of time. In this case the lift force is given by:

$$f_L = \frac{1}{2}\rho D C_L U^2 \exp[i(\int_0^t \omega t' dt' + \varepsilon)] \qquad (8.18)$$

Equation 8.17 may be solved using one of the two methods described above.

This numerical analysis of a cable was performed by Le Cunff, et al. (1999). The simulated results were compared with available experimental data conducted in a canal. The steady current in the canal was sheared with a maximum velocity located at 2m from the wall of the canal, and negative velocity near the opposite side. The model cable of diameter 2.86cm, mass 0.86kg/m and length 17.7m was tensioned horizontally to 671.6N. Due to high tension, only the low modes were excited, simulating the behavior of a riser. With the modal approach, modes 8, 9, 10, and 11 were dominant, with frequencies of 4.179Hz, 4.776Hz, 5.373Hz, 5.97Hz, 6.567Hz.

Table 8.1: Simulated and Experimental Displacement Amplitudes [Le Cunff, et al. (1999)]

Approach ↓	Displacement	
Location →	L/8	13L/16
Wave	0.28	0.20
Modal	0.45	0.32
Experimental	0.50	0.30

A comparison of the displacement at two locations along the cable between the two methods and the measurements is presented in Table 8.1. The modal approach gives a good approximation, while the wave approach underestimates the amplitude of vibration. The RMS amplitudes of

Chapter 8 Fluid Induced Vibration

displacement along the length of the cable are shown in Fig. 8.15; the most excited region of the structure corresponds to the maximum flow speed.

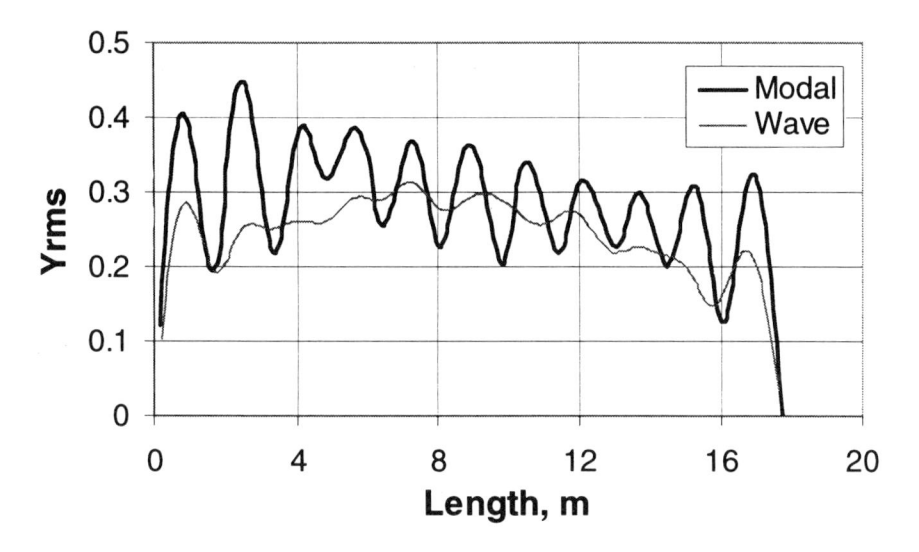

Fig. 8.15: RMS Amplitude of a Cable by Two Numerical Methods [Le Cunff, et al. (1999)]

Like a cable, VIV and associated dynamics of a long mooring line in current is a critical issue. The type of current, i.e., uniform or shear, has a profound influence on the amplitude modulation of the mooring line, and lock-in is a possibility. The vibration frequencies of the line (based on local Strouhal number) are typically quite high, whereas the amplitude of motion is small compared to the line diameter. The drag coefficient is larger ($C_D \gg 1.2$) than found for rigid members, depending on the current strength and the associated transverse vibration of the line. Therefore, the effect on the fatigue life of the mooring line is important.

Another component that behaves like a cable in water is a steel catenary riser (SCR). A catenary riser takes the shape of a catenary and is a common type of deepwater riser system. It represents a promising alternative for production and export in deepwater floating production developments. A catenary riser is different from a conventional top tensioned riser, since it experiences out-of-plane current. For TTR, the rotation rarely exceeds 10° and must be maintained below 4° for continued operation of a drilling riser.

The design of a TTR has a long history of using finite element analysis in the frequency domain. However, there are many uncertainties related to the hydrodynamics of catenary risers, such as Strouhal number, added mass and lift coefficients. A catenary riser experiences a strong nonlinear behavior close to the touchdown point caused by the varying contact condition and friction forces between riser and the seafloor. Frequency domain analysis is less common for flexible lines or for SCR, particularly for extreme response prediction, because of the geometric nonlinearity of the response and the importance of nonlinear effects such as intermittent seabed contact and friction. One of the most accurate ways of modeling its effects is to apply a nonlinear finite element model in a time domain. A conventional VIV analysis of a riser subjected to uniform current profiles is applied in a nonlinear time domain simulation.

On the other hand, the frequency domain fatigue response prediction based mainly on environment of low severity may be an attractive option. Riser dynamic equilibrium equations of motions are typically expressed in finite element matrix form as

$$\boldsymbol{M}\ddot{\boldsymbol{x}}(t) + \boldsymbol{C}\dot{\boldsymbol{x}}(t) + \boldsymbol{K}\boldsymbol{x}(t) = \boldsymbol{f}(t) \tag{8.19}$$

where the uppercase letters in bold denote matrices, lowercase bold letters denoting vectors. The major dynamic contribution to the force vector $f(t)$ comes from wave forces. The solution for the dynamic response $x(t)$ of Eq. 8.19 is found numerically in either the time or the frequency domain.

Time domain analysis rewrites Eq. 8.19 using a temporal operator at discrete times. The (unknown) displacements at the current time are solved using the (known) displacements, velocities and accelerations at the previous time step. The inputs are time histories of wave and motions of the vessel on the top of the riser; the outputs are response time histories of the riser. All the nonlinear effects can be retained in a time domain analysis, giving a more exact solution. However, the method is expensive in terms of computer time, and the level of output generated for subsequent post-processing can be significant, such as for fatigue analysis.

A frequency domain analysis assumes that the excitation and response are harmonics (or summations of harmonics). For a harmonic excitation of frequency ω, one gets

$$f(t) = f_0 \exp(i\omega t); \quad x(t) = x_0 \exp(i\omega t) \quad (8.20)$$

Substitution of Eq. 8.20 into Eq. 8.19 gives the matrix equation:

$$[-\omega_2 M + \omega C + K]x_0 = f_0 \quad (8.21)$$

Equation 8.21 can be solved for x_0 and then $x(t)$ can be computed from Eq. 8.20. The frequency domain analysis is fast and less complicated with minimal output. Moreover, there is no statistical uncertainty associated with the results from random excitation, since the input and output are the wave and the response spectra, respectively. Note that in deriving Eq. 8.21 M, C and K are assumed invariant with time. This assumption is not valid when geometric nonlinearities are important, or when intermittent effects such as seabed interaction occur. Therefore, the frequency domain analysis is generally not valid for flexible riser analysis under these circumstances.

Another major difficulty with the above derivation is that Eq. 8.20 is not valid for $f(t)$ when the Morison drag force is important. In this case, drag linearization techniques can be used to derive an approximate harmonic solution. Usually, a stochastic linearization for the random drag force is employed (Chapter 6).

If a finite element method is used for fatigue stress analysis, one may model the pipe-seafloor interaction by introducing nonlinear springs and dashpots at the element nodes with seafloor contact. This type of model will require a fully nonlinear dynamic analysis in the time domain. However, for simplicity such analyses are often replaced by linear approaches in which the springs are assumed in tension and hence prevent lift-off.

Note that the mass and damping matrices depend on the response frequency, while the response level must be known in order to calculate the external forces. Since both the frequency ω and the response x are unknown, the equation must be solved by iteration. A two-step iteration scheme may be applied. The first step determines the frequency, while the second computes the response shape and amplitude. An analysis [Lie, et al. (2001)] of the dynamic equilibrium equation having the form of Eq. 8.19 was based on empirical hydrodynamic coefficients and was solved by iteration using the frequency response method. The solutions were assumed to be in the form of Eq. 8.20. A linear finite element model was applied. The response was assumed to appear at discrete frequencies.

354   *Theory and Practice of Vibration and Hydrodynamics*

A model test was performed [Lie, et al. (2001)] to assess the VIV response of a steel catenary riser designed for tie-in of pipelines to the Statfjord C platform. This structure is located on the Norwegian continental shelf in a water depth of approximately 145 m. The riser model was made from sections of commercial rubber hose and 10 aluminum intersections specially designed to house accelerometers. The riser was fixed at both ends with universal couplings providing a boundary condition where bending is free and torsion is restricted. The tests were performed with the riser parallel to the towing direction.

Figure 8.16 presents an example of measured and predicted riser response in terms of normalized rms transverse displacements. The model test results are shown at the locations of the accelerometers. The towing speed was 0.13 m/s. The figure indicates that the 3rd mode is dominating, which was confirmed by the modal analysis. The peak frequency of 0.59 Hz was close to the computed 3rd eigenfrequency of 0.64 Hz for an added mass coefficient of one. The largest rms value of about 1/3D was slightly under-predicted.

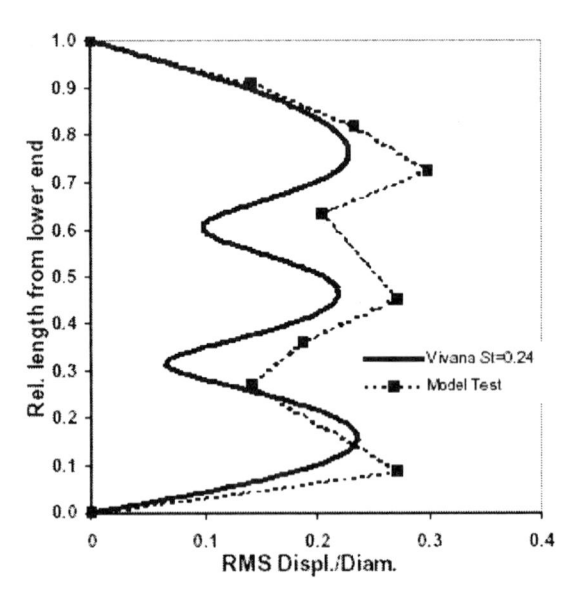

Fig. 8.16: Normalized Transverse Displacement of an SCR Model
[Lie, et al. (2001)]

It may be assumed that damping of long flexible structures is large enough to avoid significant VIV. However, data from field measurements on one such riser, a steep wave umbilical cable in use at the Foinaven field, suggested otherwise [Lyons, et al, (2000)]. The umbilical cable had a reasonably large damping, and regions with low tension. It responded at a high mode number. Furthermore, it was equipped with buoyancy modules configured in such a way as to suppress VIV. Nonetheless, significant VIV was observed under a wide variety of environmental conditions including current, waves and combined current plus waves.

It is clear by now that the largest vibration of long slender, cylindrical structures exposed to current occurs in the normal plane. The displacement in the in-line direction is smaller in magnitude. However, substantial accelerations and bending stresses can occur in the in-line direction as well. Analysis of the Auger TLP production and catenary export risers [Allen (1995)] showed that the catenary production risers had sufficient fatigue life to accommodate VIV, but the catenary export risers could be vulnerable to fatigue failure and required suppression devices.

8.5.3 VIV of Subsea Pipeline

Subsea pipelines in offshore locations are a common means of transporting oil, e.g., in the Gulf of Mexico. Because the sea floor is rough and non-uniform, pipelines need intermediate support. Suspended spans may result from the pipeline being laid on an uneven sea bottom, the pipeline resting on boulders, or the sea bottom being scoured from under the pipeline in noncohesive soil from environmental action. The suspended spans of unburied subsea pipelines may undergo vortex-induced vibration. These vibrations are low amplitude, high cycle phenomenon and as such may affect the fatigue life of the pipeline. Failure of suspended spans due to VIV is prevented by supporting them in such a manner that the maximum length between adjacent supports does not exceed the safe length of a suspended span. The safe length of a suspended span is defined as the span length that will not fail due to fatigue caused by VIV. On the other hand, too many supports increase the installation costs. The cost of installation of pipelines, especially in deep water is directly proportional to the number of these supports. Therefore, there is a need for an accurate assessment of fatigue life of such spans due to VIV so that the number of supports may be optimized. To achieve this goal, an accurate description of the amplitude and frequency response of a suspended span undergoing VIV is needed.

The proximity of a plane boundary, such as ocean floor, to a flexible pipe, changes the flow characteristics of currents or waves. The effect of wall proximity increases the flow velocity. Experiments in current [Tsahalis (1984)] showed that for a spacing between the floor and the pipe of one diameter, the transverse oscillation was reduced by a factor of about two in the reduced velocity range of $V_r = 0$ to 12. Also, the first perceptible vibration took place at a higher value of V_r. The inline vibrations were reduced in magnitude compared to the transverse, but had the same effect when compared to an isolated cylinder. In waves plus current, the reduction in the transverse oscillation experienced was reduced even further by a factor of about three.

Normal design practice for pipelines by classification guides has been to limit the length of a span in such a way that the natural frequency of the pipe is not near the vortex shedding frequency such that practically no resonant vibration would take place. For example, Det norske Veritas (Norway) Guideline No. 14 (1996) prescribes a procedure to compute VIV response, and determines unsupported spans of seabed pipelines on the basis of fatigue damage. The designed unsupported pipe spans are compared with the maximum allowable span. The maximum allowable spans are determined by limiting the length of a span such that the reduced velocity does not reach the minimum value required for the onset of VIV. Minimum reduced velocities for the onset of inline and cross flow VIV are taken as 1.7 and 4.7 respectively. Therefore, the maximum allowable span for a given pipe can be determined from the natural frequency, using the following relations:

$$\text{inline}: \quad f_n > \frac{U}{1.7D} \tag{8.22}$$

$$\text{transverse}: \quad f_n > \frac{U}{4.7D} \tag{8.23}$$

Equations 8.22 and 8.23 suggest that the design is controlled by inline VIV resulting in a short allowable span, and may be conservative. The first mode natural frequency is the lowest.

An alternative to this empirical approach is to design a pipe span based on fatigue damage caused by VIV. However, in this approach, the VIV response and consequential fatigue damage computation must be accurate with an adequate margin of safety to safeguard against the uncertainty generally associated with VIV. An alternative procedure based on simplified calculations may be stated as follows [Chaudhury (2001)]. First, obtain inline and cross flow VIV responses of the pipe span for all eigenmodes and all possible current velocities and determine the onset of VIV. Next, compute the response assuming a harmonic SDOF system. The computation may be limited to the cross flow case only as the inline VIV response is generally small and occur at a different frequency from the cross flow VIV response. Then, compute the fatigue damage for these cases and determine the maximum damage for each current velocity. Finally, based on the probability level of each current, establish the cumulative damage, which must be less than the allowable limit. This criterion establishes the acceptable pipe span. In a study following this design method [Chaudhury (2001)], the unsupported pipe span was found to be much larger than what is proposed by DnV. For a more accurate design, each span length may be verified for cumulative damage from all participating modes and current events.

8.6 Interference Effect in Vortex-Induced Vibration

The discussions so far have been limited to VIV of individual members. However, in practice the long members often appear in groups or bundle, such as tube bundle for heat exchangers, riser bundles for production platforms, tendon groups for tension leg platforms, etc. These members are in close proximity to each other and hence will experience flow interaction among them in both steady and oscillatory flows. The wake pattern and associated VIV of these groups of members are quite different from the individual members, so that the proximity should be considered in the determination of flow induced vibration. In some cases, the VIV is much more severe when the members are close to each other, such that they clash among them causing structural failure.

8.6.1 VIV between Two Closely Spaced Cylinders

When more than one bluff body is placed in a fluid flow, the resulting forces and vortex shedding pattern are completely different [Zdravkovich (1987)] from those found on a single body at the same Reynolds number. The interference between two cylinders at close proximity dramatically changes the flow around them and produces unexpected forces and pressure distributions. In addition, it intensifies or suppresses vortex shedding. It is known that vortex shedding in this case is a three dimensional phenomenon. However, as a first approximation to the problem, at low Reynolds number two-dimensional simulations may be used to give some insight about the details of the vortex dynamics and vortex impingement in the wake of a group of cylinders.

Computational fluid dynamics is attractive in solving such problems. The governing equations for a Newtonian, incompressible viscous flow, as already stated, are the conservation of mass and the conservation of momentum. The Eulerian description of the non-dimensional Navier-Stokes equations for the two cylinders are written as follows:

$$\frac{\partial u_i}{\partial x_i} = 0, i = 1,2 \tag{8.24}$$

The momentum equations for the two cylinders in the two directions are written as

$$\frac{\partial u_i}{\partial t} + u_j \frac{\partial u_i}{\partial x_j} = -\frac{\partial p}{\partial x_i} + \frac{1}{Re}\left(\frac{\partial^2 u_i}{\partial x_j \partial x_j}\right), i, j = 1,2 \tag{8.25}$$

in which the summation convention applies, and Re denotes the Reynolds number based on the free stream velocity. The outer boundary conditions for the velocity is the free stream value and for the pressure p is a prescribed value equal to p_∞. On the circular cylinder surface a no slip condition is applied, which implies that the fluid velocity is zero. Force coefficients are calculated by suitably integrating the pressure and skin friction contributions. After considering the contributions from skin friction and pressure, the force components are resolved in the inline and normal directions.

Chapter 8 Fluid Induced Vibration 359

There is an infinite number of possible arrangements of two parallel cylinders positioned at right angles to the approaching flow direction. As the spacing or gap between two circular cylinders is varied, a range of flow regimes occurs characterized by the behavior of the wake region. The interference phenomena between the two flows are highly nonlinear. At present, a complete description of such flow interference is beyond a reliable theory. Two specific orientations are presented here: two cylinders in a tandem arrangement, one behind the other at some longitudinal spacing, and a pair of cylinders facing the flow in a side by side fashion. These may serve as useful guides for the interpretation of the predicted forces on cylinder array, mainly when a qualitative approach is required.

Numerical results are presented here for flow around two cylinders at $Re = 200$ [Siqueira, et al. (1999)]. For both tandem and side by side arrangements, the distance between the centers of two identical cylinders was varied from $1.5D$, $2D$, $3D$ up to $4D$. A time history plot of the force coefficients for a spacing of $S = 1.5D$ is shown in Fig. 8.17. The drag coefficient on cylinder 1 is slightly above one, while C_D for the second cylinder is actually negative being in the shadow of the first. The average lift coefficients are near zero with a small fluctuation.

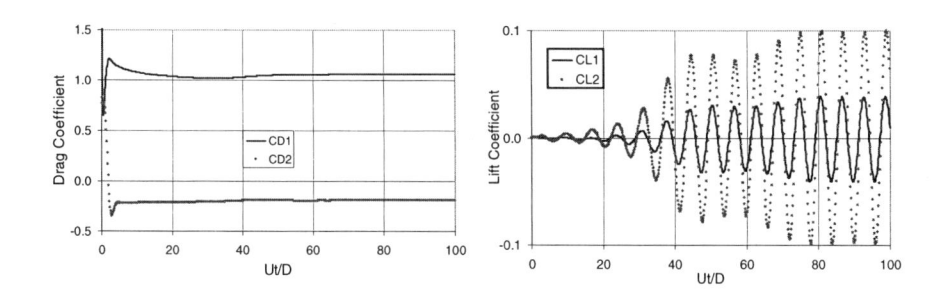

Fig. 8.17: Force Coefficients on Tandem Cylinders vs. Nondimensional Time for $S/D = 1.5$ [Siqueira, et al. (1999)]

Results on the two-cylinder interference for tandem and side by side arrangement are presented in Tables 8.2 and 8.3 from the numerical calculations of Siqueira, et al. (1999). The average drag and lift coefficients for the tandem arrangement of two cylinders are shown in Table 8.2 for a $Re = 200$. For a gap up to $3D$, the C_D values are positive for cylinder 1 and negative for 2. It clearly shows the flow interference between cylinders resulting in a repulsive force on cylinder 1. For the side-by-side arrangement, the mean drag and lift coefficients for different gaps are shown in Table 8.3. The symmetry in the flow interference is clear. For the spacing of up to $2D$ or less the drag coefficients show a decrease indicating a cancellation in the force.

Table 8.2: Average Force Coefficients for Tandem Cylinders [Siqueira, et al. (1999)]

Spacing	C_{D1}	C_{D2}	C_{L1}	C_{L2}
1.5D	1.06	-0.18	0.0	0.0
2.0D	1.03	-0.17	0.0	0.0
3.0D	1.0	-0.08	0.0	0.0
4.0D	1.18	0.38	0.0	0.0

Table 8.3: Average Force Coefficients for Side-by-Side Cylinders [Siqueira, et al. (1999)]

Spacing	C_{D1}	C_{D2}	C_{L1}	C_{L2}
1.5D	1.32	1.32	-0.40	0.40
2.0D	1.42	1.42	-0.22	0.22
3.0D	1.41	1.41	-0.10	0.10
4.0D	1.34	1.34	-0.05	0.05

The average coefficient for a dissimilar pair of cylinders in tandem [Flatschart, et al. (2000)] is shown in Table 8.4. The Strouhal number shown is based on the upstream cylinder diameter, which was twice that of the downstream cylinder. The downstream cylinder shows negative drag coefficients at smaller gaps due to the interference. The drag coefficient of the downstream cylinder is smaller in this case (than the two equal cylinders) due to higher shadow effect from the larger cylinder.

Table 8.4: Average Drag Coefficient for Tandem Dissimilar Cylinders [Flatschart, et al. (2000)]

Gap	St	C_{D1}	C_{D2}
1.5D	0.173	1.179	-0.250
2.0D	0.159	1.144	-0.300
2.5D	0.168	1.308	0.603
3.0D	0.179	1.310	0.792
3.5D	0.182	1.320	0.908

8.6.2 VIV among Multiple Cylinders

When multiple cylinders are present close to each other, a significant change is expected in the wake field from the flow interaction among cylinders. Similarly, the vortex shedding and associated vibrations are affected by the neighboring cylinders. The flow-induced interaction among multiple cylinders may be classified into two categories: (1) relative interaction among the cylinders when they are located close to one another, and (2) wake interaction when one is located in the wake of the others.

When a cylinder is placed in the wake of another, it is subjected to three different kinds of excitation forces: (1) a time averaged mean static force, which is a function of its location in the wake of the first cylinder; (2) a broad-banded force due to the oncoming turbulent flow shed from the upstream cylinder; and (3) a periodic vortex shedding force. The third type of force may cause a fluid-elastic instability, resulting in the possibility of a large amplitude vibration.

Flatschart, et al. (2000) showed numerical results for four identical circular cylinders in a square configuration. In this case, the downstream cylinders experience the presence of multiple frequencies. The implication of this analysis to tube or riser bundles is obvious.

In the case of a riser bundle, the risers generally are positioned with small gaps. Therefore, the wake interaction is dominant except immediately before an impact. The collision of risers is not an uncommon occurrence in the field. The riser collision may induce dents in the riser pipe, or damage to the appendages, which include buoyancy modules, anodes, and any strakes that may be present. This riser interference may be computed in the frequency domain based on a spectral description of the environment. The relative motions of two adjacent risers are determined from the transfer function in the usual way in terms of spectral density functions. Time series of the relative motions of the risers at a given elevation may then be generated from their spectra. Whenever the relative displacement reaches the initial riser separation distance, a riser pipe is assumed to have touched the wall of the adjacent riser. Therefore, the time history establishes the frequency of this contact and the associated velocities. The velocities at these instants are called interference velocities. The number of interference during a unit time is the interference frequency. In this approach, transparency is succinctly assumed so that adjacent risers are not affected by collision. Thus, it cannot indicate the consequence of riser interference, such as the damage caused by the riser collision. Generally, the spacing between risers is determined on the "no-interference" approach, which often results in a large spacing. For a simple approach to include impact in the analysis and compute probability of damage, refer to Li and Morrison (2000). Using their simple method, the optimum spacing among risers in order to avoid collision may be estimated.

8.7 VIV Spoilers

It should be clear by now that VIV is caused by vortices formed by the flow separation at the surface of the submerged structure, and is the principal mechanism for possible large amplitude vibration of the structure in the fluid (potentially locking in at its natural frequency). This may cause failure of the structural component. For a group or bundle of long slender members in cross flow, collision and consequential damage of the members are probable, if not properly accommodated in the design.

Therefore, if the structure is provided with some mechanism or attachment that prevents or breaks up the shedding of vortices, then the vortex-induced vibration of the structure may be reduced or entirely eliminated. There are a number of examples in the literature of devices used to suppress VIV of cylinders of circular cross-section. These VIV mitigation components, called spoilers, are designed not to interfere with the intended purpose of the structure, and are placed on or near the body of the structure.

They may take the form of a special shape for the structure such that the vortices experience difficulty in forming. Alternately, attachments may be used on the structure that break up the vortices immediately after they are formed and thus avoid the pressure fluctuations around the body created by the vortices. Zdravkovich (1981) classified the techniques for suppressing vortex shedding for both the aerodynamic and hydrodynamic industry into three categories:

- surface protrusions,
- shrouds, and
- nearwake stabilizers.

A good example of the first case is the helical strake. The second type basically surrounds the structure with a porous cap, such as a gauze, a perforated cylinder, axial slats, etc. These have found lesser applications in practice because of their interference with the intended purpose of the structure. The third type works with devices like splitter plates, close to the structure that are intended to change the vortex shedding, both avoiding synchronization and displacing the interaction further downstream.

8.7.1 Streamline Geometry

It may be possible to introduce a design in the structure that makes it forgivable to the fluid flow. In other words, the structure geometry is such that the flow past the structure does not generate frequencies in the fluid from the flow separation past the body. One such method is streamlining the cross-section of the structure. Replacing the geometry of a circular member with the shape of a fin will introduce a stagnation point at the end of the fin in steady flow, so that little separation of fluid will take place about the fin and formation of vortices may be avoided. This technique may be successfully applied to marine piers placed in the steady flow field, e.g., river. Generally, a taper ratio of 6 to 1 for a fin (Fig. 8.18) is effective. For fixed fins, the flow of fluid should be relatively constant in the direction of the streamline. If the fin is pinned (or hinged) so that it can rotate about its axis, then multiple current flow directions may be accommodated. It is, however, unlikely that this arrangement will work efficiently in an oscillatory fluid flow, such as waves.

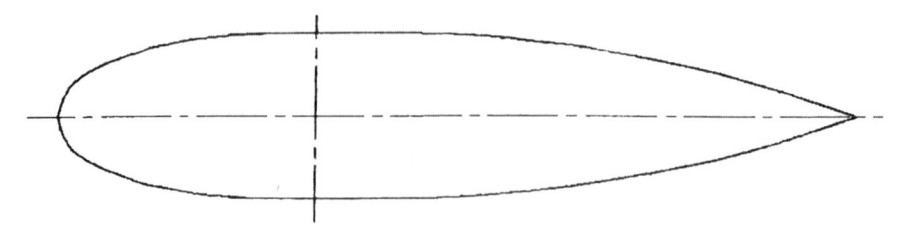

Fig. 8.18: Streamline Geometry Avoids Well-Defined Wake Field

The slender Glasgow (UK) Science tower was designed in the shape of an aerodynamic foil and has been allowed to rotate 360 deg. with the mean wind direction with the help of a turntable at its base. This minimized the movement of the top of the tower due to steady wind in order to facilitate the visitors' observation from the top of the tower. The installation of a mechanical connection for the fins poses difficulties. Moreover, fatigue of the rotating element may become a problem in this case.

8.7.2 Super Smooth Surface

As noted earlier, vortices are tripped from the surface when the body has imperfections. Super smooth surfaces on a bluff body will not allow significant flow separation or vortex formation. In the absence of well-defined vortices, the transverse force on a bluff body will be negligible. Since the principle cause of VIV is the high frequency transverse force, such smooth surfaces will eliminate or substantially reduce the vortex-induced vibration. If the surfaces of submerged bluff bodies, such as drilling and production risers, spar, or TLP tendons, can be maintained in a smooth condition in their operational mode, then these structures will theoretically have minimal VIV problems.

In a test program with cylinders in steady flow (sponsored by Shell and carried out at the US Navy's David Taylor Model Basin circular tank), a fiberglass cylinder model was built with a super smooth ground surface. The tests of these cylinder specimens in supercritical Reynolds number demonstrated the absence of VIV. Repeating the same sequence of tests with one order of magnitude rougher surface brought back the VIV. The same technique was field tested with drilling risers hung from a surface vessel offshore Trinidad. In this case, gelcoat smooth fiberglass sleeves were clamped around the drilling risers. The risers experienced a 2 knot current at the site, which would normally induce vibration and top deflection of the riser. No deflection or vibration was observed in these tests. The above examples show that smooth surfaces suppress vortex formation and associated VIV problems. Appropriate level of smoothness in this approach may be achieved with a urathane painted surface. However, marine growth is generated quickly bringing roughness onto most underwater surfaces. Therefore, in such applications, this area should be carefully evaluated.

8.7.3 External Damper

The amplitude of vibration of structures is directly related to the amount of damping present in the system. Damping is the most important element determining the amplitude of vibration. The hydrodynamic damping arises from fluid flow around the surface of the structure. The structural damping arises from the energy absorbed within the structure due to the motion or deflection of its members. In addition, the structural material introduces material damping.

One means of reducing vibrational amplitude is to make the structure very stiff or to introduce additional material damping. Certain materials have high internal damping, such as rubber or wood, that may be sandwiched in the members of the structure providing additional damping. Similarly, external damping may be introduced in certain cases, such as bumpers. Bumpers may be placed between the structure and the support system so that when the amplitude of vibration becomes large, the structure comes in contact with the bumper, which absorbs the vibration. Ships or boats are often moored with fenders, which have such effects.

The VIV may also be reduced with mechanical dampers. In this case, a dashpot, which absorbs the energy of the vibration, is attached to the structure in the direction of the vibrational motion. One such concept of mechanical damping system was conceived inside a caisson of a Tension Leg Platform [Katayama, et al. (1982)]. The buoyant TLP is typically held in place by tensioned tendons and is susceptible to high tendon loads, known as springing, from the heave motions in waves. These loads arise from the higher order frequency components due to inherent small damping in the

TLP system. In order to supplement the internal damping in this concept, an external electro-magnetic dashpot was introduced in each of the four legs attached to the tendons in the vertical direction. This damper was simulated in a model test of the TLP. The vertical tendon loads due to vibration from oscillatory waves was measured in the model with and without this external damping. Figure 8.19 presents the normalized tendon loads as functions of wave periods. The computed values are based on a single degree of freedom analysis that includes this additional mechanical damping when present. The TLP response in oscillatory waves clearly shows the reduction in the tendon loads at the natural frequency.

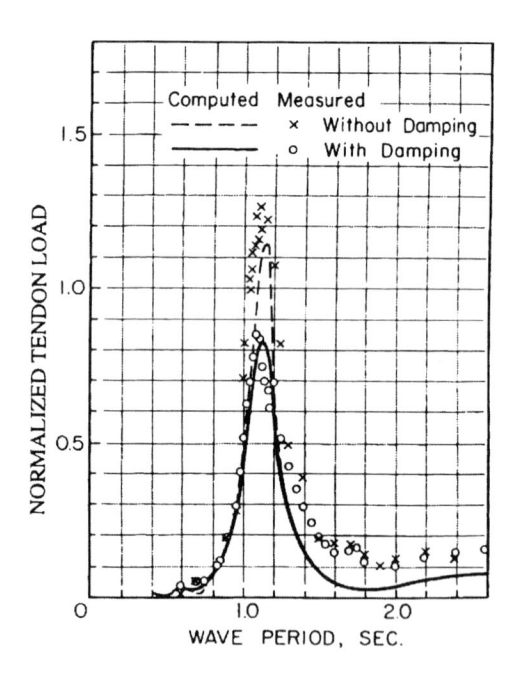

Fig. 8.19: Response of a TLP Model With and Without Mechanical Damper [Katayama, et al. (1982)]

The idea here is to damp out the natural period vibration of the structure at the vortex shedding frequency by an order of magnitude, which is technically a very difficult task. Sometimes these external devices may not be suitably accommodated to the structure. Even if it is possible to incorporate, the cost of installation may be prohibitive.

8.7.4 Pneumatic Damper

Pneumatic dampers in a floating structure can affect the overall motions of the structure due to environment. Floating structures are often equipped with open bottom tanks on their submerged hull geometry. The water columns in these tanks can flow in or out through their open bottoms. The tops of these tanks are often capped, trapping a volume of air. The trapped air acts as a spring in the heave and roll motions of the floating structure. Oscillation of water columns in these tanks generates additional damping, suppressing the vibrating oscillation of the structure due to excitation, such as from waves.

As an example, a ship may be equipped with these tanks fitted with valves along its beam, extending from above to below still water level. Opening or closing the valves can control the natural period of the floating vessel, which may be designed to avoid resonant conditions. Recently, a pneumatic platform supported by an array of open bottom cylinders was proposed for a very large floating structure, such as a floating airport. The open cylinders provide stabilization needed for operation of aircraft on the deck.

8.7.5 Damping Plate

Another method of vortex suppression is to include damping plates as a structural element. The hydrodynamic damping consists of wave radiation and viscous damping. The viscous damping is generated from the skin friction due to the tangential flow past the surface of the structure. It also comes from the separation of flow from the surface of the structure. In contrast to the radiation damping, the viscous damping is nonlinear so that the damping magnitude increases with the motion, which limits the motion more effectively. Therefore, a method in which the viscous effect through flow separation is incorporated may be an effective vibration control system. Bilge keels as flat plates on a longitudinal floating structure, such as ship, is a simple and effective way to reduce the ship roll. Similarly, plates of different size and shape may be used to reduce the motion or vibration of submerged structures. One of these structures is the spar having a submerged cylindrical caisson with a truss structure beneath it. In one design, horizontal plates are introduced in the latticed structure, which enhances flow separation from the structure in the heave direction. The heave plates, as they are commonly known, introduce velocity dependent forces providing heave damping to the vibration of the structure. The plates, however, also add inertia force to the overall inertia of the structure making the net benefit somewhat less efficient.

An alternative to these solid plates is the use of perforated plates. Use of the perforated plates in absorbing energy is not new. They have been used as breakwaters in coastal structures. One of the offshore applications of this concept is the circular wall around the Ekofisk gravity production platform in the North Sea. The purpose of this wall was to break up the waves prior to reaching the solid walls of the gravity structure. Another application has been in the design of artificial beaches in the laboratory to break up the incident wave energy using a series of vertical plates of different porosity (e.g., the wave basin facility at Offshore Technology Research Center, Texas A&M University, Texas). This technique was also introduced as a slotted stabilizer in the design of the Roseau tower in the North Sea for offshore application. The stabilizer reduced the structural vibration due to fluid flow. Downie, et al. (2000) tested a spar model with the upper buoyant part and the lower truss composed of four bays of small structural members. The plates were introduced in the bottom bay and were perforated in the upper part and solid in the lower portion (Fig. 8.20). Four different geometry of the plates were tested, some solid and some perforated. The heave response of the structure modified with the heave plates was investigated. A typical result on the heave RAO from the experiment is given in Fig. 8.21. The four cases include the cases of solid and perforated plates. The individual natural periods for these cases were different as shown by the vertical lines. The response for the spar model with the perforated plates is greatly reduced at the respective natural periods.

Chapter 8 Fluid Induced Vibration

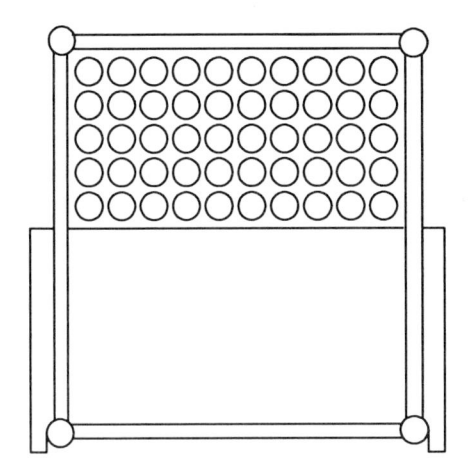

Fig. 8.20: Small Perforated and Large Solid Plates in a Spar Model [from Downie (2000), et al.]

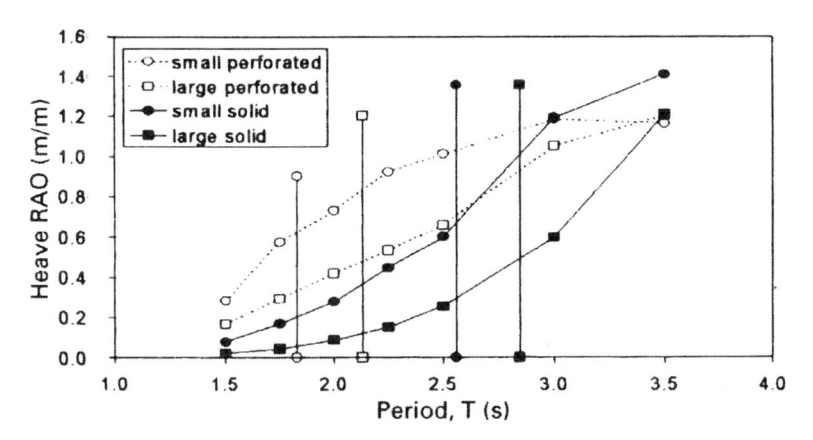

Fig. 8.21: Comparison of Heave Response of Spar with Solid and Perforated Plates [Downie (2000), et al.]

8.7.6 Helical Strakes

A common method of spoiling the formation of vortices is to introduce helical strakes around cylindrical members. This technique has worked well for a long time on large chimneys or stacks exposed to high wind flows. Strakes may be equally effective in steady and oscillatory flows. They are a very popular method for large caisson deep draft spar platforms used in the offshore industry today. The helical strake is a continuous thin plate of defined width that is wound spirally at a specified gap around the caisson (Fig. 8.22). The width of the strake is typically about 10 percent of the cylinder diameter. Larger widths have also been designed. These strakes act in the fluid in such a manner that the synchronization between excitation and lateral motion, a necessary condition for the lock-in, is avoided. The purpose is to break up the wake region of the member, which will disrupt the formation of regular vortices behind the member.

A disadvantage of the strakes is that it increases the overall drag force on the structure. They may obstruct or hamper the operation of the member as well, such as, in the application of drilling risers. Moreover, the strakes are large, difficult to construct and difficult to handle during the transportation and installation process. However, they have been found to be very effective in reducing VIV.

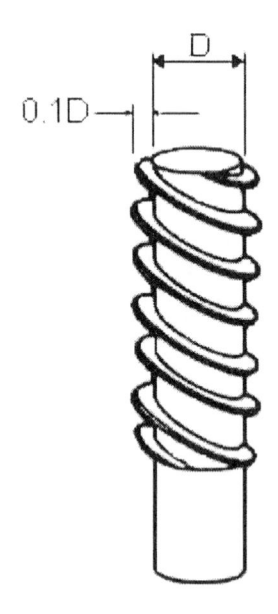

Fig. 8.22: Helical Strake on an Offshore Spar Platform

These strakes are also installed on flexible members, for example, marine risers, with some success. For steel catenary risers used in the production of oil in deepwater offshore floating platforms, some VIV suppression device appears inevitable. In a model experiment [Allen (2001)], one method that proved effective is an ultra-short fairing. Since the fairings are small, they are not subject to appreciable torsional moments or rotation, and may be effective in reducing marine growth. Huse and Saether (2001) tested a marine riser model with strakes of four different heights – $0.025D$, $0.05D$, $0.1D$ and $0.15D$. They found the highest reduction of transverse amplitude with the $0.15D$. The amplitude of motion was about one diameter for the smooth riser model, which reduced to about $0.5D$ for the strake height of $0.05D$ and only $0.1D$ for the strake height of $0.15D$.

Plastic ribbons woven on marine cables are an effective way to break up vortices in the flow and reduce vibration. Typically, the width of the ribbon is about the size of the cable diameter and extends approximately four diameters.

8.7.7 Wavy Edge

The introduction of waviness in a structure was observed to reduce the drag force. Waviness in the leading edge flow is effective in modifying the flow, producing a near wake that varies in thickness across the span. It has been determined experimentally that vortex shedding from bluff bodies can be weakened, and in some cases suppressed, when the flow separation lines are forced to be sinuous. For rectangular cross-section bodies with a wavy front face, and circular cross-section bodies with a sinuous axis, vortex shedding was suppressed for the wave height/wavelength ratio above a critical value. The effectiveness was greatest when the wavelength was close to around 4 to 5 cylinder diameters. The introduction of relatively small degrees of spanwise waviness to the flow separation lines had a large effect on Von Kármán vortex shedding. In the case of the circular cross-section body, for example, reductions in the drag coefficient of up to 47% have been observed. Also, vortex shedding has been completely suppressed at high Reynolds numbers for the rectangular and circular cross section bluff bodies.

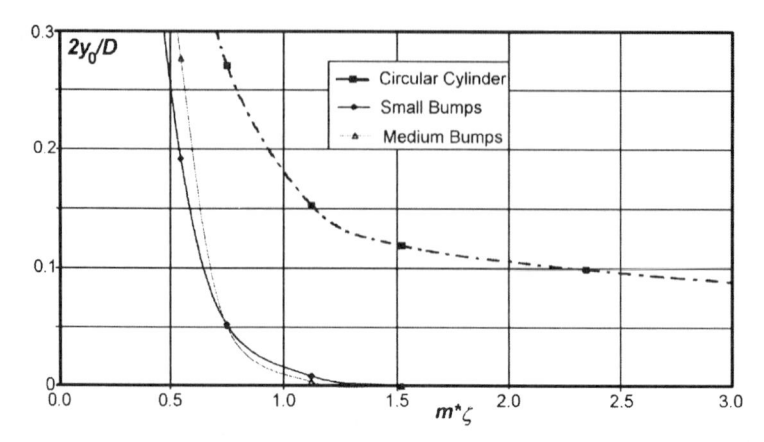

372   *Theory and Practice of Vibration and Hydrodynamics*

Fig. 8.23: Amplitude of Vibration for a Smooth vs. Bumped Cylinder
[Bearman (2000)]

It is not practical to have a wavy tube for a structure, such as a riser. Therefore, the underlying concept is to produce a body shape with reduced drag and no detectable vortex shedding. It was found [Bearman (2000)] that these objectives could be met by attaching small hemispherical bumps to the leading edge of a circular cylinder. In order to make the suppression effective in all flow directions, bumps were applied in a staggered arrangement at a regular angular and spanwise position. At Reynolds numbers up to about 10^5 it was found that the application of the bumps at an axial separation of 7 diameters reduced the drag coefficient from 1.2 to 0.9. This is illustrated in Fig. 8.23, which plots the amplitude of vibration as a function of mass damping coefficient $m^*\zeta$ where the value of m^* is 14.3. The Reynolds number for this example was 4000.

8.7.8 External Disturbance

One of the effective means of reducing vortex-induced vibration is the introduction of disturbances on or near the structure itself, which will interact with the fluid to eliminate or substantially reduce the formation of vortices from the structure's surface. There are several different ways this can be accomplished:

Chapter 8 Fluid Induced Vibration 373

- Interfering with the vortex formation in the wake of the structure,
- Minimizing the adverse pressure gradient fore and aft of the structure, thus reducing the vortex mechanism, or
- Disrupting or reducing the spanwise coherence of vortex shedding.

These disturbing agents may take the form of fairings such as hair, fringe, or ribbons in helical or spiral form. This is illustrated in Fig. 8.24 with a few examples. The hair or ribbons are attached to the body, and will interfere with the flow past the body and inhibit the formation of vortices by breaking them up. The difficulty in their introduction is that they may interfere with the intended operation of the structure or with other members. Moreover, their active life span in the underwater operation may be limited.

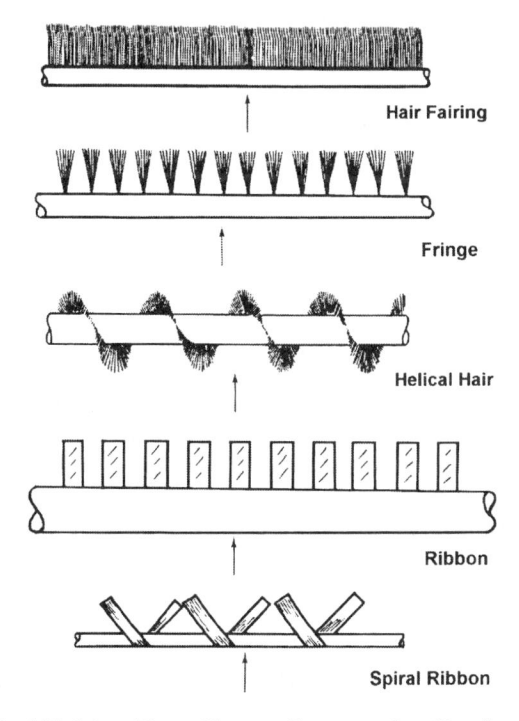

Fig. 8.24: Typical Fairing Type Vortex Suppression Devices [Hafen, et al. (1976)]

Studs and rings have been used to influence boundary layer separation and vortex formation. Similarly, splitter plates have been employed to prevent vortex interaction. Herringbones have been used to cut down the spanwise coherence. Perforation has been used in spoiling the wake formation. These are pictorially illustrated in Fig. 8.25.

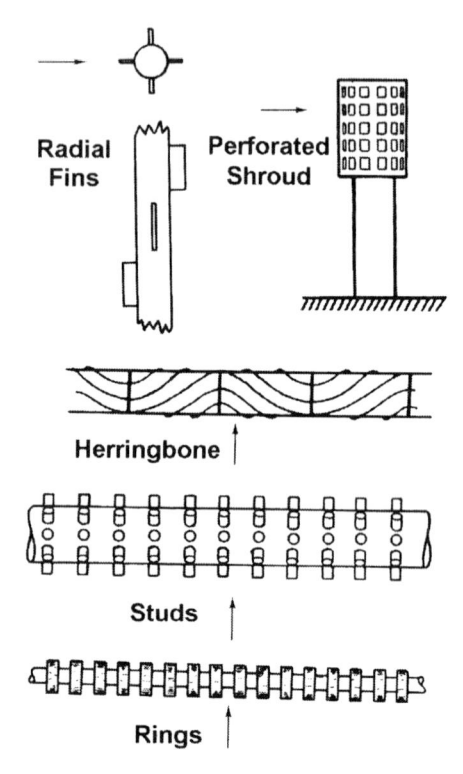

Fig. 8.25: Strumming Vortex Suppression Devices [Hafen, et al. (1976)]

8.7.9 Porosity

In order to disturb the vortices, controlled porosity may be introduced in the structure that allows the flow enter automatically and leave the structure. In this case, openings or perforations are designed into the structure that allows

Chapter 8 Fluid Induced Vibration

the incoming flow to pass through and re-enter into the fluid stream where the vortices are present, thus disturbing their formation. This is commonly known as bleeding. In the aeronautical industry, the bleeding technique is not new. External fluid, such air, may also be introduced in the original fluid flow with the intention of changing the vortex formation. The external fluid may be injected from within the structure. Limited experimental results indicate that this concept is effective in reducing VIV. Zdravkovich (1981) suggested almost a complete longitudinal slit in the direction of the flow in a cylindrical structure. Table 8.5 indicates the typical drag coefficient (C_D) for some of these cases. Note the reduced value of C_D for the slotted cylinder compared to the straked cylinder. There has been a renewed interest in the guided porosity. Fernandes, et al. (2000) proposed to arrange ducts across a cylinder in such a manner as to collect the flow from high pressure regions (such as the stagnation point) and discharge it to the low pressure regions. A conceptual arrangement is shown in Fig. 8.26.

Table 8.5: Drag Coefficients for Modified Cylinder (Zdravkovich, 1981)

Case	Helical Strake	Perforated Shroud	Slit along Cylinder
	1.4	0.6	0.8

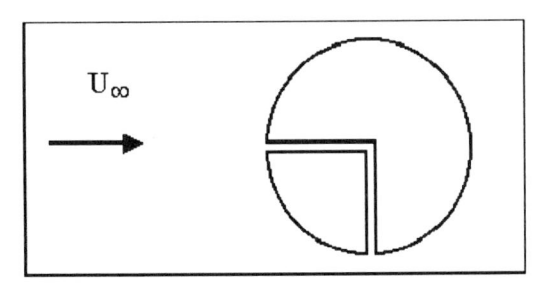

Fig. 8.26: Porous Cylinder with 90 deg. Duct [Fernandes, et al. (2000)]

Implementation of porosity in a real structure appears possible only if the dimension of the structure is large. In the case of large diameter spar, it is possible to introduce porosity in the column without compromising the operation of the structure. Additionally, the drag coefficient for the porous spar is likely to be smaller than for a spar with helical strakes. For risers (and stack chimneys) this is more difficult to implement, since the integrity of the internal duct needs to be maintained.

8.7.10 Tension Control

The VIV interference of riser bundles and the problem of riser clashing is usually avoided in a design by increasing the spacing between the risers at the seabed and/or at the floater, as well as increasing the riser pretension. Pretension of the individual risers reduces the interference or clashing, but does not eliminate it. Spacer frames at different levels along the risers is another possibility. However, such solutions may not be very attractive from an economic point of view.

Another possible solution is to control the riser tensioners in such a way that they give equal payout for all risers in the array. This will force the similar risers in the bundle to have approximately equal static deflection in current flow, and thus avoid clashing [Huse (2000)]. For instance, for a spar buoy, all risers may be connected to a common buoyancy tank providing common tension control, instead of pretensioning them individually by separate buoyancy tanks. In order to verify this principle of payout, a model test [Huse (2000)] was carried out in which all risers in the group were connected to a common frame at the top end. The frame, in turn, was connected to the towing carriage by springs. No clashing was observed, and the distance between individual risers at the mid-level was found to be equal to the spacing at top and bottom, even at the largest realistic current velocities.

8.8 Exercises

Exercise 1

Consider the case for a cylinder where $KC = 34.7$ and the rms value of $C_L = 1.59$. The value of the Strouhal number is assumed to be its steady flow value of $St = 0.2$. Compute the lift force using the formula in Eq. 8.4.

Exercise 2

Construct the lift force time history for a vertical cylinder unit section at the mean water surface using Fig. 8.6 for a $KC = 12$ and a wave of period 2s and amplitude of 6 inches (152 mm).

Exercise 3

Using the above excitation, find the transverse oscillation of the floating cylinder section above for a 10 percent damping factor and a natural period of 3s. Use Eq. 8.2 and assume linear superposition applies.

Exercise 4

Construct a conceptual design for a mechanical damper attached to the tendon in the leg of a TLP (see Section 8.7.3). Explain how the system will reduce response near the natural period of the structure.

8.9 References

1. Achenbach, E., & Heineche, E., "On Vortex Shedding from Smooth and Rough Cylinders in the Range of Reynolds Numbers 6×10^3 to 5×10^6", Journal of Fluid Mechanics, 1981, Vol. 109, pp. 239-251.

2. Allen, D.W., "Vortex-Induced Vibration Analysis of the Auger Production and Steel Catenary Export Risers", Offshore Technology Conference, OTC 7821, Houston, 1995.

3. Allen, D. W., and Henning, D. L., "Prototype Vortex-Induced Vibration Tests for Production Risers", Offshore Technology Conference, Houston, TX, OTC 13114, May 2001.

4. Allen, D. W., and Henning, D. L., "Surface Roughness Effects on Vortex-Induced Vibration of Cylindrical Structures at Critical and Supercritical Reynolds Numbers", Offshore Technology Conference, Houston, TX, OTC 13302, May 2001.

5. Bearman, P.W., "Vortex Shedding from Oscillating Bluff Bodies", Annual Reviews of Fluid Mechanics, Vol. 16, 1984, pp. 195-222.

6. Bearman, P.W., "Developments in Vortex Shedding Research", Workshop on Vortex-Induced Vibrations of Offshore Structures, Sao Paulo, Brazil, 2000.

7. Bearman, P.W., Graham, J.M.R., Lin, X.W., and Meneghini, J.R., "Numerical Simulation of Flow-Induced Vibration of a Circular Cylinder in Uniform and Oscillatory Flow", Flow Induced Vibration, Bearman (Editor), A.A. Balkema, Rotterdam , 1995, pp. 231-240.

8. Chaudhury, G., "Vortex-Induced Vibration and Design of Pipelines in Deepwater", Offshore Technology Conference, Houston, TX, OTC 13018, 2001.

9. Davies, M.E., "The Effects of Turbulent Shear Flow on the Critical Reynolds Number of a Circular Cylinder," National Physical Laboratory, UK, NPL Report Marine Science R151, Jan. 1976.

10. Davis, J.T., Hover, F.S., Landolt, A., Triantafyllou, M.S., "Vortex-Induced Vibrations of Rigid and Flexible Cylinders", Workshop on Vortex-Induced Vibrations of Offshore Structures, Sao Paulo, Brazil, 2000.

11. Det norske Veritas, "Rules for Submarine Pipeline Systems", DnV Guideline No 14, 1996.

12. Downie, M.J., Graham, J.M.R., Hall, C., Incecik, A., Nygaard, I., "An Experimental Investigation of Motion Control Devices for Truss Spars", Marine Structures, Vol. 13, 2000, pp.75-90.

13. Fernandes, A.C., Esperança, P.T.T., Sphaier, S.H. and Silva, R.M.C.; "VIV Mitigation: Why not Porosity", Proceedings on 19th International Symposium on Offshore Mechanics and Arctic Engineering, New Orleans, USA, Feb., 2000.

14. Flatschart, R. B., Meneghini, J. R., Saltara, F., Siqueira, C. R., Ferrari Jr., J. A., "Numerical Simulation of Flow Interference Between Two Circular Cylinders in Tandem", Proceedings on 19th International Conference on Offshore Mechanics and Arctic Engineering, OFT-4081, ASME, Feb., 2000.

15. Flatschart, R.B., Meneghini, J.R., Saltara, F., "Numerical Simulation of Flow Interference Between Four Circular Cylinders", IUTAM Symposium on Bluff Body Wakes and Vortex-Induced Vibration, France, June, 2000.

16. Gerrard, J.H., "The Mechanics of the Formation Region of Vortices Behind Bluff Bodies", Journal of Fluid Mechanics, Vol. 25, Part 2, 1966, pp. 401-413.

17. Hafen, B.E., Meggett, D.J., and Liu, F.C., "Strumming Suppression – An Annotated Bibliography", Civil Engineering Lab., Technical Report N-1456, Port Hueneme, CA, 1976.

18. Halse, K.H., "Norwegian Deepwater Program: Improved Predictions of Vortex-Induced Vibrations", Proceedings of Offshore Technology Conference, Houston, TX, OTC11996, 2000.

19. Huse, E., "Flow Interference and Clashing in Deep Sea Riser Arrays", Workshop on Vortex-Induced Vibrations of Offshore Structures, Sao Paulo, Brazil, 2000.

20. Huse, E., Saether, L.K., "VIV Excitation and Damping of Straked Risers", Proceedings of 20th International Conference on Offshore Mechanics and Arctic Engineering, OMAE2001/OFT-1363, Rio de Janeiro, Brazil, ASME, June, 2001.

21. ISSC, "Structural Design of Pipeline, Riser and Subsea Systems", Report from Committee V5, International Ship and Offshore Structures Congress, Nagasaki, Japan, 2000.

22. Iwagaki, Y., and Ishida, H., "Flow Separation, Wake Vortices and Pressure Distribution around a Circular Cylinder under Oscillatory Waves", Proceedings of 15th Conference on Coastal Engineering, ASCE, Vol. 3, 1976, pp. 2341-2356.

23. Kaasen, K.E., Lie, H., Solaas, F., and Vandiver, J.K., "Norwegian Deepwater Program: Analysis of Vortex-Induced Vibrations of Marine Risers Based on Full-Scale Measurements", Proceedings of Offshore Technology Conference, Houston, TX, OTC 11997, 2000.

24. Katayama, M., Unoki, K., and Miya, E., "Response Analysis of Tension Leg Platform with Mechanical Damping System in Waves", Proceedings of Behaviour of Offshore Structures, MIT, Boston, Vol.2, 1982, pp. 497-522.

25. Larsen, C.M., "Empirical VIV Models", Workshop on Vortex-Induced Vibrations of Offshore Structures, Sao Paulo, Brazil, 2000.

26. Le Cunff, C., Biolley, F., and Durand, A., "Prediction of the Response of a Structure to Vortex-Induced Vibrations: Comparison of a Modal and a Wave Approach", Proceedings on 18th International Conference on Offshore Mechanics and Arctic Engineering, ASME, OFT 4072, 1999.

27. Li, Y., and Morrison, D.G., "The 'Colliding Participating Mass': a Novel Technique to Quantify Riser Collisions", Proceedings of ETCE/OMAE2000 Joint Conference, ASME, OFT-4132, Feb. 2000.

28. Lie, H., Larsen, C.M., and Tveit, Ø., "Vortex Induced Vibration Analysis of Catenary Risers", Proceedings of Offshore Technology Conference, Houston, TX, OTC 13115, May, 2001.

29. Lyons, G.J., Vandiver, J. K., Larsen, C.M., & Ascombe, G.T., "Vortex-Induced Vibrations Measured in Service in the Foinaven Dynamic Umbilical, and Comparison with Predictions", Proceedings on Flow-Induced Vibration 2000 Conference, Lucerne, Switzerland, June 2000.

30. Meneghini, J.R., and Bearman, P.W. "Numerical Simulation of High Amplitude Oscillatory Flow About a Circular Cylinder" Journal of Fluids and Structures, Academic Press, Vol.9, pp. 435-455, 1995.

31. Mørk, K.J., Fyrileiv, O., Verley, R., Bryndeum, M., Bruchi, R., "Introduction to the DnV Guidelines for Free Spanning Pipelines", 17th International Conference on Offshore Mechanics and Arctic Engineering, Lisbon, Portugal, ASME, June, 1998.

32. Peltzer, R.D., "Vortex Shedding from a Vibrating Cable with Attached Spherical Bodies in a Linear Shear Flow," Naval Research Laboratory Memorandum Report 4940, Oct. 1982.

33. Peltzer, R.D. and Rooney, D.M., "Effect of Upstream Shear and Surface Roughness on the Vortex Shedding Patterns and Pressure Distributions Around a Circular Cylinder in Transitional *Re* Flows," Virginia Polytechnic Institute and State University, Report No. VPI-Aero-110, Apr. 1980.

34. Price, S.J., "A Review of Theoretical Models for Fluidelastic Instability of Cylinder Arrays in Cross-Flow", Journal of Fluids and Structures, Vol. 9, 1995, pp. 463-518.

35. Sarpkaya, T., "Computational Methods with Vortices", Journal of Fluids Engineering, Vol. 111, No. 1, March 1989.

36. Schulz K. and Kallinderis Y., "Numerical Prediction of Vortex Induced Vibrations", Proceedings of 17th International Conference on Offshore Mechanics and Arctic Engineering, ASME, OMAE98-0362, 1998.

37. Siqueira, C.L.R., Meneghini, J.R., Saltara F., Ferrari Jr., J.A., "Numerical Simulation of Flow Interference Between Two Circular Cylinders In Tandem and Side By Side Arrangements", Proceedings of 18th International Conference On Offshore Mechanics and Arctic Engineering, ASME, OMAE99/OFT-4080, July, 1999.

38. Tsahalis, D.T., "Vortex-Induced Vibrations of Flexible Cylinder Near a Plane Boundary Exposed to Steady and Wave-Induced Currents", Journal of Energy Resources Technology, ASME, Vol. 106, June, 1984, pp. 206-213.

39. Vaicaitis, R., "Cross-flow Response of Piles Due to Ocean Waves", Journal of Engineering Mechanics Division, Trans. ASCE, Vol. 102, No. EM1, Feb., 1976, pp. 121-134.

40. Yamamoto, C.T., Meneghini, J.R., Saltara, F., Ferrari Jr., J.A., Martins, C.A., "Hydroelastic Response of Offshore Risers Using CFD", Proceedings of 20th International Conference on Offshore Mechanics and Arctic Engineering, ASME, OMAE 01-1052, June, 2001.

41. Zdravkovich, M.M., "Review and Classification of Various Aerodynamic and Hydrodynamic Means for Suppressing Vortex Shedding", Journal of Wind Engineering and Industrial Aerodynamics, Vol. 7, 1981, pp 145-189.

42. Zdravkovich, M.M., "The Effects of Interference Between Circular Cylinders in Cross Flow", Journal of Fluids and Structures, No. 1, 1987, pp. 239-261.

43. Zienkiewicz, O.C., and Morgan, K., <u>Finite Elements and Approximation</u>, Wiley, New York, 1983.

Chapter 9

Practical and Design Case Studies

9.1 Introduction

This chapter is dedicated to the practical application of the theory outlined in the previous chapters. Several practical applications of structures and structural concepts have been selected to demonstrate how vibration and hydrodynamic theories may be combined to solve the vibration problem of structures that are subjected to fluid flow. The practical cases include a mix of concept design studies, model tests of practical structures, and field tests of installed structures. The examples are chosen to include structures experiencing vibration in steady fluid flow as well as in waves in the ocean, and include full-scale experience as well as model experiments in a controlled environment. It is noted that the design of these structures often uses simple calculations without resorting to complicated analysis. This is particularly true in the initial phases of design. In many cases, simplified theory is used, recognizing that the methodology may not be mathematically consistent and correct, but the results are of practical importance. The examples demonstrate how some of the experience-based, overly simplified analyses may be able to explain the physical phenomenon, and provide reasonable results suitable for design and analysis of the system under consideration.

9.2 Damping Experiment of a Tethered TLP Caisson

Determination of an appropriate damping value to apply in a structural design is often a difficult problem. Due to inadequate theory in computing this damping, experiments are carried out to determine the damping coefficients subjected to fluid flow. This is particularly true for the vertical oscillatory motion of a caisson type structure with sharp bottom edges. One such structure is the tension leg platform (TLP) tethered vertically to the ocean floor. The first-order wave-induced vertical motion of a TLP does not induce excessive resonant motions in the system, primarily due to the high tension in the tethers. However, the tethers experience high resonance loads, known as springing loads at the second-order (sum frequency) components in both heave and pitch. The tether load is a critical concern in the design of a TLP system. The damping is the limiting factor in the amplitude of tether loads.

As with other floating structures, the TLP system experiences both material and hydrodynamic damping. The material damping appears from the tendon hysteresis and their attachments to the TLP as well as to the seafloor. The subsea template also provides some damping. Hydrodynamic damping appears in the form of radiation damping and viscous damping. The radiation damping is quite small for the small vertical motion of the caisson. The viscous damping consists of skin friction and pressure drag. The latter arises from the flow separation and vortex formation at the sharp edge of the bottom of the caisson.

Here, two experimental projects will be described, each of which tested one leg of a TLP in the form of a circular caisson in vertical motion to determine the damping coefficient. These tests were carefully designed so that the heave damping, which is quite small for a taut moored large caisson, may be accurately determined. An example of an erroneous test setup to derive damping, and the subsequent erroneous results is highlighted.

9.2.1 Caisson with Leaf Springs

For a floating caisson attached vertically to pretensioned springs of high stiffness characteristics, the system will have a high natural frequency. Since the damping coefficient at a high frequency is extremely small, care must be taken in setting up the test so that accurate information may be obtained from the test. In a test with a TLP column model [Huse (1990)], heave-damping coefficients were obtained. The model represented a TLP column of 25m diameter and 37.5m draft at a scale of 1:20. The steel model had a sharp corner at the lower edge with a corner radius of less than 0.1mm. The model was suspended from the middle of two 6m long horizontal steel beams. The beams acted as leaf springs, which represented the stiff tethers providing restoring forces to the oscillatory system in the vertical direction. The ends of the beam were welded to rigid structures to avoid additional damping that would be introduced in the system from any mechanical fasteners. The spring constant of the suspension was high (1.414 x 10^6 N/m).

The model was quite heavy (3,905 kg) when ballasted with steel and water to achieve the required draft. The steel ballast inside was secured to the hull. The vertical and angular motions were measured by accelerometers. The tests were carried out as free oscillation tests in heave. The heave amplitudes of the model ranged from 0.1 to 2.0 mm. The angular motions were analyzed to ensure pure heave motions of the column. The frequency of vertical oscillation in water was 3.03 Hz corresponding to a prototype period of ($\sqrt{20}/3.03$ =) 1.48s. The KC and Re numbers were in the range of 0.0005 to 0.01 and 0.2 to 70 respectively.

In order to initiate the test, the model was pulled upward from equilibrium by a crane and the steel wire connecting the crane to the model was cut. The initial displacement of the model was controlled by an electronic balance. The subsequent heave oscillation was recorded. The damping in the system was computed from the free decayed oscillation measured by the vertical accelerometer. The equation of motion governing the oscillation involves a second-order nonlinear equation including the linear and quadratic terms (see Eq. 3.67). The quadratic damping term was linearized and the total damping coefficient was obtained from the formula:

$$c = c_1 + c_2 \frac{16T}{3} X \qquad (9.1)$$

where X and T are the amplitudes of the decayed oscillation and its period. The damping was obtained by fitting a straight line through this linear equation. The slope of the line yields the second order term c_2 in Eq. 9.1, while the intercept gives the linear damping term c_1.

The total damping includes tmaterial damping of the system including the beam support. However, the interest here is in the hydrodynamic part of the damping only, not that of the model support system. Therefore, tests were carried out in the dry to determine the material damping of the system. Then, the hydrodynamic damping alone was computed by subtraction of the material damping from the total. The slope of the fitted line was nearly zero for all test runs. Thus, the hydrodynamic damping of the tethered caisson was found to be mainly linear, the quadratic damping being much less than

10 percent of the total. An average value of the damping factor (ζ) from these tests was about 0.08 percent.

Additional tests were performed [Huse (1994)] with a smaller model in which the sharp edge at the bottom was rounded off to reduce the effect of vortex shedding due to vertical motion. The setup with the beams was similarly welded as in the earlier tests and provided low mechanical damping to the overall system. In this case, the resonant frequency was 2.75 Hz. For the purpose of representing a TLP caisson, a scale factor of 1:40 may be assumed for this setup so that the equivalent prototype heave period becomes 2.3 seconds.

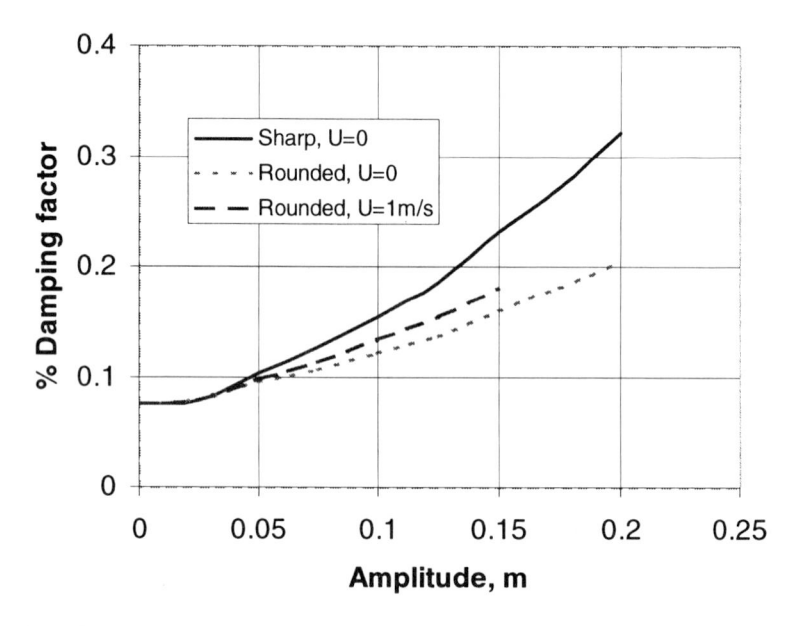

Fig. 9.1: Heave Damping Factor in Still Water and Current [Huse (1994)]

These tests were conducted in current as well as waves. Additionally, the bottom edge was rounded to a radius of 0.8m. For the rounded edge, the flow was more streamlined and the scale effect in the model was expected to be larger. The (percent) damping factor in still water and in current as a function of (prototype) heave amplitude is shown in Fig. 9.1. It is clear that the rounded edge reduced the damping factor, since the flow would not produce substantial vortex shedding. It may also be seen that the presence of current increased the heave damping, the amount of increase being directly related to the current speed. In waves, the damping increased even further. The damping was found to be linear for the (prototype) heave amplitudes up to 0.05m. For amplitudes near 0.13m the second-order damping had contributions comparable to the linear damping. At higher amplitudes, the higher order damping dominated. For the rounded edge, the higher order contribution was about 50 percent of that for the sharp edge. In general, the springing damping of the TLP caisson was found to vary from 0.1 to 0.5 percent of the critical damping.

9.2.2 Caisson with Coil Springs

In another test with a floating vertical caisson [Chakrabarti and Hanna (1991)], the added mass and damping coefficients in heave were evaluated. The model consisted of a single vertical cylinder scaled from a vertical column of a typical TLP (scale factor 1:20). A 2-1/2 ft. (0.76m) diameter model was selected for this test, representing a 30-ft (9.15m) caisson. The draft of the model (about 4 ft. or 1.22m) also represented the typical draft of a TLP. The model surface was painted to a smooth finish. The bottom edges were sharp. Tests were run in still water and in waves. The cylinder was held in place vertically in the water by two sets of linear springs pretensioned to represent tendons (Fig. 9.2). Load cells were attached to the ends of the springs to record the loads (and hence, motion). The cylinder was displaced vertically from its equilibrium position and released, and the free decaying oscillation was recorded.

In order to ensure pure heave and to examine the presence of any simultaneous pitching motion of the caisson, in a first series of tests, the motions of the caisson were recorded by using two potentiometers placed at two diametrically opposite points on the top face. While no appreciable pitch of the caisson was introduced, these potentiometers were found to introduce substantial material damping, especially at the high frequencies, and the results were totally erroneous due to their presence. In the subsequent tests [Chakrabarti and Hanna (1991)], their removal and reduced damping in the spring sets assured that the total material damping in the test set-up was indeed small, and most of the damping in still water was attributed to hydrodynamic effects.

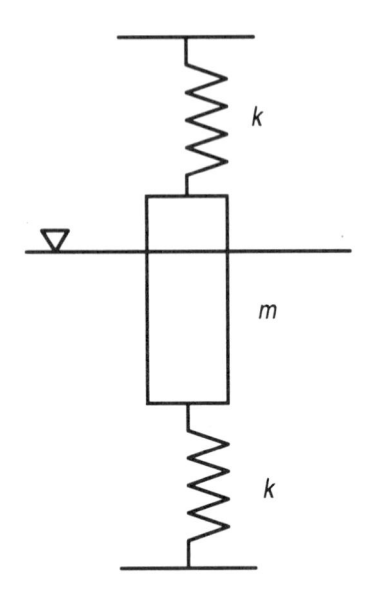

Fig. 9.2: Free Vibration Test Setup of a Vertical Cylinder

Since hydrodynamic damping was of importance for this experiment, the material damping present in the pair of springs used in the tests was extracted. In order to derive this damping, the spring sets were hung in air with an equivalent dead weight representing the pretension in the springs. The weight was displaced vertically an amount equivalent to the displacement of the caisson in water. The loads in the spring system were measured with a load cell at the fixed end of the setup. The damping in air was determined from this measurement using the logarithmic decrement technique described in Chapter 3. The material damping was generally quite small compared to the hydrodynamic damping, although not negligible. This damping was subtracted from the total damping. The results are shown in Fig. 9.3 for various spring sets. The results obtained from the erroneous setup are also shown for comparison. The effect of the additional potentiometer lines in the earlier series of tests is found to be quite large, especially near the low periods.

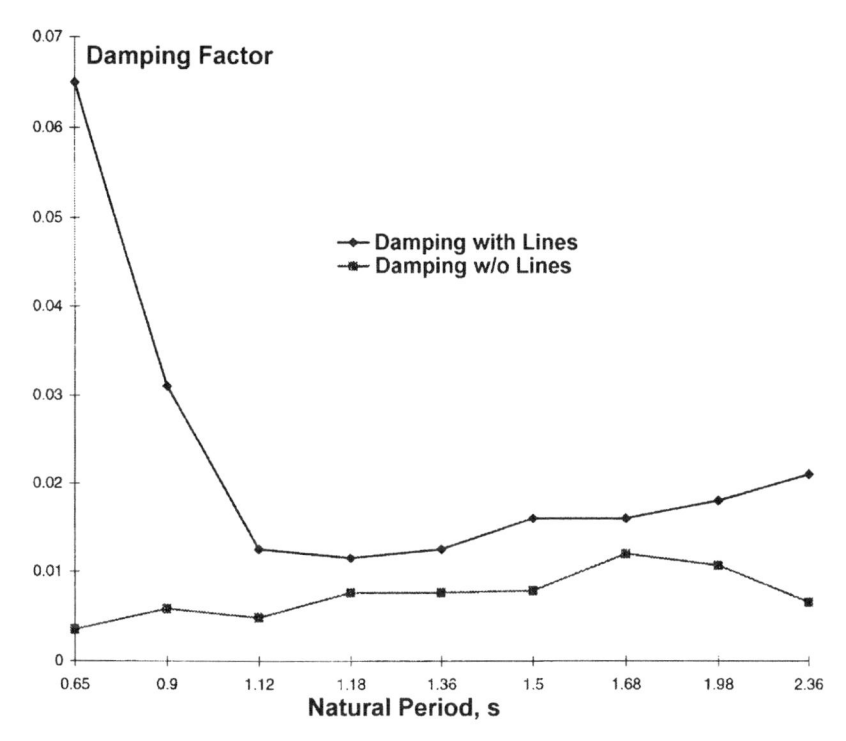

Fig. 9.3: Hydrodynamic Damping of the Vertical Caisson Model

An interesting observation of the test results is that, as before, the heave damping for the vertical cylinder was linearly dependent on the cylinder velocity in low amplitude heave. For larger amplitude of oscillation, there was some evidence of nonlinearity. Note, however, that for a TLP hull, the heave amplitude is expected to be small. The hydrodynamic damping was computed to be on the order of 1 percent of critical. However, it decreased even further at higher natural frequencies. At a frequency near 2 Hz, the damping was about 0.3 percent. At a higher frequency (3Hz) [Huse (1990)], the damping factor in free oscillation was shown to be even lower.

This test series demonstrates that the damping factor for a TLP in the heave motion is expected to be less than 0.005. It is linear at low heave amplitudes and constitutes linear and nonlinear components when the amplitude of motion is large. Because of this low damping, the tether vibration loads are expected to be quite high, and should be considered carefully in a design of the TLP system. This is why the concepts of external damping device described in Chapter 8 were deemed important.

9.3 Field Experiment of Vibration of a Riser

Deep-sea vertical risers are subject to strong currents prevalent in many offshore locations. Risers are an essential element of any platforms used for offshore exploration and production. Since the exploration of petroleum products is moving into much deeper water, the risers are prone to several technical difficulties including high current forces, VIV, and possible collision within the riser bundle, and subsequent damage or even failure. Therefore, it is essential that the fluid flow effect on these deep-sea risers be known. Unfortunately, the ultradeep water exceeding 1500m prevents small-scale experiments, as no suitable testing facility exists for such experiments.

9.3.1 Collision Within Riser Group

This section describes a case study of a model test of risers in a deep-water fjord. The purpose of this project was to investigate riser collision and generate a data set that might be used to verify a numerical model to predict riser behavior under the action of current [Huse (1996)]. The test site was chosen at Skarnsund, 100 km north of Trondheim. The sound has a water depth of 190 m, and tidal currents well above 2 knots. An existing bridge spanning the sound was used as the work platform. A set of riser models were suspended from a surface catamaran and had a weight attached to their bottom end, but supported by a pulley system to introduce the desired tension in the risers. The weights were arranged in a bottom frame such that the risers maintained their modeled spacing in equilibrium. The catamaran as well as the bottom framework was moored to the existing bridge in such a way as to maintain the horizontal positions of the risers.

Chapter 9 Practical and Design Case Studies 391

The riser group consisted of an array of risers in a 3x4 rectangular arrangement (Fig. 9.4) with equal spacing. One riser in the middle of the array represented a drilling riser while the others were smaller diameter production risers. The array represented a riser system for a Tension Leg Platform. The spacing at the top and bottom end among the risers were maintained at equal distances in the inline and transverse directions.

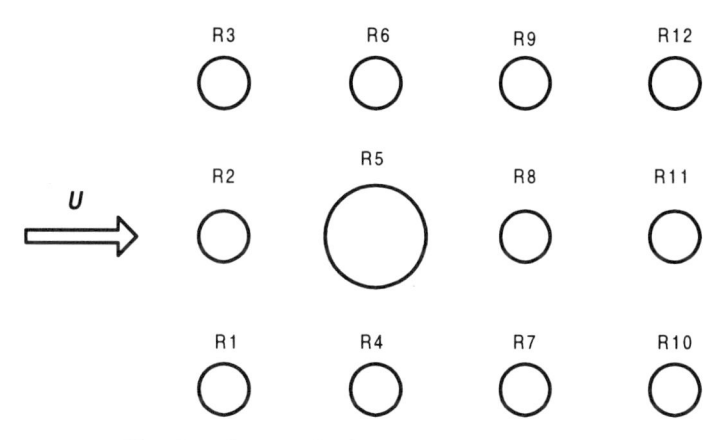

Fig. 9.4: Setup of Riser Array in the Fjord

An interesting scaling law was used. Froude scaling was not adopted since the wave effect was not of concern. Instead, the current velocity was scaled 1:1. Since VIV was an important consideration, the bending stiffness was considered important and modeled appropriately. The material of the model riser was steel, the same as the prototype. This resulted in scaling the riser pipe thickness according to the scale factor, λ. The scaling parameter for various quantities was chosen as shown in Table 9.1. For the purposes of discussion, a scale factor of 1:30 was selected for the riser model.

Vortex induced vibration increases the mean inline drag force, causing large static deflection in the middle of the risers. This, in turn, may induce collisions between the neighboring risers. During the testing, accelerometers were fitted inside the riser model. In addition, a collision detecting system between the risers was installed to detect collisions. Moreover, the collision generated a loud audible noise indicating a collision between the risers.

Table 9.1: Scaling for Deep Sea Riser Model Testing

Parameter	Scaling Factor
Length	λ
Velocity	1
Time	λ
Acceleration	λ^{-1}
Frequency	λ^{-1}
Force	λ^2
Mass	λ^3
Modulus of Elasticity	1

The drilling riser had a pretension of about 1205 kN while that of the production risers was varied from 412 kN to 862 kN for two test conditions. Several tests were performed at different current velocities and profiles encountered at the site. At low current velocities, no collision of risers was observed. As the current velocity increased, the collision between neighboring risers was initiated and the frequency of collision increased with the increase in the magnitude of current velocity. The current profile at the site was shear type, decreasing in magnitude with depth. Additionally, the phases of the maximum velocity at different elevations were different.

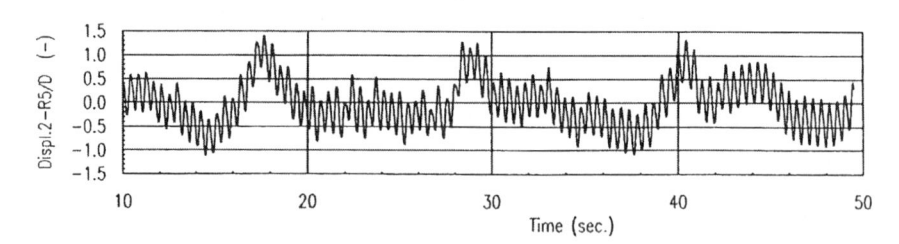

Fig. 9.5: Displacement Time History of Drilling Riser [Huse (1996)]

The displacement time history shown in Fig. 9.5 was computed from the accelerometer measurement. The time history shows that the drilling riser experienced clear evidence of lock-in vibration at its natural frequency. The VIV amplitude was about half the riser diameter. Additionally, the risers experienced an irregular low frequency inline oscillation of large magnitude, almost of a chaotic nature. The peak-to-peak amplitudes of these motions were as much as 30 to 40 diameters. Typically, the far upstream risers remained stationary. The next riser collided with the upstream riser and then moved far downstream in a slow motion before returning upstream and colliding again with the upstream riser. This nature of motion is physically difficult to explain and is probably due to instability arising from the hydrodynamic interaction. This situation arose at or above the collision velocity of current. In a practical design, of course, it is undesirable to have collisions and they should be avoided in a design. Thus, the low frequency oscillation of the intermediate risers should not arise in a properly designed spacing of a riser system. However, the observation was unexpected and is of particular interest in the study of fluid induced vibration. A theoretical explanation for such an observed riser motion in shear current is not known.

9.3.2 Towing of Riser

Additional tests were performed on risers in the fjord at a scale of 1:15 [Huse, et al. (1998)]. The test was conducted at a floating quay made with two barges spanning 180m. The water depth was 97m. These tests were controlled, rather than being dependent on the environment present at the site. The tests consisted of mechanically towing the riser group along the quay, extinction tests by a mechanical impact on the riser group, and surface current tests induced by a propeller. This last series of tests was found to be difficult to perform and points to the difficulty in controlling large-scale tests in a natural environment. The results were not satisfactory. The towing tests were performed by a moving catamaran holding the top end of the risers. The risers were pre-tensioned with the help of buoyancy tanks connected with kevlar at the bottom end of the risers going over pulleys. A triangular equivalent current profile was encountered due to the setup. Tests with a single riser varied the towing speeds up to a maximum of 2 m/s. During a two-riser test, the spacing between the risers was varied from 1m to 2m at the top end. The bottom end spacing was maintained at 2m.

The model pipe riser was 30mm in outer diameter and 2mm thick. At a scale of $\lambda = 15$, this represents a riser length of 1350m and an outer diameter of 0.45m. One riser was composed of 6m sections joined by flanges. The other had 2.9m long elements with 0.1m long bending moment transducers in between. Thus, the bending moment on the instrumented riser was measured every 3m along its length. Some transducers failed, leaving about 50 good measurements. The angles of the risers were also measured.

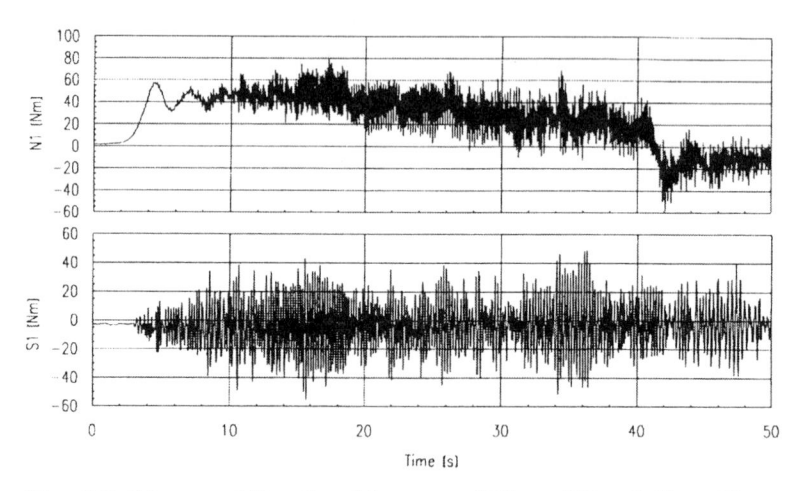

Fig. 9.6: Measured Bending Moment of Riser (N1 = Inline; S1 = Transverse) [Huse, et al. (1998)]

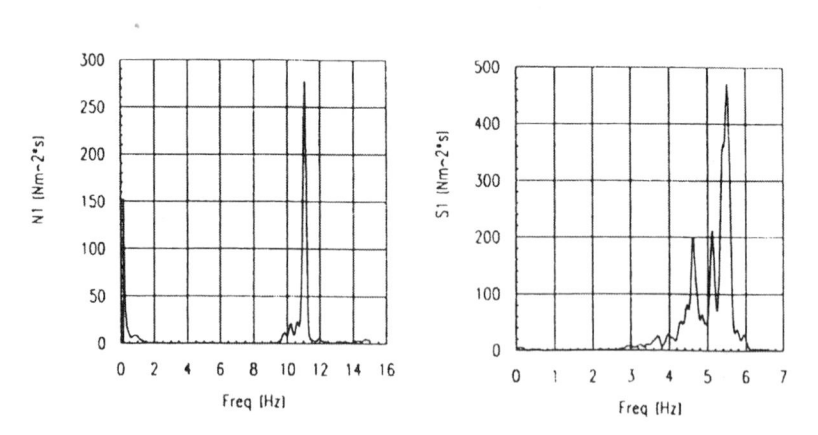

Fig. 9.7: Spectra of Inline and Transverse Bending Moment of Riser [Huse, et al. (1998)]

Chapter 9 Practical and Design Case Studies

An example of the measured bending moment at 3m from the top end of riser is shown in Fig. 9.6. The corresponding spectra of the measured moment at a towing speed of 1.39 m/s are given in Fig. 9.7. The VIV about the transverse direction is seen at a frequency of 5.5 Hz, while the VIV about the inline axis is around 11 Hz. Thus, the inline oscillation frequency is twice the transverse vibration frequency. In addition, the riser tension was found to vary from zero to twice the mean tension found at twice the transverse VIV frequency. This high frequency oscillating riser tension might pose a problem with the fatigue life of the riser.

A simple analysis is presented here to show the effect of the axial vibration of the riser. It follows the work of Huse, et al. (1998). A vertical or near vertical riser is considered, which is pinned at both ends. The VIV transverse oscillation is assumed to have a single mode at a high frequency ω_n. The displacement along the horizontal axis x of the riser is a function of the vertical coordinate y and time t, and has the form of a standing wave:

$$x(y,t) = x_0 \sin(ky)\sin(\omega_n t) \tag{9.2}$$

where k is the wave number given by $2\pi/L$, and L is the wave length. The length along the riser is given by

$$l(y,t) = \int_0^y \sqrt{dx^2 + d\zeta^2} = \int_0^y \sqrt{1+\left(\frac{dx}{d\zeta}\right)^2}\, d\zeta \tag{9.3}$$

Considering the slope to be small, the integrand is expanded in an infinite series and the first two terms are retained:

$$l(y,t) = \int_0^y \left[1 + \frac{1}{2}\left(\frac{dx}{d\zeta}\right)^2\right] d\zeta \tag{9.4}$$

Substituting Eq. 9.2 for x, this equation becomes

$$l(y,t) = \int_0^y \left[1 + \frac{1}{2} x_0^2 k^2 \sin^2(\omega_n t)\cos^2(k\zeta)\right] d\zeta \tag{9.5}$$

The term $\cos^2 k\zeta$ is replaced by its average value of ½ under the present assumption. Then,

$$l(y,t) = \left[1 + \frac{1}{4}x_0^2 k^2 \sin^2(\omega_n t)\right] y \qquad (9.6)$$

Noting that $\sin^2 a = [1 - \cos(2a)]/2$ and replacing y by the length of the riser, the relative elongation of the riser is computed. Thus, the dynamic stress due to axial tension fluctuation is obtained as

$$\sigma_t = \frac{1}{8} E x_0^2 k^2 \cos(2\omega_n t) \qquad (9.7)$$

The dynamic bending stress due to the transverse vibration is

$$\sigma_b = \frac{DE}{2} \frac{\partial^2 x}{\partial y^2} \qquad (9.8)$$

where D is the riser outer diameter. Substituting Eq. 9.2, one gets

$$\sigma_b = \frac{1}{2} D E x_0 k^2 \sin(ky) \sin(\omega_n t) \qquad (9.9)$$

The amplitude ratio of axial to bending stress of the riser becomes

$$\left(\frac{\sigma_t}{\sigma_b}\right)_0 = \frac{x_0}{4D} \qquad (9.10)$$

If the amplitude of the transverse oscillation is on the order of the riser diameter D, then the VIV-induced axial tension results in a 25 per cent increase in the fatigue stress in the riser. Moreover, the number of stress cycles increase, as the frequency of the tensile stress is twice that of the bending stress. This frequency relationship in the analytical development was found to be consistent with the experimental results from the model riser shown above.

The field experiment and a simple analysis with deep-water risers demonstrate that there is a potential increase in the probability of fatigue damage to the riser resulting from its axial vibration and associated tension. Therefore, this case study indicates that the axial tension should be included in the overall design of risers, along with the bending stresses.

Chapter 9 Practical and Design Case Studies 397

9.4 Impact Loading on a Fixed Platform

Many offshore platforms built in the 1960's are still in service after more than 30 years in the Gulf of Mexico. These aged platforms are currently being evaluated for serviceability and additional life expectancy. One of the critical areas of investigation is the environmental load experienced by the platform during a 100-year hurricane. In particular, the effect of current and breaking waves on these platforms and the associated vibrations are of importance. This area is presented with the support of experimental studies of two existing structures in the Gulf of Mexico – one is a space frame structure and the other is a shallow water caisson.

In designing space-frame structures, the 20th Edition of API RP2A 20th Edition Guideline (1992) outlines the procedure for evaluating the static wave forces. For a space frame structure, the wave kinematics and current will suffer shielding and blockage through the members of the space-frame. The API RP2A guideline for wave force evaluation specifies a wave kinematics factor and a current blockage factor to be used. However, the methodology (and coefficients) outlined is mainly for typical modern jacket structures, which are more transparent to waves. Most of the platforms designed and constructed in the 1950's have a large number of legs and densely arranged bracing members, and the API procedure may be conservative in predicting wave forces due to the more significant blockage and shielding effects in these structures.

This case study examines an experimental investigation [Chakrabarti and Gu (1996)] undertaken to document extreme wave-current loading on a typical Gulf of Mexico platform constructed in the 1950's. It illustrates the effect of impact loading of fluid on a structure. The Vermilion 46A platform, located approximately 30 miles offshore from New Orleans, was chosen as the prototype for the model testing.

The prototype structure that was modeled consisted of a 40-leg vertical pile jacket structure with a deck. The testing was performed with a 1:20 scale model of the jacket platform (Fig. 9.8) at a wave/current generating facility. The model platform consisted of two identical 20-leg jackets connected together at the base. For convenience in the test setup each 20-leg jacket was individually instrumented. This permitted testing of a single 20-leg platform separately from the complete 40-leg model. Each platform was instrumented at its base to measure the two-component global shear and moment. The deck of each platform was separately instrumented to measure the two-component horizontal impact load due to waves. The stiffness of the model was considered to be within the range typical of jackets of this kind, with the first mode of oscillation of the 20-leg model of about 4.5 Hz.

Fig. 9.8: Jacket Model Setup for Load Measurement

The setup of the base and deck of the jacket model in the testing facility is shown in Fig. 9.8. The two units are shown in end-on orientation. The individual decks were floated on load cells with a small gap between the decks and the vertical jacket members. Three component fluid particle velocities were measured using an acoustic Doppler current meter at 12 inches (305 mm) below the still water surface at a fixed location transverse to the platform model.

9.4.1 Current Effect on Jacket

The steady current loading on the platform model was obtained for current speeds ranging from 0.4 ft/s to 1.4 ft/s (0.12 to 0.43 m/s). The test results were used to develop the coefficient representing a steady current drag coefficient multiplied by the projected area and blockage factor:

$$AC_D F_b = \frac{2F_D}{\rho U^2} \qquad (9.11)$$

where C_D = drag coefficient, F_b = blockage factor defined in API RP2A 20th Edition, F_D = load due to drag, ρ = mass density of water, A = projected area of model to loading, and U = current velocity.

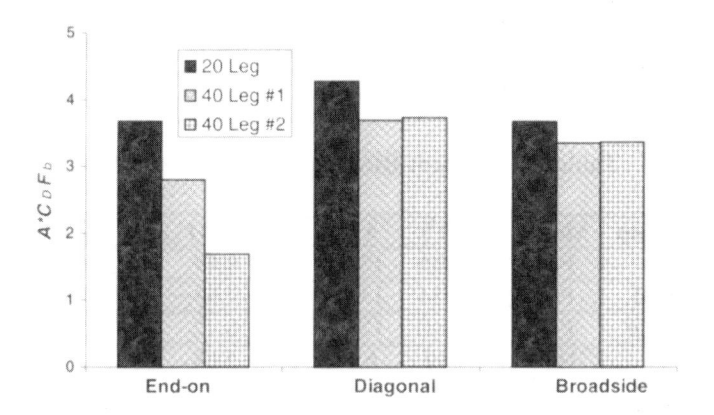

Fig. 9.9: Drag Coefficient of Jacket Model in Current

Figure 9.9 shows the results of this coefficient for each model orientation. For the 20-leg model, the diagonal orientation produced a higher area-drag coefficient than the end-on orientation. This result is to be expected due to the increased number of members that was influenced by the current as well as lower blockage. For the 40-leg model, the broadside and diagonal orientation showed little to no effect of blockage. The area-drag coefficients for the diagonal and broadside orientations for load cell #1 and load cell #2 were virtually identical. The coefficients for the stand-alone module are higher in these cases. The effect of blockage can be clearly seen in the end-on orientation. The coefficients for the #2 half model in this case are about two-thirds of those for the #1.

The blockage factors were calculated using a drag coefficient C_D = 1.05, which is the value specified in API RP2A for rough cylinders which was considered appropriate. Table 9.2 lists the results for the (20 leg) half and (40 leg) full jacket model. It clearly shows the blockage effect by densely arranged members and concludes that the blockage for the end-on orientation is quite high. The difference in the blockage effect between the broadside and diagonal cases for both the full and the half model is not significant. The main text of API RP2A (1992) recommended the values of the blockage factors for 8 piles. In its commentary, an extension method for structures with more than 8 piles is recommended. Using this method, the calculated blockage factors are shown in the last column of the table and are close to those obtained in the test. Thus, the current blockage test conducted for the densely populated jacket verifies reasonably well the recommended method of API for such structures.

Table 9.2: Current Blockage Factors for Jacket

Orientation	Model	A (ft^2)	$A*C_D$	$A*C_D*(F_b)^2$	F_b	API F_b
End-on	Full	13.286	13.950	4.49	0.567	0.617
Diagonal	Full	12.554	13.182	7.43	0.751	0.770
Broadside	Full	12.140	12.747	6.72	0.726	0.760
End-on	Half	6.589	6.918	3.68	0.729	0.767
Diagonal	Half	6.209	6.519	4.27	0.809	0.813

9.4.2 Jacket Structure in Regular Waves

The jacket was subjected to regular waves having heights and periods conforming to the extreme waves at the Gulf of Mexico site. For each wave period, a non-breaking steep wave maximized for height, and a breaking wave was chosen.

Since the deck was suspended from the aluminum rods that transfer the jacket loading to the base load cells (Fig. 9.8), the deck load was removed from the measured base load, in order to determine the loading on the jacket members. The base loads on the individual section indicated the effect of shielding and blockage.

The platform model was not designed to scale the structural stiffness of the prototype. However, the stiffness introduced by the model material and load cells introduced a high frequency damped oscillation in the load from the wave impact. The natural frequency of the model was measured in the range of 4 - 5 Hz. This oscillation frequency present in the measured load was filtered digitally before analyzing the shielding effect on the loads.

The forces on the jacket due to regular waves are given by the Morison equation (see Eq. 2.49). Assuming linear wave theory, the water particle

Chapter 9 Practical and Design Case Studies 401

acceleration is related to the particle velocity by $\dot{u}_0 = \omega u_o$, where the subscript zero refers to the amplitude. The normalized force is written as:

$$\frac{f}{1/2\rho D u_o^2} = C_M \frac{\pi^2}{2} \frac{1}{KC} \sin \omega t + C_D |\cos \omega t| \cos \omega t \qquad (9.12)$$

Thus, for linear waves, the normalized force (f / u_o^2) is a function of the Keulegan-Carpenter number KC. The dependence of the coefficients on the Reynolds number is weak, especially for low range of Re, which is typical for wave tank tests. Therefore, if the measured force amplitudes on the 20-leg and 40-leg jackets are normalized with the square of the measured velocity amplitudes, the results should show the qualitative effect of shielding.

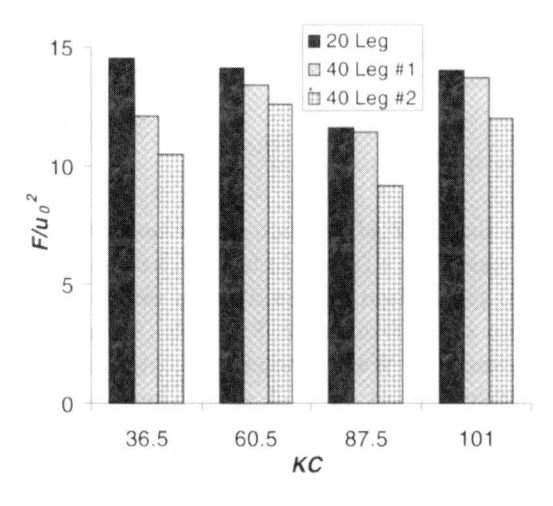

Fig. 9.10: Normalized Force on Jacket – End-on

The normalized force amplitudes are shown in Fig. 9.10 for the end-on model orientation, which should show the most significant effect from shielding. The KC number was defined for this purpose in terms of the individual diameters of the vertical legs of the jacket structure. This definition was assumed to be reasonable since all vertical members were of equal diameter and the measured force on the structure principally depended on these members. The test runs having close *KC* numbers in each model case were chosen in plotting the bar charts. The four *KC* numbers correspond to four wave periods.

For the end-on case (Fig. 9.10), the 20 leg model showed the highest normalized load [lbs/(ft/s)^2] in all cases (except for *KC* = 80, where #1 jacket module measured higher load). This is followed by the #1 module. The loads on the #2 module are the lowest. The waves were breaking or near-breaking around *KC* = 80 and the exception in this group is probably due to the breaking effect of waves. The case of *KC* = 101 was a breaking wave. Hence, the load on the #2 module is substantially lower, once the wave broke on the #1 module.

The shielding effect from waves on the downstream jacket module by the upstream module is very clear. It shows that the stand-alone module has the highest forces, as compared to the two half modules in the full model tests. This is because the fluid velocities through the stand-alone module were always higher than those past the full model for similar wave conditions, due to shielding effects.

9.4.3 Pile Structure in Random Waves

A pile situated in the Gulf of Mexico was chosen to show the effect of breaking random waves on a fixed platform [Chakrabarti, et al. (1997)]. A 36 inch (914 mm) production pile in a water depth of 50 ft (5.2 m) was modeled at a scale of 1:20 using a 1.8 inch (46 mm) diameter smooth stainless steel cylinder. Wave forces and overturning moments were measured with strain gauges located at the base of the cylinder. The vibrational natural frequency of the pile was also scaled. In the prototype, the pile was estimated to have a natural frequency of approximated 2 Hz (natural period of 0.5 s) in the sway mode. The design of the strain gauge assembly gave the pile model a natural frequency of 8.5 Hz (natural period of 0.12 s), which is close to that desired, according to Froude model.

A random wave having a maximum height of 40 ft (12.2m) and period of 13s was modeled. Simultaneously, currents up to 2.1 knots (1.1 m/s) were generated in model scale. Several individual waves within the random wave profile reached the breaking wave limit, and several waves broke directly on the pile. These waves excited significant vibration of the model pile with a dynamic amplification of forces, just as the wave crest passed the

Chapter 9 Practical and Design Case Studies

pile. This is illustrated in Fig. 9.11 [Kriebel, et al. (1996)] showing the measured force time history at a large wave crest.

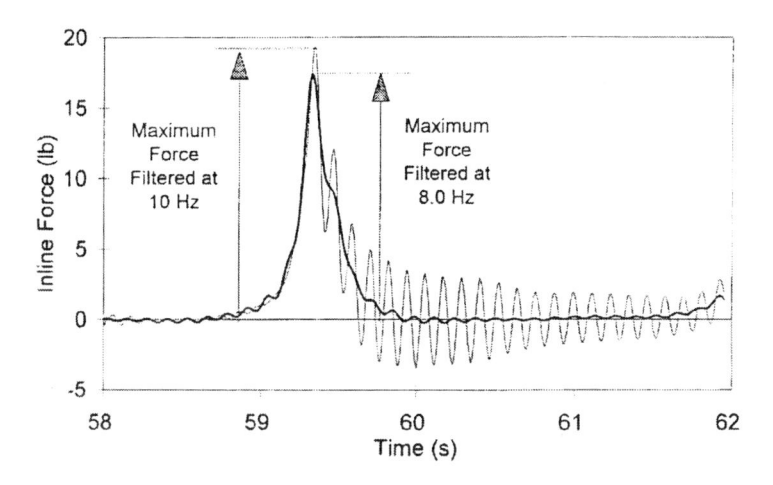

Fig. 9.11: Unfiltered and Filtered Measured Forces on Pile

A simplified calculation was carried out to show the effect of this dynamic amplification in the following way. First, the force time series was filtered to remove the dynamic effects. Note that this method is sensitive to the cutoff frequency used. The cut-off frequency for the digital filter was chosen as 8 Hz to remove the dynamic frequency of 8.5 Hz. Then, the ratio of the filtered and unfiltered maximum force gives the experimental dynamic amplification factor (DAF):

$$DAF = \frac{F_{unfiltered}}{F_{filtered}} \qquad (9.13)$$

This expression approximately represents the ratio of maximum dynamic base shear force from the vibrating structure relative to maximum static base shear force on an equivalent rigid pile.

An approximate theoretical dynamic amplification factor (DAF) for the system was estimated assuming a single degree of freedom system. The impulse loading was considered to apply during the passage of the wave crest, which was taken as a half-sine wave. The crest period was taken as twice the "crest front period" T_F, which was defined as the time from the zero-crossing of the wave, where wave loading was negligible to the wave crest where wave loading was generally largest. This crest period was considered to have a harmonic form for the analysis. The equation of motion for a simple spring-mass oscillator was adopted as

$$m\ddot{x} + kx = F_0 \sin \omega_F t, \qquad 0 < t < 2T_F \qquad (9.14)$$

where x is the characteristic displacement of the structure, m is the system mass, k is the system stiffness and F_0 is the maximum applied load. The frequency of the half sine wave impulse load, ω_F, is given by $\omega_F = 2\pi/(4T_F)$. In Eq. 9.14, the effect of damping was neglected, since damping is expected to have little impact during the short duration of the impulse load.

The dynamic motions were solved from Eq. 9.14 by applying initial conditions that the structure displacement and velocity would be zero at the beginning of the impulse. If the static response of such a system is defined by $x_s = F_0/k$ and the natural frequency is $\omega_n = 2\pi/T_n$ (T_n = natural period), then the dynamic response during the impulse load is given by

$$\frac{x(t)}{x_s} = \frac{1}{1-\beta^2}(\sin \omega_F t - \beta \sin \omega_n t) \qquad (9.15)$$

The first term on the right hand side is the forced response at the impulse frequency, and the second term is the free response at the natural frequency of the structure. The parameter β is the ratio:

$$\beta = \frac{\omega_F}{\omega_n} = \frac{T_n}{4T_F} \qquad (9.16)$$

The theoretical dynamic amplification factor was defined from the maximum value of the response given by Eq. 9.15.

The computed DAF is shown in Fig. 9.12 as a solid line as a function of the normalized impulse duration $2T_F$. For a value of $2T_F/T_n = 1$, the theoretical values approach infinity in the absence of any damping. The measured values are shown by different symbols on the figure for the three current velocities of 0, 0.8 and 1.2 ft/s (0, 0.24, and 0.37 m/s respectively) present with the waves. A value of DAF = 1 implies no amplification. The theoretical DAF ranged in value from 1.05 to 1.16 with a mean value of about 1.09. Measured values ranged from 1.0 to 1.15 with an average value of 1.05.

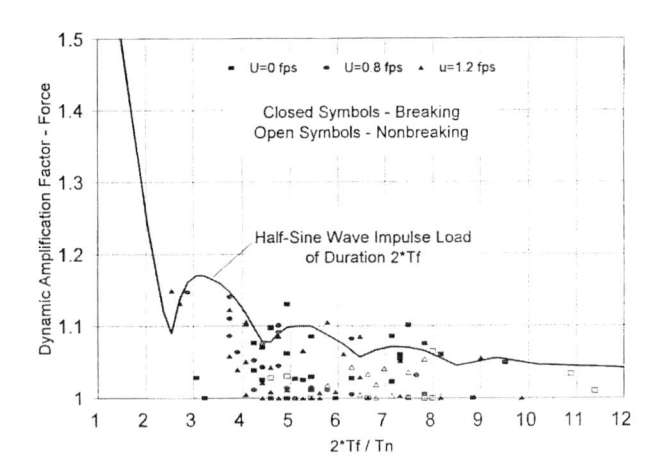

Fig. 9.12: Dynamic Amplification for Impulse Loading --Measured vs. Theoretical

It was observed that the waves did not excite strong vibration of the cylinder despite the short crest front period. For breaking waves, the degree of dynamic amplification was dependent on whether the waves broke directly on the pile (larger amplification) or whether the waves broke slightly ahead of the pile (smaller amplification). The theoretical DAF generally acted as an upper bound of the measured dynamic amplification. The case study shows that the simple theory of impulsive vibration of a structure from breaking waves and the resulting vibration may be used to explain the amplitude of dynamic amplification.

9.5 Sloshing in a Hydroelastic Vertical Caisson

Liquid sloshing inside a partially filled storage tank subjected to environmental forces is a well-known phenomenon [Abramson (1966)]. The standing wave in a partially filled tank, or "sloshing" of liquid as it is commonly known, involves the transfer of liquid from one side of the storage tank to the other in the form of a wave. Sloshing is a natural period phenomenon caused by the vibration of the container from external sources. The energy is transferred to the fluid through the motion or deformation of the wall of the storage tank. There are multiple natural periods in which the sloshing may be excited. Usually, the fundamental mode of motion is the most critical one.

Simple calculations provide these natural periods for a rigid rectangular or circular cylindrical tank. For example, the free longitudinal oscillation of fluid in a rectangular tank of length l is given in terms of the wave length L of the standing wave for the nth mode of oscillation as

$$L = \frac{2l}{n}, \qquad n = 1,2,3... \qquad (9.17)$$

The natural period of the standing wave may then be computed if a shallow depth is assumed for the tank fluid. For a shallow depth d, the natural period is given by

$$T_n = \frac{L}{\sqrt{gd}} \qquad (9.18)$$

For example, for a rectangular tank of 100 ft (30.5 m) length and 10 ft (3.1 m) depth, the first natural period of sloshing is 11.14 s. The 2nd period of sloshing is half that, i.e., 5.57 s.

If the storage tank is in motion, such as liquid transportation tanks on road or on rail, then the force generated on the tank wall by this motion may be severe and should be included in a design. For a liquid transport barge, the sloshing in the contained liquid from the motion of the barge in waves introduces similar force on the liquid storage tank. The phenomenon of sloshing of liquids in partially filled cargo and ballast ship tanks has been known to cause severe damage to ship structures.

Here, a case study of a large offshore oil storage tank is considered. The cylindrical tank was designed as a gravity structure placed on the ocean floor (Fig. 9.13) to store crude oil from a nearby offshore oil production platform. For such systems, the energy is transferred to the contained liquid, either through the wall of the tank from the waves or through the foundation due to the motion of the foundation from waves or earthquake. The foundation acts as a stiff spring for the tank. Thus, the coupled tank-foundation system may be treated as a damped spring-mass system.

Assuming that the tank is fairly elastic, the dynamic interaction between the motion of the liquid and the foundation and the elastic deformation of the tank wall is quite important. The associated standing waves in the free-surface liquid may pose a structural problem.

A simplified analysis and model test of the oil storage tank was conducted to investigate the sloshing waves due to translational and rotational motion of the tank. The purpose of this exercise was to determine the sloshing of free-surface liquid in the elastic container, and evaluate if the generated standing waves of the confined liquid were a threat to the structural design of the vessel.

Fig. 9.13: Cylindrical Gravity Storage Tank Model in Waves

9.5.1 Simplified Sloshing Theory

The sloshing theory was based on an elastic shell structure seated on an elastic foundation, which was simulated as a linear spring. The problem was separated into two parts. The first part considered only the deflection of the fixed elastic shell. The second part included solely the motion of the rigid shell on the elastic foundation. The two solutions were then combined on the assumption of small amplitude wave theory. It was recognized that the foundation as well as the shell wall would introduce damping in the system. Therefore, a damping factor was introduced to account for the total damping present in the system.

First consider the theory of lateral sloshing due to elastic walls. Assume that the wall is elastic, the interaction of the tank wall and standing waves is negligible for small waves, and flow is irrotational. Since the tank has a cylindrical shape, it is convenient to introduce cylindrical polar coordinates, r, θ and y at the bottom center of the cylinder. Therefore, the velocity potential, $\Phi = \Phi(r, \theta, y, t)$, is written as the Laplace equation (Section 2.2.4):

$$\frac{\partial^2 \Phi}{\partial r^2} + \frac{1}{r}\frac{\partial \Phi}{\partial r} + \frac{1}{r^2}\frac{\partial^2 \Phi}{\partial \theta^2} + \frac{\partial^2 \Phi}{\partial y^2} = 0 \qquad (9.19)$$

The boundary conditions at the tank floor and the free surface (see Eq. 2.42) of the standing wave are

$$\frac{\partial \Phi}{\partial y} = 0, \qquad \text{at } y = 0 \qquad (9.20)$$

$$\frac{\partial^2 \Phi}{\partial r^2} + g\frac{\partial \Phi}{\partial y} = 0, \qquad \text{at } y = h \qquad (9.21)$$

where h is the fluid height inside the tank. The boundary condition at the tank wall (the velocity of the water particle at the wall set equal to the wall velocity at that point) is:

$$\frac{\partial \Phi(y,\theta,t)}{\partial r} = \dot{\varepsilon}(y,\theta,t), \qquad \text{at } r = a \qquad (9.22)$$

where a is the tank radius, $\varepsilon(y,\theta,t)$ is the displacement of the wall which is a function of time t, elevation y, and angular position θ. The shell deflection arises from the external pressure acting on the shell wall resulting from the incident wave. The relationship between this external wave and the shell deflection is complex and cannot be expressed in simple mathematical terms. However, that is exactly what is needed at this point in order to proceed further. For this study, assume that the deflection and outside pressure at a point on the shell are related linearly with a constant of proportionality:

$$\varepsilon(y,\theta,t) = \alpha p(y,\theta,t), \qquad \text{at } r = a \qquad (9.23)$$

where the constant of proportionality α is known at all points on $r = a$. In practice, α is not a constant with an external pressure gradient, even though

Chapter 9 Practical and Design Case Studies

the circumferential deflection of the tank wall at a given elevation conforms closely to the circumferential pressure distribution. Therefore, Eq. 9.23 is considered a simplification. Applying linear wave theory, the pressure due to diffraction of waves from the outside wall of the cylindrical surface is obtained from the MacCamy and Fuchs theory [see Chakrabarti (1990)]:

$$p(y,\theta,t) = \frac{\rho g H}{\pi k a} \frac{\cosh ky}{\cosh kd} \sum_{n=0}^{\infty} \gamma_n X_n \cos n\theta \sin(\omega t - \delta_n - n\pi/2) \qquad (9.24)$$

where $\gamma_0 = 1$ and $\gamma_n = 2$ for $n > 0$, ρ = mass density of water, g = acceleration due to gravity, k = wave number, d = water depth, H = incident wave height, ω = circular wave frequency, δ_n = phase angle of the nth component given by

$$\delta_n = \tan^{-1}\left[\frac{Y'_n(ka)}{J'_n(ka)}\right] \qquad (9.25)$$

and the amplitudes are

$$X_n(ka) = \frac{1}{\{[J'_n(ka)]^2 + [Y'_n(ka)]^2\}^{1/2}} \qquad (9.26)$$

The quantities J_n and Y_n are the Bessel functions of the first and second kind of the nth order and the prime denotes derivative with respect to their arguments. The above pressure (Eq. 9.24) is valid for a rigid cylinder. Assuming a small deflection of the tank wall, the external pressure distribution is considered applicable to the present case.

The solution of the differential equation (Eq. 9.19) may be obtained by the separation of variable technique. Including all the boundary conditions except the body surface condition, the solution for the velocity potential is written as

$$\Phi_n(y,r,\theta,t) = [A_n J_n(kr) + B_n Y_n(kr)]\frac{\cosh ky}{\cosh kd}\cos n\theta \cos(\omega t - \delta_n - n\pi/2) \qquad (9.27)$$

The relationship between the frequency and wave number is obtained as the dispersion relation (Eq. 2.45). Noting that Y_n approaches infinity at $r = 0$ (which would yield $B_n = 0$), and applying the boundary condition at the wall (Eq. 9.22), the expression for A_n becomes:

$$A_n = \frac{\alpha\omega}{kJ'_n(ka)} \frac{\rho g H}{\pi k a} \gamma_n X_n \tag{9.28}$$

Then the potential function describing the standing wave inside the tank becomes:

$$\Phi_n(y,r,\theta,t) = \frac{\rho g H}{\pi k a} \frac{\alpha\omega}{k} \frac{J_n(kr)}{J'_n(ka)} \frac{\cosh ky}{\cosh kd} \gamma_n X_n \cos n\theta \cos(\omega t - \delta_n - n\pi/2) \tag{9.29}$$

The water surface elevation inside the tank due to the lateral sloshing is obtained from the first term of the Bernoulli equation (Section 2.3):

$$\eta(r,\theta,t) = \frac{\rho g H}{\pi k a} \frac{\alpha\omega^2}{k} \sum_n \frac{J_n(kr)}{J'_n(ka)} \gamma_n X_n \cos n\theta \cos(\omega t - \delta_n - n\pi/2) \tag{9.30}$$

As an example, the wave amplitude inside the cylinder at a point given by $r = a$ and $\theta = 180$ deg (i.e., the leading edge) is the absolute value of

$$\zeta_i = \frac{\rho g H}{\pi k a} \frac{\alpha\omega^2}{k} \sum_{n=0}^{\infty} \frac{J_n(ka)}{J'_n(ka)} \gamma_n X_n (-1)^n \tag{9.31}$$

The natural period of sloshing in this case is obtained from the relationship:

$$J'_n(ka) = 0 \tag{9.32}$$

which represents the zeros of the derivatives of the Bessel function of first kind of order n. Thus a series of natural periods may be computed for a given tank radius from the relationship

$$ka = j_n, \quad n = 0,1,2\ldots \tag{9.33}$$

where j_n are the zeros of the Bessel function of the first kind.

Next, consider the walls of the cylindrical tank to be rigid while considering the translational and rotational motion of the tank foundation due to waves. In the lateral sloshing due to the translation, the tank base is allowed to move harmonically in a forced oscillation, assuming small amplitude theory. Then the boundary condition at the tank wall is written as

$$\frac{\partial \Phi}{\partial r} = i\omega x_0 \cos\theta \exp(i\omega t) \quad at\ r = a \tag{9.34}$$

Chapter 9 Practical and Design Case Studies 411

where x = tank displacement in the direction of wave, and the subscript zero indicates its amplitude. The total potential is divided into two parts: potential due to the motion of the container, ϕ_c, and the potential of the liquid moving relative to the container, ϕ_l. Thus

$$\Phi = (\phi_c + \phi_l)\exp(i\omega t) \tag{9.35}$$

The potential due to the movement of the container is computed by the integration of Eq. 9.34 with respect to r:

$$\phi_c = i\omega x_0 r \cos\theta \tag{9.36}$$

In this case the constant of integration may be combined with that of ϕ_l.

The boundary conditions for ϕ_l are similar to the external wave shown previously (Chapter 2), except for the height h of the internal liquid in the tank. The solution technique is similar by separation of variable. The series solution is written as

$$\phi_l = \sum_{m=0}^{\infty}\sum_{n=0}^{\infty} A_{mn} \frac{\cosh\xi_{mn}\bar{y}}{\cosh\xi_{mn}\bar{h}} J_m(\xi_{mn}\bar{r})\cos m\theta \tag{9.37}$$

in which the quantities with the upper bar are normalized by the tank radius, a. The arguments ξ_{mn} are derived from the free surface condition. The constants A_{mn} are 0 for $m \neq 1$, so that only the terms with $\cos\theta$ appear in the series expression and A_{mn} may be expressed simply by A_n.

The free surface boundary condition yields the expression for the natural period:

$$\omega_n^2 = \frac{g\xi_n}{a}\tanh\xi_n\bar{h} \tag{9.38}$$

Then the quantity A_n becomes:

$$A_n = \frac{i\omega x_0}{\left(\frac{\omega_n^2}{\omega^2}-1\right)}b_n \quad ; \quad b_n = \frac{2a}{\left(\xi_n^2-1\right)J_1(\xi_n)} \tag{9.39}$$

After substituting these values, the velocity potential becomes

$$\phi_1(\bar{y},\bar{r},\theta) = i\omega x_0 \left[\bar{r} + \sum_{n=0}^{\infty} \frac{2}{\left(\frac{\omega_n^2}{\omega^2}-1\right)(\xi_n^2-1)J_1(\xi_n)} \frac{\cosh\xi_n \bar{y}}{\cosh\xi_n \bar{h}} J_1(\xi_n \bar{r}) \right] \cos\theta \qquad (9.40)$$

and the free surface wave profile inside the tank at $y = h$ is obtained as

$$\eta(\bar{r},\theta,t) = \frac{\omega^2 x_0}{g/a} \left[\bar{r} + 2\sum_{n=0}^{\infty} \frac{J_1(\xi_n \bar{r})}{\left(\frac{\omega_n^2}{\omega^2}-1\right)(\xi_n^2-1)J_1(\xi_n)} \right] \cos\theta \exp(i\omega t) \qquad (9.41)$$

The wave amplitude at $\theta = 0$ and $r = a$, including a linear damping factor ζ is given by

$$\delta = \frac{\omega^2 x_0}{g/a} \left[1 + 2\sum_{n=0}^{\infty} \frac{1}{\sqrt{\left(\frac{\omega_n^2}{\omega^2}-1\right)^2 + \left(2\zeta\frac{\omega_n}{\omega}\right)^2} (\xi_n^2-1)} \right] \qquad (9.42)$$

Now consider the lateral sloshing due to a pure rotation of the tank. Assume that the tank rotates about the transverse (z) axis in a harmonic oscillation:

$$\psi = \psi_0 \exp(i\omega t) \qquad (9.43)$$

where ψ = angle of rotation. The boundary condition at the tank wall has a form similar to that in translation:

$$\frac{\partial \phi}{\partial r} = i\omega y \psi_0 \exp(i\omega t)\cos\theta \qquad \text{at } r = a \qquad (9.44)$$

As in the case of translation, the total velocity potential is divided into two parts. The potential due to the movement of the container is obtained by integrating Eq. 9.44 with respect to r:

$$\phi_c = i\omega y \psi_0 r \cos\theta \qquad (9.45)$$

Chapter 9 Practical and Design Case Studies

in which the constant of integration is absorbed in ϕ_l.

The solution for the spatial part of ϕ_l is assumed to be of the form

$$\phi_l(y,r,\theta) = \sum_{m=0}^{\infty}\sum_{n=0}^{\infty} A_{mn}(y) J_m(\xi_{mn} \bar{r}) \cos m\theta \tag{9.46}$$

Upon substituting this expression in the Laplace equation and replacing the quantity A_{mn} using A_n by the same reasoning as in the previous case,

$$A_n''(y) - \frac{\xi_n^2}{a^2} A_n(y) = 0 \tag{9.47}$$

which is valid for all non-negative n. Upon applying the bottom boundary condition,

$$A_n'(0) = -i\omega \psi_0 b_n \tag{9.48}$$

Similarly, upon applying the free surface boundary condition:

$$g A_n'(h) - \omega^2 A_n(h) = (\omega^2 h - g) i\omega \psi_0 b_n \tag{9.49}$$

Considering the boundary value problem including the differential equation (Eq. 9.19) and the boundary conditions discussed above, the general solution is assumed to be of the form:

$$A_n(y) = B_n \cosh(\xi_n y/a) + C_n \sinh(\xi_n y/a) \tag{9.50}$$

Equation 9.48 gives

$$C_n = -i\omega \psi_0 \frac{ab_n}{\xi_n} \tag{9.51}$$

Writing the eigenfrequency as

$$\omega_n^2 = g_0 \frac{\xi_n}{a} \tanh(\frac{\xi_n h}{a}) \tag{9.52}$$

the expression for B_n becomes

$$B_n = \frac{i\omega\psi_0 b_n \left(-\frac{\omega^2 a}{\xi_n}\sinh(\xi_n h/a) + g\cosh(\xi_n h/a) + \omega^2 h - g\right)}{(\omega_n^2 - \omega^2)\cosh(\xi_n h/a)} \quad (9.53)$$

Noting that the expression for $A_n(y)$ is given by Eq. 9.50 and B_n and C_n are shown in Eqs. 9.51-52, the solution for the total potential, Φ, becomes:

$$\Phi = [A_n(y)J_1(\xi_n r/a) + i\omega y r \psi_0]\cos\theta \exp(i\omega t) \quad (9.54)$$

The wave profile at $y = h$ is obtained from the formula:

$$\eta = i\omega/g[A_n(h)J_1(\xi_n r/a) + i\omega h r \psi_0]\cos\theta \exp(i\omega t) \quad (9.55)$$

Substituting the value of $A_n(h)$ from Eq. 9.51, using the eigenfrequency relationship (Eq. 9.52) introducing a linear damping factor and normalizing by the radius a, the standing wave amplitude at a point given by $\theta = 0$ and $r = a$ reduces to the form:

$$\delta = \omega^2 a \psi_0 \left[\frac{h}{g} + 2\sum_n \frac{\left\{\frac{1}{\omega^2 \cosh(\xi_n h/a)} + \frac{h}{g} - \frac{1}{\omega^2}\right\}}{\sqrt{\left(\frac{\omega_n^2}{\omega^2} - 1\right)^2 + \left(2\beta\frac{\omega_n}{\omega}\right)^2}(\xi_n^2 - 1)} \right] \quad (9.56)$$

Combining the translation and rotational oscillation of the tank, the lateral sloshing is obtained by the linear superposition (with an arbitrary phase shift given by μ):

$$\delta = \sum_n \frac{2a\left[\left\{\frac{1}{\cosh(\xi_n h)} - 1 + \frac{\omega_n^2 h}{g}\right\}\psi_0 \exp(i\mu) + \omega_n^2 x_0/g\right]}{\sqrt{\left(\frac{\omega_n^2}{\omega^2} - 1\right)^2 + \left(2\beta\frac{\omega_n}{\omega}\right)^2}(\xi_n^2 - 1)} \quad (9.57)$$

This theoretical analysis enables the reader to evaluate sloshing in a tank due to waves.

9.5.2 Elastic Model Design

From geometric similitude, the ratio of the shell deflection between a model and the prototype linearly follows the scale factor. For an elastic structure, the Cauchy number must be satisfied. The modulus of elasticity for the model scales directly as λ (Chapter 4). Both axial and bending stiffness can not be modeled at the same time if the elastic modulus does not scale as λ. Bending of the shell to produce an oval shape appears to be the most likely source of radial shell deflection and associated standing waves.

A scale of 1:48 was selected. For a prototype steel structure and a scale factor of $\lambda = 48$, the model E should be 625,000 psi. This value of E is higher than most plastics and lower than common metals. After studying a tabulation of common plastic materials, their strengths, moduli of elasticity and available sizes and thickness ranges, Lexan unfilled polycarbonate thermoplastic was selected. This material exhibited the most isotropic properties with the narrowest range of published values. For this material, the modulus of elasticity in tension is 3.5×10^5 psi, and in bending, the flexural modulus is 3.4×10^5 psi. With this material and proper modeling of the bending stiffness, the scaling ratio between axial and bending was 1.49. In other words, the model shell was 1-1/2 times stiffer in axial tension than the actual scaled down Cauchy model.

Figure 9.14 shows a plot of the comparison of the simulated deflection of a radial shell at 0 deg. azimuth between the model and scaled down prototype when subjected to a regular wave. The shapes of the deflected curves are similar, but the model shell simulation does not show the circumferential growth that the prototype simulation does. Much of this error can be attributed to the distorted axial stiffness of the model design.

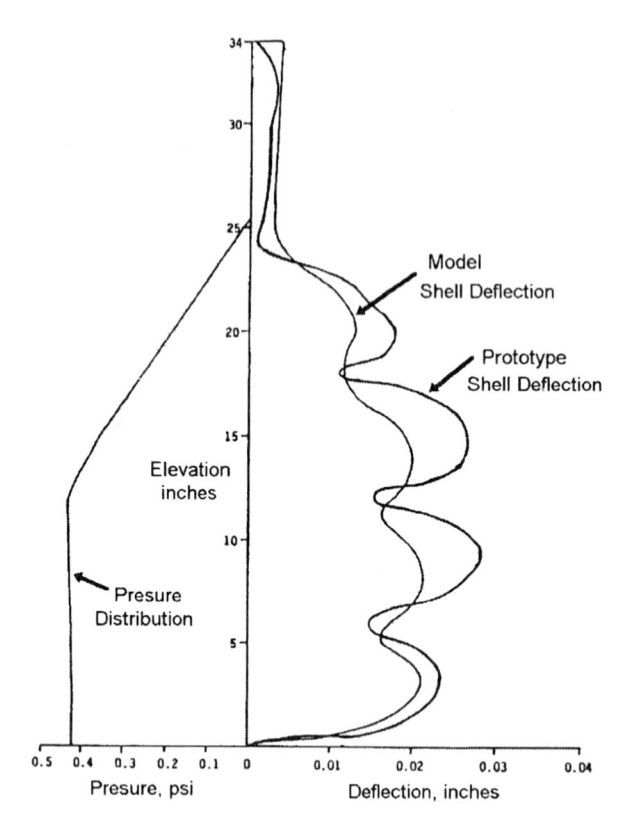

Fig. 9.14: Tank Shell Deflection Under External Pressures – Prototype vs. Model

9.5.3 Elastic Model Tests

The model (Fig.15) in the wave tank was supported by the floor of the tank and was subjected to regular incident waves. The first set of tests (Case 1) included a fixed base, which restricted the model's motion except for the deformation of its wall. No standing waves of significant amplitude were generated during any of the test runs for the elastic shell mounted on a rigid foundation. No runs generated a standing wave amplitude greater than 6.3 mm (0.25 in) and generally, it was less than 3.1 mm (0.12 in).

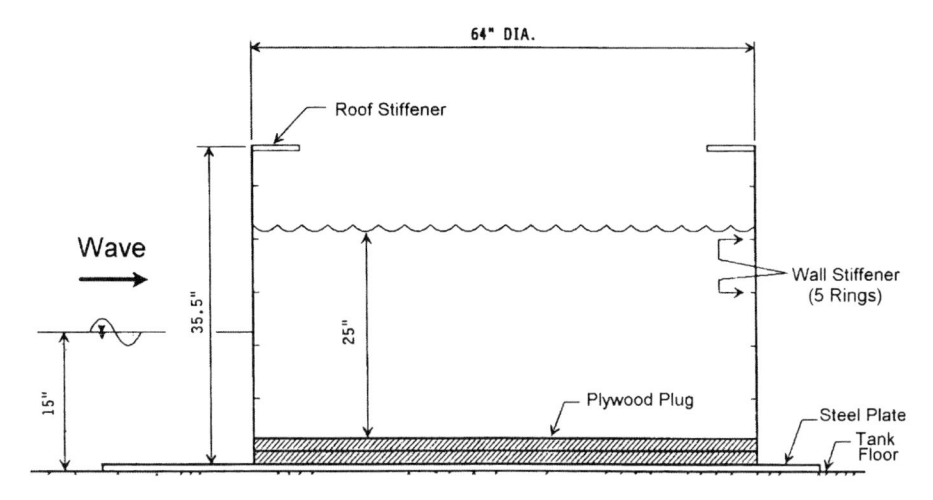

Chapter 9 Practical and Design Case Studies 417

Fig. 9.15: Fluid Filled Elastic Model Subjected to External Waves

Next, the elastic model was placed on different foundations. In the second series of tests (Case 2), the model was placed on horizontal (Fig. 9.16) and vertical (Fig. 9.17) springs. These springs were subsequently replaced with softer springs in both directions (Case 3). In the final phase, the horizontal springs were removed, leaving only the vertical springs (Case 4). In this case, the horizontal movement of the model was unrestrained.

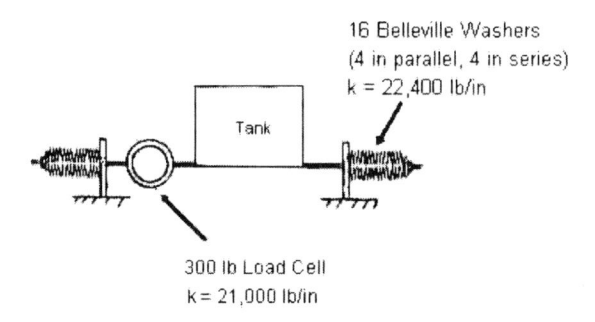

Fig. 9.16: Simulated Horizontal Soil Spring

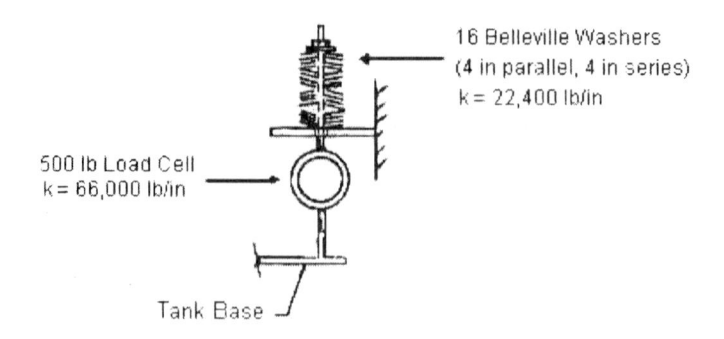

Fig. 9.17: Simulated Vertical Soil Spring

For the prototype soil foundation at the selected site for the structure, the equivalent linear spring stiffness in the vertical, horizontal and rotational directions were computed as follows:

Vertical soil stiffness   9.916E09 N/m (39,500 kips/in)
Horizontal soil stiffness   2.172E10 N/m (86,500 kips/in)
Rocking soil stiffness   2.905E14 N-m /rad (1.66 x 10^{11} kips-in/rad)

The base of the model was suspended from stacked Belleville washers, representing the foundation in the vertical direction, and was attached to horizontal springs fore and aft. Combination of load cells and stacked Belleville washers provided the scaled stiffness in the model setup (Figs. 9.16 and 9.17). The softer springs were achieved by removing Belleville washers in the vertical and horizontal springs.

In order to compare the simplified theory with the measured values, the constant of proportionality (α in Eq. 9.23) between shell deflection and external pressures was determined by computer simulation. The horizontal displacement of the tank was measured at 0 deg in the tests. The rotational displacement was not directly measured, but was computed from the measured overturning moment and spring stiffness. Two different cases were chosen for comparison with the approximate theory:

    Case 1   Incident Wave: $T = 1.40$ s, $H = 0.25$ ft, $\theta = 0$ deg
                  Average deflection = 0.00289 in (along vertical panels 2-4)
                  Average pressure = 0.0221 in (along vertical panels 2-4)

Chapter 9 Practical and Design Case Studies

        Average $\alpha = 0.000075$

Case 2   Incident Wave: $T = 1.70$ s, $H = 0.29$ ft, $\theta = 0$ deg
            Average deflection = 0.00124 in (along vertical panels 2-4)
            Average pressure = 0.00256 in (along vertical panels 2-4)
            Average $\alpha = 0.00028$

The average values of α were used to compute the sloshing waves in the tank. A typical damping factor for a smooth circular cylinder was chosen as 5 per cent [given by Abramson (1966)]. Four response curves for the two α values (undamped and damped for each) were generated. The standing wave amplitude at $\theta = 0$ deg was normalized with the incident wave amplitude and is shown in Fig. 9.18. Most of the data points fell within the range of the two theoretical curves. The scatter in the test data was due to the nonlinear nature of the incident waves in shallow water. The α value corresponding to $T = 1.40$ s (which was the natural period of the standing waves) appeared to be the more appropriate value. This example included only the elastic deformation of the wall.

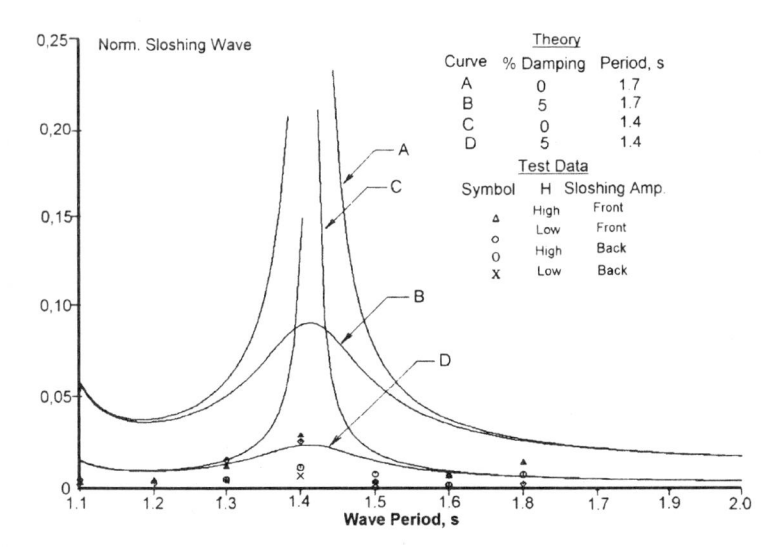

Fig. 9.18: Comparison of Sloshing Waves for Different α and ζ Values

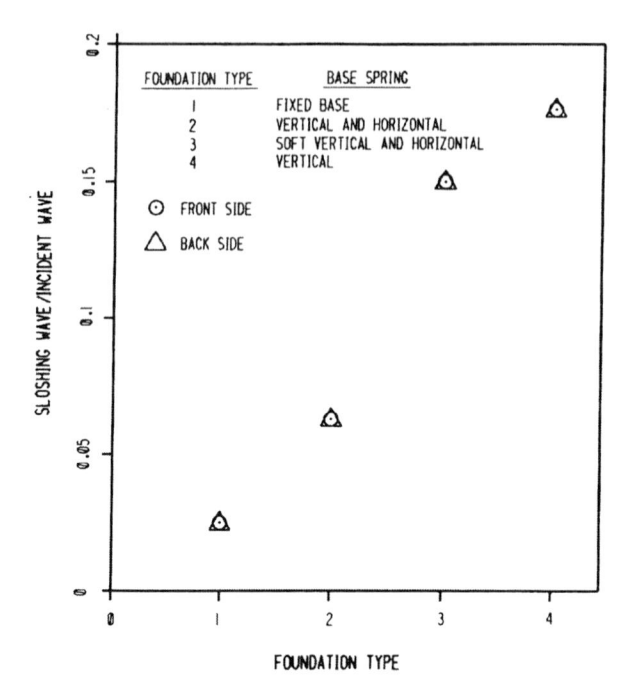

Fig. 9.19: Normalized Sloshing Wave Amplitudes for Various Foundations

In comparing the results, larger standing waves were observed when the tank model was supported on foundation springs than when it was mounted on a rigid foundation. As the foundation stiffness decreased from Case 1 through Case 4, the sloshing wave amplitudes were found to increase. This is illustrated in Fig. 9.19 for the fundamental natural sloshing period of 1.40 s. It is clear that the higher wave amplitudes observed as the foundation was softened were in response to the rigid body movement of the tank at its foundation to a much greater extent than they were due to the flexing of the tank shell.

The normalized data for foundation type 2 and 3 are compared with the corresponding computed values in Fig. 9.20. The theoretical curves were generated from the deflection data at 1.40 and 1.70 s and damping factors of 0 and 0.05. The figure clearly shows that the natural period is near 1.40 s. The nonlinearity of the data is also clear. The α value at 1.70 s provides a

Chapter 9 Practical and Design Case Studies   421

higher response. The approximate theory appears to provide reasonable results in describing the standing wave generated in the tank.

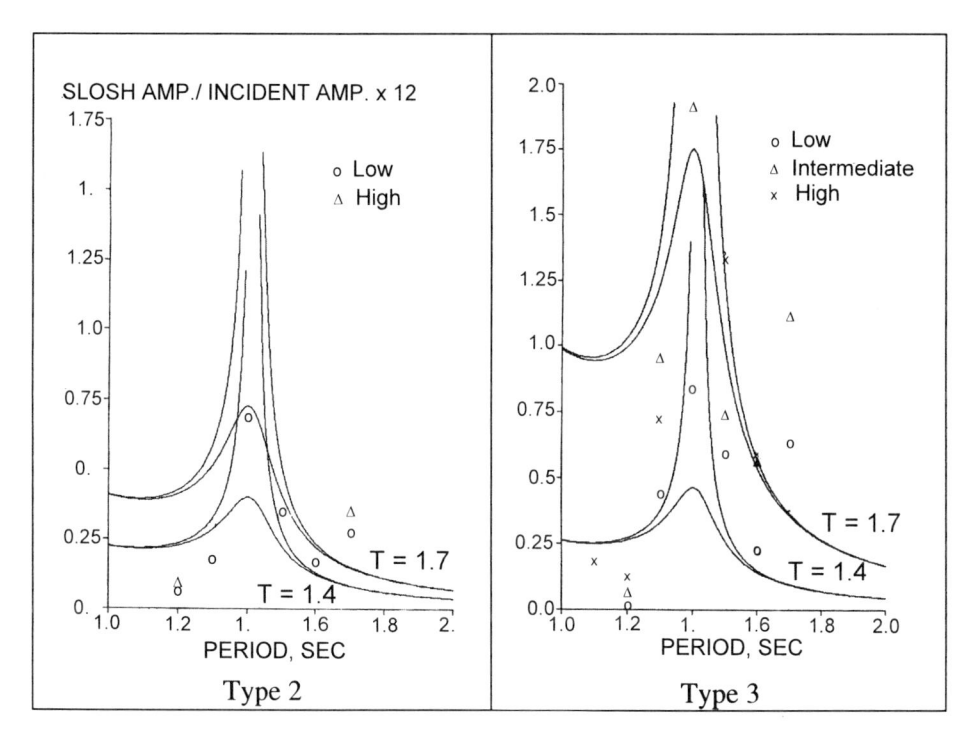

Fig. 9.20: Comparison of Measured Sloshing Wave Amplitude with Theory

This case study illustrates the difficulty of modeling large elastic structures having both axial and bending stiffness. The model test showed that the standing waves in cylindrical tanks can be significant near sloshing period, where damping is usually small. While the example shows that the elasticity was not an important cause of sloshing compared to the foundation movement, it is not suggested that such scaling is not necessary. In such cases, the scaling should be explored carefully during planning a structural model test of this type.

9.6 Articulated Tower Instability in Waves

Small moving structures subjected to fluid loading and having relatively small damping values may experience global aperiodic behavior. This is particularly true when the fluid introduces vortex shedding past the structure and the structure is free in multiple degrees of freedom. This case study considers an illustrative example of such a system where such aperiodic behavior was observed.

One of the common means of off-loading crude oil offshore is the placement of floating structures that can be used to moor large shuttle tankers. These structures are attached to the ocean floor and are designed to withstand the environment experienced at the site. Several such off-loading towers have been installed in different parts of the world.

The example study considers a simple innovative articulated tower, which was considered as a possible means for transferring crude oil from an offshore field. A single leg inclined mooring (SLIM) tower was conceptually developed to moor tankers with a single hawser in shallow water. The idea was that the initial inclination of the tower would reduce the dynamic loads on the hawser and the bottom joint of the tower. However, it was discovered that the dynamic characteristics of the tower under certain wave actions were such that near-chaotic motion was experienced. The study includes experimental results that discovered the erratic behavior of the tower, and a theory to explain the observed behavior of the tower in waves. Based on this study, the concept was abandoned from further development.

9.6.1 Theory of Inclined Tower

The inertia of an articulated tower in oscillation is counteracted by the restoring moment of the tower. For an upright articulated tower, most of the restoring moment arises from the buoyancy tank near the surface. This restoring moment is usually linear, and the nonlinear contribution is generally small. Therefore, such towers are stable in waves [Chakrabarti and Cotter (1979, 1980)]. However, when the tower is inclined in its equilibrium position, the restoring moment becomes highly nonlinear due to the buoyancy increase that results from the displacement.

Chapter 9 Practical and Design Case Studies

The SLIM tower naturally assumed an inclined position under equilibrium (Fig. 9.21) and thus exhibited a circle of equilibrium. Inside the circle, the tower was unstable. The equation of motion applicable to a statically stable upright articulated tower is generally not applicable for SLIM. Considering that the initial SLIM inclination was large and that SLIM oscillated about this inclined position, it would produce nonlinear terms in the restoring moment that would not be negligible.

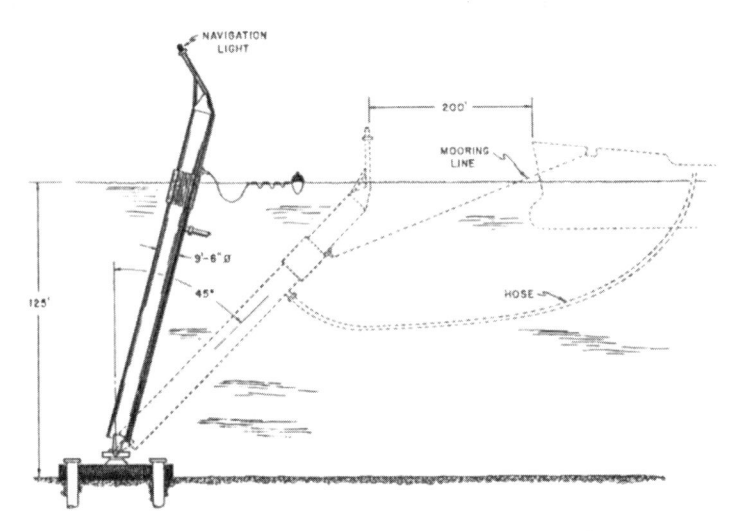

Fig. 9.21: SLIM Tower in Equilibrium in Still Water

The hinge-swivel type pivoted connection for the SLIM design was a two degree of freedom joint with one axis of rotation being vertical, and the other horizontal and normal to the axis of the tower. The tower was thus restrained from rotating about a second horizontal axis lying in the plane of inclination.

The motion of the top of the tower consisted of oscillation at the frequency of the incident wave, and precession at the natural frequency of the tower and at various harmonics of the incident wave. These additional harmonics in the transverse direction were created by the lift force on the tower transverse to the wave direction. Note that the tower had no restoring force in precession near its equilibrium position and its natural period was long. Therefore, the tower would have a precession angle slowly moving transversely along with the wave-induced inline oscillation.

In order to include various nonlinearities in their exact form, a time domain solution was sought. To describe the equation of motion of the tower, Newton's second law was applied (see Chapter 3 for details) about the tower bottom joint, in which the left-hand side represented the inertia terms with the structure mass and moment of inertia in the respective directions. The right-hand side represented all of the resultant forces and moments on the vessels and the tower in the respective coordinate directions. The internal loads were the buoyancy and hydrostatic loads, while the external loads consisted of the wind, waves and current loads, and hydrodynamic added mass, damping and nonlinear drag loads.

9.6.2 Model Test of Inclined Tower

A wave tank test of a model of SLIM was performed to demonstrate the instability problem. A model of a SLIM tower designed for a water depth of 130 ft (39.6 m) was chosen for the test. The tower was a buoyant circular cylinder having an outside diameter of 9.6 ft (2.93 m), with a boat bumper at the water surface. The tower had a no-load tilt angle of about 20 degrees, and rested on a 360-deg swivel joint. A scale factor of 1:33 was chosen for this test. At this scale, the diameter of the tower was 3.5 inches (89 mm), while the water depth was 47.25 inches (1200 mm).

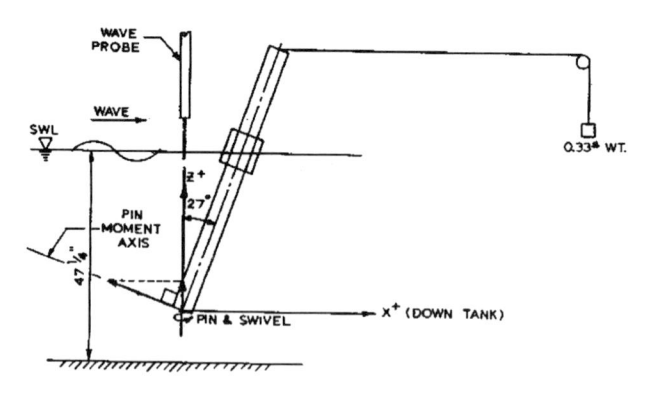

Fig. 9.22: Test Setup for SLIM

The tower was subjected to steady loads at 0 degrees, i.e., inline with the waves and at 90 degrees, i.e., normal to the waves. The steady current loads were simulated by using weights suspended over a pulley. The test setup for the tower alone at 0 deg. is shown in Fig. 9.22. The angular rotation of the tower as well as the load at the joint was measured.

The dynamics of an articulated tower in waves depend on the added mass and damping characteristics of the tower. If the frequency of the exciting force is such that these towers experience resonance, then the damping coefficients become very important in the design of such towers. This is particularly true for a tower that is inclined from vertical in its equilibrium position. Extinction tests of the SLIM model provided the results for the added mass and damping coefficients at various angles to the vertical in water. These values of the hydrodynamic coefficients were used in the time domain solution.

9.6.3 Analysis of SLIM Response

Comparison of the amplitudes of measured inclination of the tower in waves and current was first made with a simplified analytical frequency domain method (modified Eq. 3.76 including relative velocity drag damping). Figure 9.23 shows the measured inline response amplitudes (normalized with respect to wave amplitudes) of the tower and the corresponding analytical results. The test results are marked with open symbols for different wave heights. At lower wave periods, nonlinear effects are expected to be small. Therefore, the experimental results agree fairly well with the linearized frequency domain results for the tower inclination. At higher wave periods, there is a large spread in the experimental data at a given frequency, and the frequency domain solution is not reliable. A time domain solution was sought for this system based on the method described in Chapter 6. In this case, several nonlinear terms arising from the higher waves were included in the analysis. In particular, large angle oscillation, nonlinear relative-velocity drag, convective inertia, nonlinear restoring force, and nonlinear lift force were included. The tower was assumed to have two degrees of freedom. The amplitudes of oscillation from the time domain solution due to regular waves are shown as solid symbols corresponding to the open symbol from the tests in Fig. 9.23. The trend in the experimentally measured responses with higher waves is clearly demonstrated by the time domain solution. The response is found to be a strong function of the wave amplitudes at a given wave period, which is a clear indication of strong nonlinearity.

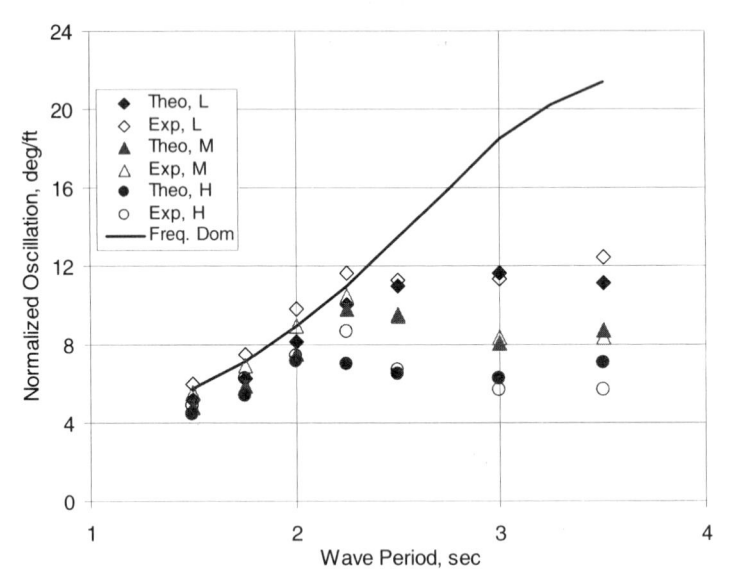

Fig. 9.23: Response of SLIM in Regular Waves

Unlike a vertical tower, an inclined (SLIM) tower with a pin and swivel connection had no lateral stiffness due to buoyancy. Consequently, in the absence of an external restoring force, the tower would swing out of the plane of the wave, under the action of steady transverse current as well as forces due to flow separation and vortex shedding. This behavior was seen clearly in those tests where the steady inline forces, serving to fix the tower's heading in the inline direction, were not sufficiently large. In fact, for the tower with no applied steady inline load, the movement of the tower transverse to the wave direction was pronounced. As the wave amplitude increased, the steady offset tended to be obscured by large transverse motions at the natural frequency of the tower in precession. In addition, higher harmonics were evident in the transverse motion during larger wave

Chapter 9 Practical and Design Case Studies   427

amplitudes. Figure 9.24a represents near chaotic displacement of the top of the tower during a thirty-second interval for a test with a wave period of 2.75 s at a large wave amplitude. Figure 9.24b represents the numerical displacement as calculated by the time domain analysis for the same environmental conditions. The experimental results included a slowly varying transverse motion at the natural frequency in precession. The nonlinear damping term was included in the numerical analysis to restrict this response. The numerical and experimental results in Fig. 9.24 exhibit quite similar behavior in magnitude and form.

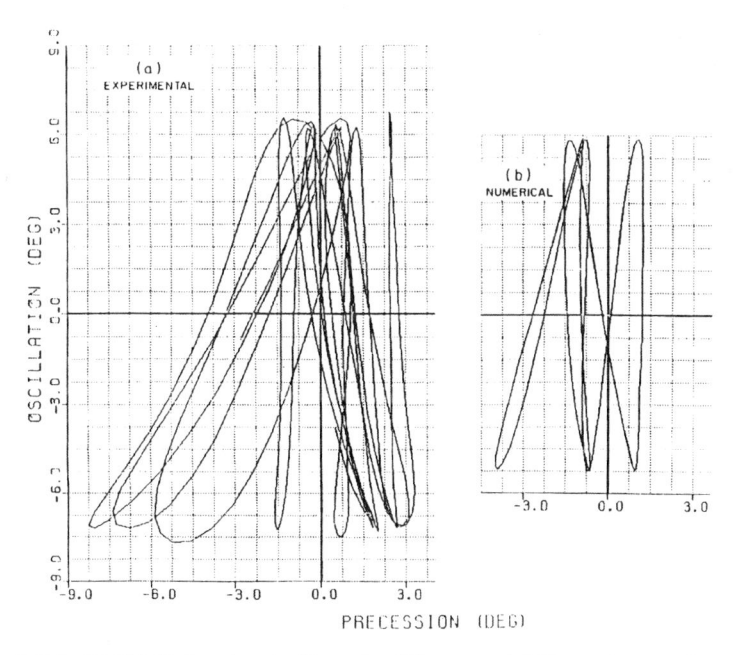

Fig. 9.24: SLIM Behavior in Regular Waves – (a) Experimental vs. (b) Numerical

This example demonstrated that the SLIM design was vulnerable to large transverse motion in high waves. Since the tower bottom connection was a hinged joint, the rotation of the tower about the wave direction had the potential of very large loads on the hinge at the bottom. Since the tower behaved like a fixed tower in this orientation after 90-deg rotation, this large load made the design of the bottom joint highly expensive, and removed the perceived advantage of the concept.

9.7 Hydroelasticity of a Multi-Moduled Structure

Structural deflections under fluid loading are always present to some degree in any structure. However, the structural deflections of many long floating ship-shaped structures are insignificant from the hydrodynamic perspective, when compared to the first-order displacements, regardless of hull form. In these cases, the fluid acts strongly on the structure, and the structure reacts only weakly on the fluid. Hence, the state of deformation of the structure does not affect the applied pressure or force field to any significant extent. This does not preclude the possibility of dynamic amplification or rigid body resonant response of the structure. If the rigid body natural periods of the structure approach those of the excitation periods, resonant response can develop. Also, the nature of the excitation may cause elastic dynamic overshoot of the structure, resulting in a dynamic amplification of the response. This area has already been covered in Chapters 3 and 7.

From the structural load perspective, even small deflections and curvatures may equate to large internal loads. However, even if the vessel exhibits small bending deflections, or local deformation due to wave slamming, the vessel is defined as completely rigid with six degrees of freedom, for hydrodynamic and "global" dynamic purposes. If estimates of "local" responses, such as internal wave-induced loads are desired, then that analysis is usually done independently of the global response analysis in a subsequent structural analysis program.

As the structure becomes long horizontally, the flexibility of the structure in relation to the rigid body motion may become significant. One important consideration in the design of very large floating structures is their elastic response. When such structures are conceptually designed, care must be taken in the rigidity and mass distribution of the structure so that their desirable hydroelastic response characteristics are achieved. There are several such structures in the development stage. An example of this is the floating airport being designed for placement offshore Japan, and a smaller prototype already exists. The analysis of such a structure, called Mobile Offshore Base, will be described here.

9.7.1 Mobile Offshore Base

The Mobile Offshore Base (MOB) is a very long floating structure that may be placed in shallow to deep water in any part of the ocean. The size of these large floating structures range up to a mile (1.6 km) in length and hundreds of feet in width. Typical dimensions of the MOB are: 5000 ft (1524 m) long, 400 ft (122 m) wide and 100 ft (30.5 m) deep. The mission of the MOB concept was to serve as an alternative to existing land bases for military operations such as, delivery of critical tactical equipment, personnel and amenities. The MOB concepts, that were analyzed, consisted of a modular design in which each module had a flat deck connected to submerged buoyant members by some articulation and/or flexible means. The individual modules were connected to each other by springs and mechanical connectors. The concept allowed the subsurface buoys to move in response to the environment and to conform to the shape of the surface waves while the deck on top remained relatively motionless. In this manner, a stable platform was provided, allowing military operations to be conducted. The mere size of a MOB makes it a unique structure and nothing exists or operates today that comes close to it in its size, uniqueness and complexity.

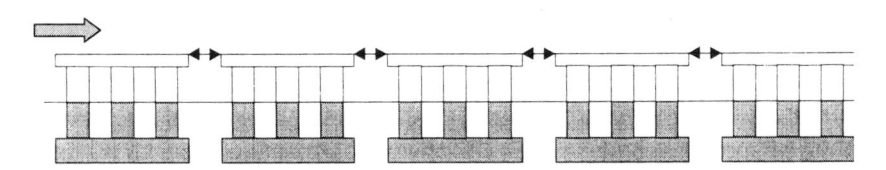

Fig. 9.25: Mobile Offshore Base Concept

The MOB structure is characterized by the elastic response of multiple-connected modules (Fig. 9.25) that differentiates it from single conventional rigid floating structures. Whether or not a prototype MOB will ever become viable or not remains to be seen. However, it is an important case study to consider the hydroelastic vibrational behavior of long structures subjected to fluid forces and the associated response.

9.7.2 Analysis of a MOB

Substantial work has been carried out in the last couple of decades concerning the development of hydroelastic methods of analyses. The most capable of these analyses include three-dimensional models of a flexible structure and an associated three dimensional model of the surrounding fluid. The MOB structure being modular, a simpler approach was taken here. The structure was divided into its individual modules, which were treated as rigid and were connected to their adjacent modules by elastic connectors with appropriate bending rigidity (Fig. 9.25). The hydrodynamic and elastic responses of the structure are simultaneously included in the analysis.

An improved efficient method based on the linear diffraction theory (Chapter 7) was applied to the MOB on the assumption that the individual modules are relatively rigid and the flexing (such as, in bending) occurs at the connectors. The analysis developed a load generator interface to efficiently transfer accurate hydrodynamic (surface pressure) forces from an arbitrary hull discretization into an arbitrary finite element structural discretization for structural analysis. The conventional diffraction theory, while applicable for hydrodynamic pressure, would require prohibitively excessive memory and execution time, and would have limited accuracy such that the results would be unreliable.

The method made use of the diffraction analysis coupled with a multiple scattering technique. In this method, the diffraction/ radiation coefficients were computed by the diffraction theory on each isolated module in the absence of the other modules in the structure. When multiple modules within the structure had the same geometry, this method had an additional distinct advantage of using the diffraction analysis only once for the identical modules. The interaction effect among modules was accounted for in the next step during the application of multiple scattering. Combining the features of the multiple scattering technique and the direct matrix method using the linear diffraction theory, an "exact" linear hydrodynamic interaction theory [see Chakrabarti (2000) for details] was developed for a group of modules of arbitrary three-dimensional geometry and locations.

The analytical method of solution may be explained in the following way. Assume N modules subjected to an incident wave. The ambient

Chapter 9 Practical and Design Case Studies 431

waves incident upon a particular module will be diffracted from this module. These waves will then be re-reflected from the other modules in the array (Fig. 9.26). The multiple interaction method is based on isolating a selected module within the array. This module will experience the ambient incident wave as well as all the scattered waves from the other modules in the array. These waves are described by their velocity potentials at a field point in the neighborhood of the selected module. By using the plane wave description, the total potential at the field point may be expressed due to the ambient incident wave, and the scattered and radiated waves from all the other modules. These latter waves are not known beforehand and are, therefore, given in terms of the unknown scattered coefficients.

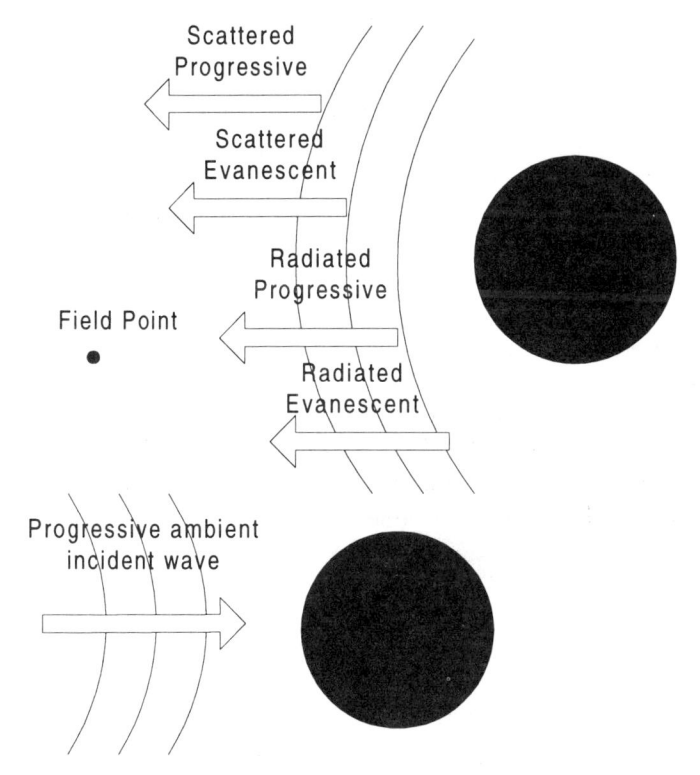

Fig. 9.26: Schematics of Numerical Method

This total potential may be treated as the incident wave on the selected module. If the diffraction coefficients for the isolated selected module are known (which may be computed independently from the 3-D diffraction theory), then the above potential may be related to the scattered potential of the selected module, including these diffraction coefficients. This gives rise to a set of linear equations in a matrix form in terms of the unknown coefficients. The inversion of the matrix provides the solution for the linear set of equations. Once these coefficients are known, then the forces on each module may be computed. The multiple interaction is included by selectively choosing each module of the structure in developing the complete set of equations. Since the formulation selects each module in an array, multiple scattering is ensured. The motions of the floating modules may subsequently be determined by simultaneously solving a set of algebraic equations describing the 6-DOF motions of each module (similar to Eq. 7.15).

Thus, this method is amenable to conveniently separating the two problems. Since the diffraction coefficients are determined first based on the individual modules, the execution time as well as the memory requirement for this computation is substantially reduced. Moreover, once the coefficients are known for a given module, they need not be computed again. The second part, namely the multiple scattering problem, can then produce results for various configurations, spacings, and mooring/ connecting systems. Therefore, it is conceivable to compute the responses of a structure composed literally of 100 identical modules in a reasonable execution time.

The equations of motion are written in the usual way with the forcing function on the right-hand side. The mass and stiffness terms of each module are included in the equations of motion. In addition, the added mass and damping coefficients corresponding to each isolated module enter in the equation for the unknown amplitudes.

$$\sum_{l=1}^{6} \left[-\omega^2 \left(m_{kl}^j + M_{kl}^{ji} \right) - i\omega N_{kl}^{ji} + C_{kl}^{ji} \right] \eta_{li} = f_k^j ; k = 1,2,..6 \quad (9.58)$$

in which the subscripts i and j refer to N modules and η_{li} (l=1,2,...6) represents the complex amplitude of motion of module i in the l-th direction.

The quantity m_k^j is the mass or moment of inertia of module j in the k-th mode, M_{lk}^{ji}, N_{lk}^{ji}, and C_{lk}^{ji}, respectively are the 6x6 added mass, damping and restoring force coefficients for the module j acting in the k-th direction. The matrix C includes hydrostatic as well as the external stiffness terms. The added mass and damping coefficients as well as the exciting forces are obtained on the isolated module j. The multiple interaction effect on the added mass and damping (i.e., the effect due to motions of the adjacent modules) are included in the coefficient matrix.

The right hand side of Eq. 9.58 includes the exciting forces on the isolated module. The external force on the module j in the k-th degree of freedom is obtained from the pressure integration over the submerged surface of the module (see Eq. 7.8). It is the sum of three distinct components that consider the multiple interactions among modules. The first term in the exciting force is the incident partial wave components, which is multiplied by the total isolated body force for each partial wave component. The second term arises from the interaction of diffraction terms from other modules, and the third term is the interaction term due to motions of the other modules. The radiation forces are equivalent to the hydrodynamic coefficients included in Eq. 9.58.

The multiple module interaction effect is demonstrated with a couple of examples. First, the effect of a small module in the neighborhood of a large one is investigated with a supply vessel next to a TLP. The TLP is a 4-column TLP of overall dimension 305 ft x 305 ft x 115 ft (93m x 93m x 35m). The small vessel is representative of a typical supply vessel of rectangular section with dimensions 50 ft x 120 ft x 25 ft (15.2m x 36.6m x 7.6m) placed beam to the waves. The center to center distance between the two modules is chosen as 210 ft (64m). The multiple interaction results are compared with the results for the isolated modules in Fig. 9.27. As one would expect, the sway motion of the TLP with and without the presence of the supply vessel shows little difference. However, the supply vessel being much smaller in size, compared to the TLP, shows the influence of the TLP in its sway motion. The natural period of the free TLP in heave is 20 s, which is why the coupled motion of the supply vessel near 20 s is high.

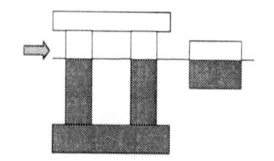

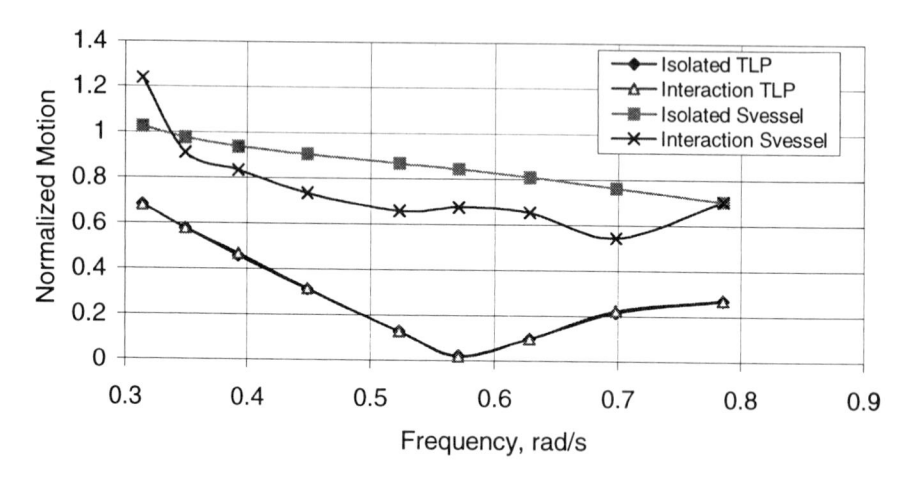

Fig. 9.27: Sway Motion of Supply Vessel at the Lee of a TLP

A second example is a MOB concept with five semisubmersible type modules (Fig. 9.25). The overall length of the MOB was 2750 ft (838.4m). The modules were connected in tandem, attached with linear springs only in the longitudinal direction (surge). The major dimensions for the MOB in this example are as follows:

- module overall dimension = 510 ft x 400 ft x 60 ft draft (155.5m x 122m x 18.3m)
- pontoons = 510ft x 150 ft 30 x ft deep (155.5m x 45.7m x 9.1m)
- columns = 90 ft dia x 30 ft long (27.4m x 9.1m)
- center spacing = 560 ft or 170.7m (Gap = 50 ft or 15.2m)
- spring constant = 4.0E6 lbs./ft or 5.84E3 kN/m

Chapter 9 Practical and Design Case Studies 435

In the first example, the modules are considered to be freely floating longitudinally with the specified gap in head seas. The surge and heave motions of the center module in the five-module arrangement are shown in Fig. 9.28 as functions of wave frequency. These coupled motions are compared with the corresponding motion of a single module to illustrate the effect of interaction. The interaction effect is quite evident in the results.

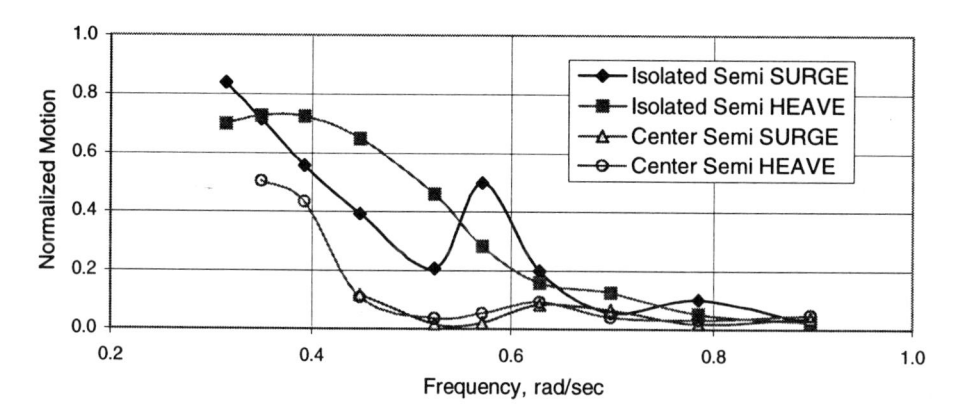

Fig. 9.28: Motions of Isolated and Center Module for a MOB in Head Sea

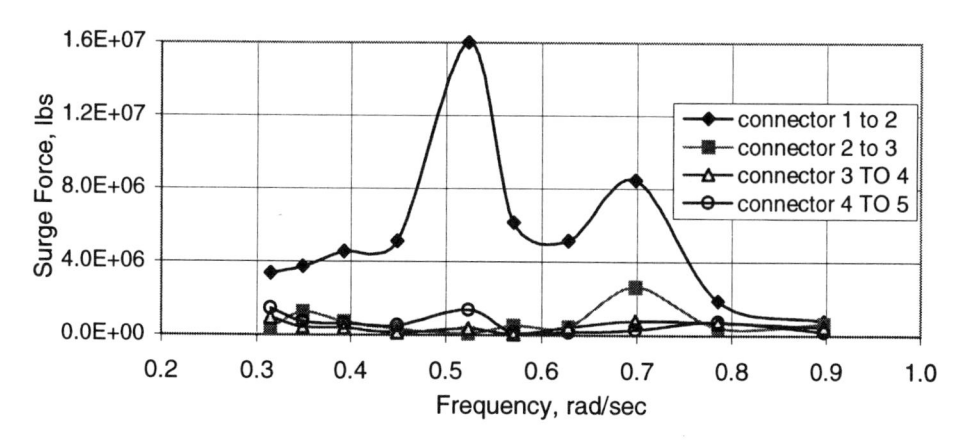

Fig. 9.29: Connector Surge Loads in a Five Module MOB in Head Sea

In the next step, the modules are connected in tandem by identical linear springs of magnitudes given above. The modules are connected off center so that a pitch moment is simultaneously applied. No rotational springs are used here. The connector loads in surge for the springs between the module pairs are shown in Fig. 9.29. Note that the loads are highest for the connector between modules 1 and 2, and the symmetry in the connector loads does not exist due to the interaction.

9.8 Dynamics of a Flexible Jack-Up Unit

A structural dynamic analysis and design method for structures subjected to oscillatory fluid forces, such as those experienced from wave action, is illustrated with a jack-up drilling rig. A jack-up rig generally consists of a platform supported on three vertical legs made of framework (Fig. 9.30). The platform is allowed to move on these bottom-supported legs for various operations and jacked up above water during drilling. In this orientation, the legs of the rig are subjected to the fluid forces. Because of the size and length of these framework legs, they are subject to dynamic deformation. A jack-up rig is a dynamically sensitive structure, especially when it operates in deeper water.

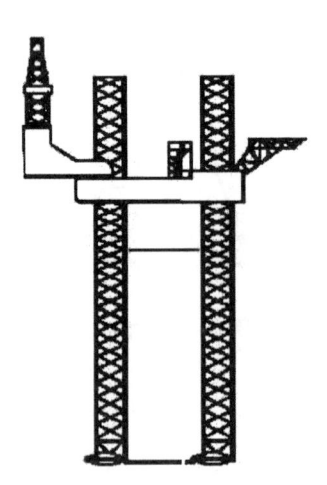

Fig. 9.30: Typical Jack-up Rig

For dynamically sensitive structures, such as jack-up rigs, compliant towers etc., the dynamic analysis assesses the safety of the unit during design storms. The dynamic analysis model becomes complicated for several reasons, primarily, nonlinearity of the drag force, random nature of the wave, and structural or foundation nonlinearity.

The current practice of analyzing such systems varies over a wide range. It starts with the simplest method of idealizing the structure as a single degree of freedom system (SDOF) for the evaluation of the dynamic amplification factor (DAF), and moves to the most complete nonlinear time domain method using random waves. Moreover, for the time domain simulation, a time series is realized, and statistical analysis is made for the design values.

The random time domain approach is considered most accurate although most complex, whereas the SDOF method is the simplest but is an approximation. Frequency domain methods are generally applied to simple linearized models. Even for a regular wave, higher harmonic components are present in the wave forces due to the nonlinearities stated above. The SDOF method severely overestimates the inertia forces near the natural period, while possibly underestimating the dynamic response further away.

In view of these limitations, a practical frequency-domain dynamic analysis method, referred as "Wave Response Analysis" (WAVRES), was developed [Chakrabarti, et al. (1999)]. WAVRES is a frequency-domain method that is less complex and computationally less intensive than the time domain solution, but produces results much better than the SDOF method, and often produces results that are close to the random time domain method. It leads to comparable results on extreme values as are produced by the random time domain analysis. The method is based on mode superposition without the assumption of linearization and considers the effect of higher harmonics, which reduces the overall resonance effects predicted by the SDOF method. This method is described here with some examples.

9.8.1 Hydrodynamic Jack-Up Analysis

For structures of interest, the members are slender and prismatic. The wave forces on this type of members are calculated by the Morison equation as described in Section 2.7.5. In the presence of current, the modified form of Morison equation is written as

$$F(t) = \frac{1}{2}\rho C_D D |u(t)+U|(u(t)+U) + \rho C_M \frac{\pi}{4} D^2 \dot{u}(t) \qquad (9.59)$$

where C_M, C_D are the inertia and drag coefficients, ρ is the mass density of water, D is the member diameter, t is time, U is the steady current and $u(t)$ and $\dot{u}(t)$ are the horizontal water particle velocity and acceleration.

The drag force introduces nonlinearity to the excitation force. For a sinusoidal wave in the presence of current, the drag term can be expressed in a series form as:

$$|u(t)+U|(u(t)+U) = u_0^2 \{C_o + C_1 \sin\omega t + C_2 \cos 2\omega t + C_3 \sin 3\omega t + C_4 \cos 4\omega t + ...\} \qquad (9.60)$$

where $u(t) = u_0 \sin \omega t$, ω is the circular wave frequency, u_0 is the amplitude of wave particle velocity and C_i (I = 0,1,2...) are functions of U/u_0.

The above expression (Eq. 9.60) is the basis of most of the linearization techniques, in which higher terms are dropped as being insignificant for mathematical simplification. The equivalent linearized drag and inertia components are evaluated and used for wave load computation and subsequent structural dynamic analysis.

The higher order terms in the dynamic analysis occur not only because of the velocity-squared term (Eq. 9.60), but also due to nonlinear waves, and varying submergence of members near the free surface. As the current strength increases with respect to the wave particle velocity, the forcing terms with frequencies of ω, and 2ω become significant, while the terms with frequencies of 3ω and 5ω approach zero.

The linearization technique in Eq. 9.60 can also be applied to random Gaussian waves. Non-Gaussian wave excitation, however, introduces additional complexities.

9.8.2 Jack-Up Structural Analysis

The dynamic displacement equation for a jack-up can be written in a matrix form in terms of the structural displacement, $\{v\}$ as:

$$[M]\{\ddot{v}(t)\} + [C]\{\dot{v}(t)\} + [K]\{v(t)\} = \{F(t)\} \qquad (9.61)$$

where [M] is the mass matrix including added mass effects for submerged members, [C] is the damping matrix, [K] is the linear stiffness matrix, and $\{F(t)\}$ is the excitation vector.

For linear systems, the above equations can be decoupled using mode shapes. For the uncoupled set of equations, the response for a steady-state harmonic excitation is harmonic, and it is fairly simple to express the solution in closed form (see Chapter 3). However, structures such as jack-ups experience large drag force. Even in regular waves, the hydrodynamic forces include harmonics higher than the wave frequency, as well as a steady force due to current (Eq. 9.60). Thus, the contribution to the total force on the structure is distributed over a range of discrete harmonics.

An irregular sea-state is represented by a wave spectrum, which can be viewed as a superposition of harmonic waves of different frequencies, amplitudes and phases. Due to the nonlinearity of the drag force, a frequency domain approach normally fails to represent the true statistical properties of excitation and response, unless special attention is given to the higher harmonics. Therefore, it is common to perform time domain calculations for the structural response. In this method, the wave train is simulated by the superposition of a large number of linear waves of different frequencies, the amplitudes of which are selected from a description of the wave spectrum, and phases are selected as random variables. Care is taken so that the resulting wave train has the same statistical properties as the original spectrum. Such time domain calculations are performed for a sufficiently long duration, and several realizations are analyzed to achieve stable statistics for the response [SNAME TR-5-5A (1997)].

9.8.3 Wave Response Method

In the present wave response method of analysis WAVRES, the force is not linearized, but rather the hydrodynamic excitation is expressed as a superposition of several harmonics. When regular waves are considered, the following steps are used:

- A linear or appropriate higher order wave theory is used and the wave force on the structure is calculated, stepping the wave one wavelength through the structure in small increments.

- The required eigenmodes of the structure are extracted so that the desired mass participation is achieved.
- Fourier components of the generalized loading for each mode are evaluated. In general, several higher harmonics (multiples of wave frequency) are present. The number of Fourier components is chosen based on the modal frequencies and the accuracy of the representation desired.
- Steady-state modal responses to these individual Fourier components due to the selected structural modes are evaluated.
- For the final structural analysis, the chosen modes are superimposed for the dynamic response with due regard to phase relationships among all components.

Since the method depends on modal superposition, the structural system should essentially be linear. In certain cases involving limited nonlinear elements, such as the foundations, linearized properties can be used in the structural model. Other improvements such as isolating static and dynamic response for enhanced recovery of the static response can also be achieved. For regular waves, this method produces results almost identical to the time domain analysis.

For random waves, the sea state is represented by a wave spectrum $S(\omega)$. Spectral analysis for any linear system involves constructing the transfer function of the response $R(\omega)$, such as base shear, lateral deflection, etc. The response spectrum $S_R(\omega)$ is computed using the following expression:

$$S_R(\omega) = S(\omega) R^2(\omega) \qquad (9.62)$$

The wave spectrum is normally given for a site, and often, standard forms of the spectrum, such as Pierson-Moskowitz, or JONSWAP, are used. For the dynamic wave analysis, the construction of the structural transfer function involves analyzing the structure for a range of frequencies at unit wave amplitude. For a zero-mean Gaussian process, once the response spectrum is evaluated, the statistical properties of the response, such as the standard deviation, zero-crossing period etc., are found by standard spectral analysis techniques. Extreme values are predicted, assuming the peaks to be Rayleigh distributed. However, when the process involves nonlinearity such

Chapter 9 Practical and Design Case Studies 441

as that due to drag, the evaluation of response and its statistics is not as straightforward. The difficulties arise because

- The transfer function itself is not linear, i.e., it is dependent on the amplitude of the regular wave selected, and
- The process is not Gaussian, and the evaluation of statistical extreme is more complex.

9.8.4 Response in Regular Waves

In order to demonstrate the relative merits of the various analysis techniques above, an independent latticed leg jack-up rig was chosen as an example. The particulars of the 3-leg jack-up model are shown in Table 9.3.

Table 9.3: Example Jack-up Rig Particulars

Item	Value	Metric
Leg length	477 ft	145 m
Total weight	25,153 kips	1.12E8 N
Elevated weight	15,100 kips	6.72E7 N
Water depth	300 ft	91 m
Penetration of spudcan	27 ft	8.2 m
Air gap	50 ft	15.2 m
Equivalent leg diameter	87 in	2210 mm
Effective drag coefficient C_D	4.9	4.9
Inertia coefficient C_M	2.0	2.0

All calculations for time-domain as well as frequency-domain wave response analysis were performed with standard structural design software. Two cases were considered: (a) pinned base, and (b) fixed foundation spudcan.

The behavior of the spudcan that penetrates the soil is nonlinear due to the soil-structure interaction. In this case, spudcan springs were simulated with three translational and two horizontal rotational directions. These springs represented linearized springs for the nonlinear spudcan behavior. It was assumed that the global inertial load on the structure could be predicted well by using a time-averaged spring stiffness of the foundation, as long as failure of the soil did not occur. The fundamental natural period of vibration for this model when the spudcans were pinned was 11 s, whereas with fixed spudcan, it was 7.3 s. Due to the wide variation of the natural period for the two cases, the bottom conditions analyzed represent two completely different dynamic system characteristics.

Ten structural modes were included in the calculations. A 7 percent damping factor was chosen for all modes. For equivalent lattice leg segments, the hydrodynamic parameters, C_D and C_M, were evaluated based on procedures given in SNAME 5-5A (1997).

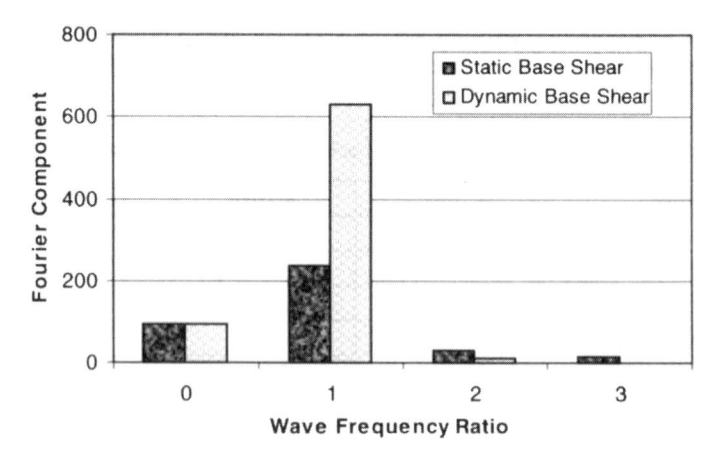

Fig. 9.31: Higher Harmonics in Regular Wave Force

To demonstrate the effect of higher harmonics on the dynamic response, the structure pinned to the bottom was analyzed using the frequency-domain wave response method for a regular wave height of 32.7 ft (10 m) and period of 13 s and a small current of 0.9 knots (0.46 m/s). A Fourier decomposition of the static and dynamic base shear is plotted in Fig. 9.31. The Fourier components for static base shear show the presence of higher harmonics in the time history. The corresponding dynamic components get amplified or de-amplified, depending on the proximity of the natural frequencies of the structure.

The design force acting on the structure is often defined as the sum of the static force obtained from the drag-inertia parameter, assuming the wave is regular and the inertia force is obtained from either a spectral or a random time domain analysis. This definition was used to compute the total design force for the pinned spudcan, as well as when rigidly fixed using the WAVRES method and in a more elaborate time domain. The results are compared in Fig. 9.32. For the wave response method, the inertia force (dynamic minus static) was calculated using Rayleigh distribution, and was added to the static maximum force. As shown in the figure, the random time domain and spectral wave response calculation lead to results very close to each other.

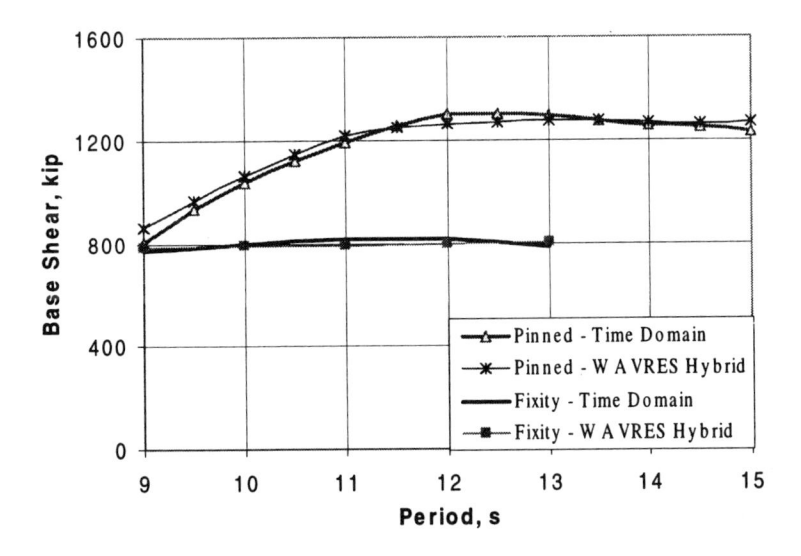

Fig. 9.32: Comparison of Design Force on a Jack-up

In view of the above study, the hybrid method of spectral wave response analysis is shown to be a practical alternative to the random time domain method. This frequency domain method is efficient and easy to apply, compared to the time domain analysis. As demonstrated in this example, in many situations, this method can provide acceptable results.

9.9 Impact of a Submarine with a Jacket Structure

In general, the resonant vibration of jacket structures in the North Sea conditions had not been an important consideration in their design. This is primarily because the natural frequencies of the jacket structures had traditionally been far removed from the wave frequencies. However, a deep water jacket may have natural periods above 3 s approaching the wave energy zone, and the resonant response is important in the design from the point of view of fatigue life and extreme value.

A jacket type structure called the Osberg jacket was built in the North Sea in the summer of 1987. The Osberg jacket represented a drilling and wellhead platform situated in a water depth of 110 m. It was an eight-legged platform housing 48 conductors braced at five horizontal bracing levels. Within the year of its installation, the jacket was struck accidentally by a German submarine. The submarine had a displacement of about 500 metric tons and was traveling about 25 m below the mean water level (MWL) at the time of the accident. The hit was in the form of an impact load of short duration. The impact caused a serious damage to one of the diagonal bracing members located between 14 m and 43 m below the MWL [Thuestad and Nielsen (1990)]. The tube wall experienced a dent of about ten times the wall thickness. The damage is shown in Fig. 9.33. It was important to re-analyze the structure to determine its reserve strength. It was also necessary to comply with the design code subsequent to the accident and the damage to the bracing member.

The above jacket was instrumented to record its inclination in two orthogonal directions, the deck acceleration in three directions, accelerations at two corners of the platform, and nominal strains at numerous bracings. Several modes of vibration of the structure were discovered in the recorded data from impact, depending on the location of the instrument. The data analysis from these time history records showed that the excitation caused vibration of the structure in its first three modes. In addition, the impact caused a decayed oscillation of the jacket structure at its natural frequencies.

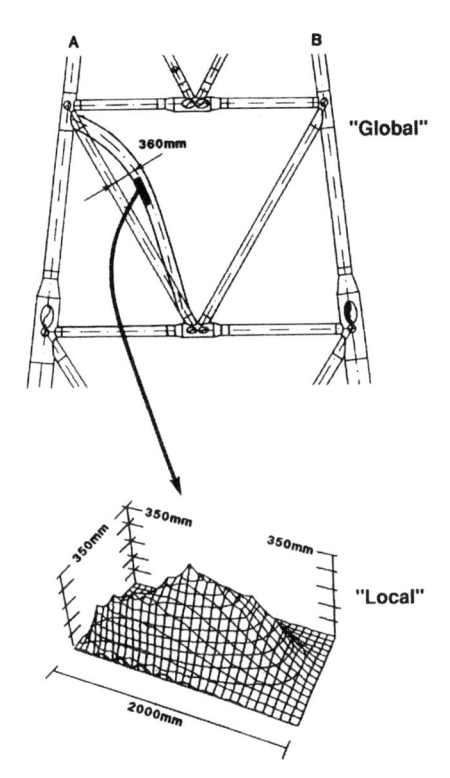

Chapter 9 Practical and Design Case Studies 445

Fig. 9.33: Damage to Jacket Bracing from Impact [Thuestad and Nielsen (1990)]

9.9.1 Safety Check on Residual Strength

Safety checks of the structure were performed [Thuestad and Nielsen (1990)], in which the finite element structural software was coupled with progressive collapse analysis. The environmental load from waves approaching from the north was incremented in steps until the ultimate capacity of the structure was reached. The analysis was performed for the intact and the damaged platform. The results are shown in Fig. 9.34. The vertical axis represents a load factor, l_{LF}, which is defined as the ratio of the actual load to the design environmental load. The reserve strength of the structure is characterized by the ability of the undamaged structure to withstand loads exceeding the design load. The results in Fig. 9.34 showed that the intact jacket had a reserve strength factor of 3.8.

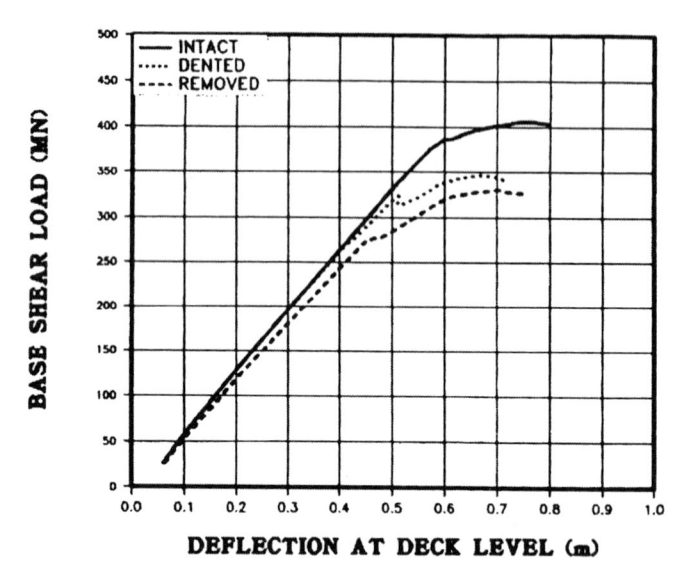

Fig. 9.34: Reserve Strength of Jacket Before and After Collision [Thuestad and Nielsen (1990)]

Even when the damaged member was removed in the analysis, the reserve strength showed a factor above 3. The residual strength factor for the damaged structure was about 90%, while it was about 85% when the bracing member was removed. The jacket was considered safe for the 100-year environmental storm, even with the damaged member. In order to satisfy the code requirements of the Norwegian Petroleum Directorate, however, the bracing member was replaced.

9.9.2 Structural Damping

Since the structure was instrumented, the impact from the submarine permitted investigation of the amount of damping present in the jacket structure. It was particularly important for at least two reasons. Deep water

Chapter 9 Practical and Design Case Studies

jacket structures may be susceptible to resonance from the environmental loads in which damping plays an important design consideration. Secondly, determination of the damping parameter in a model scale may be difficult, particularly if the damping turns out to be nonlinear. Therefore, the unfortunate accident provided the opportunity to determine the damping of the full-scale structure without any scale effect. A typical recording of the acceleration time history measured by a linear servo accelerometer located at the lower deck level following the impact is shown in Fig. 9.35.

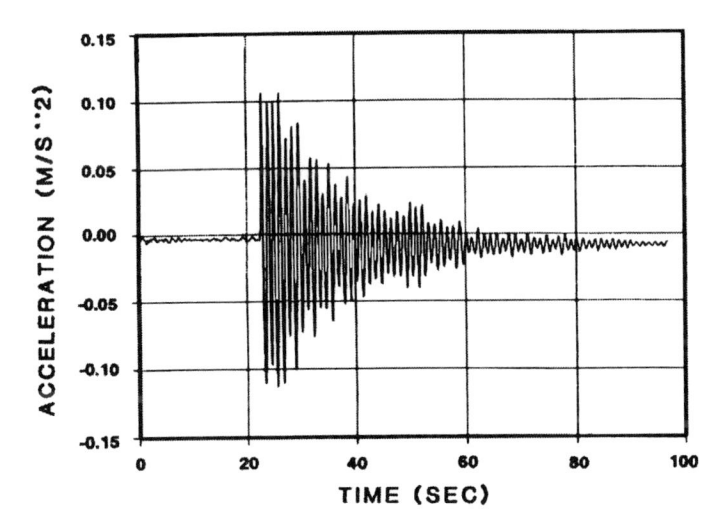

Fig. 9.35: Recorded Acceleration Time History Near the Damage [Thuestad and Nielsen (1990)]

The significant wave height at the time of impact was low, having a value of 1.8 m. No significant wave-induced response could be observed in the acceleration measurement. It was concluded that the impact oscillation corresponded only to the eigenmodes of the jacket structure. A spectral representation of this time series obtained by the Fast Fourier Transform (FFT) is shown in Fig. 9.36. The peaks in the frequency spectrum at 0.65 Hz and 0.90 Hz correspond to the second (i.e., sway in the North-South direction) and third (i.e., rotation) modes of vibration of the structure. The first mode of oscillation of the jacket was 0.55 Hz (corresponding to the lowest natural period in sway of 1.8s), which had a direction perpendicular to the second mode. Thus, the decaying oscillation corresponded to both the second and third mode of oscillation of the jacket.

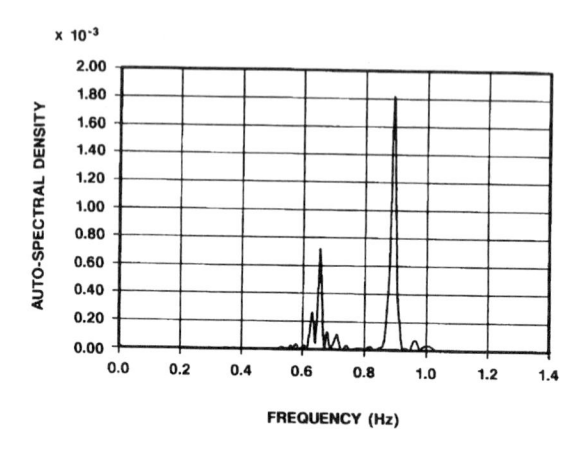

Fig. 9.36: Spectral Density of the Recorded Acceleration Above [Thuestad and Nielsen (1990)]

In order to examine the hydrodynamic and structural damping of the jacket, the first task was to separate the two frequencies in the acceleration data. This was accomplished with the help of a band pass digital filter. It was assumed that the different modes of vibration were uncoupled with no interaction among them. This allowed the above decay curve to be treated as a SDOF system. The method of computing the damping coefficient from a decay curve was discussed in Section 3.3.1.

The nature of the total damping in the jacket structure was investigated by fitting the amplitude of the decaying displacement individually with linear, Coulomb and drag type damping. Each natural frequency mode was analyzed independently. The Coulomb model provided the best fit of the measured data. A comparison of the measured decay and the exponential fit is shown in Fig. 9.37. The natural period of the free oscillation in this case corresponds to the third mode of the jacket structure.

As shown in Eq. 3.81, the Coulomb damping factor depends on the amplitude of oscillation. The best-fit damping estimate based on Coulomb damping is shown in Fig. 9.37. The measured data is shown as open circles

Chapter 9 Practical and Design Case Studies 449

in the figure. The vertical line around 0.2 corresponds to a damping factor of infinity. At this point the motion of the structure stops. The damping was highly nonlinear, with a damping factor varying from 0.5 to 3 percent over the displacement amplitude. The damping steadily decreased with the increase in the motion amplitude. Similar ranges were found in other cases of frequency as well [Thuestad and Nielsen (1990)].

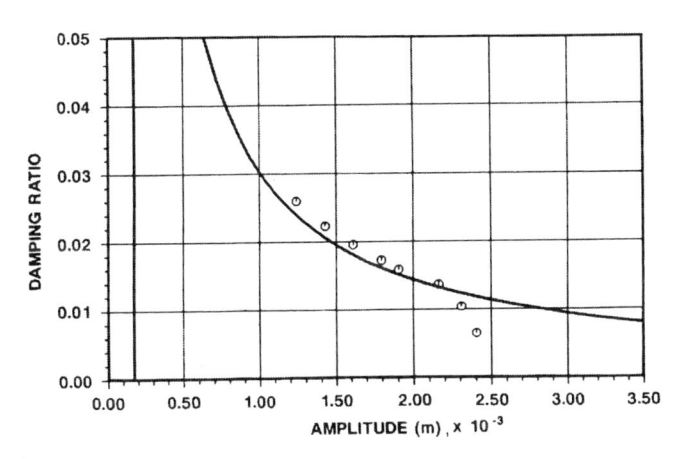

Fig. 9.37: Damping Factor for the Jacket Structure [Thuestad and Nielsen (1990)]

9.10 References

1. Abramson, H. N., "The Dynamic Behavior of Liquids in Moving Containers With Application To Space Vehicle Technology", NASA SP-106, 1966.

2. American Petroleum Institute, "Recommended Practice for Planning, Designing and Constructing Fixed Offshore Platforms", API-RP2A, 20th Edition, 1992.

3. Chakrabarti, P., Chakrabarti, S. K., and Mukkamala, A., "A Practical Frequency-Domain Method for Random-Wave Analysis and Its

Application to Jack-up Units", Proceedings of Offshore Technology Conference, Houston, TX, OTC 10795, 1999.

4. Chakrabarti, S.K. and Cotter, D.C., "Motion Analysis of an Articulated Tower", Journal of the Waterway, Port, Coastal and Ocean Division, ASCE, Vol. 105, August 1979.

5. Chakrabarti, S.K. and Cotter, D.C., "Transverse Motion of an Articulated Tower", Journal of the Waterway, Port, Coastal and Ocean Division, ASCE, Vol. 106, November 1979 and February 1980.

6. Chakrabarti, S.K. and Gu, G., "Environmental Effects on a Structurally Dense Jacket Platform Model", Proceedings of the Offshore Mechanics and Arctic Engineering Symposium, Vol. 1, Part A, June, 1996.

7. Chakrabarti, S.K. and Hanna, S.Y., "High Frequency Hydrodynamic Damping of a TLP Leg", Proceedings of the Offshore Mechanics and Arctic Engineering Symposium, Vol. I, Part A, June 1991, pp. 147-152.

8. Chakrabarti, S.K., Kriebel, D., and Berek, E.P., "Forces on a Single Pile Caisson in Breaking Waves and Current", Applied Ocean Research Vol. 19, 1997.

9. Chakrabarti, S.K., "Internal Waves in a Large Offshore Storage Tank," Journal of Energy Resources Technology, ASME, Vol. 115, 1993, pp 133-141.

10. Chakrabarti, S.K., "Response Due To Moored Multiple Structure Interaction", Marine Structures, Vol. 14, Nos. 2, 2001, pp. 231-254.

11. Chakrabarti, S.K., "Numerical Simulation of Multiple Floating Structures with Nonlinear Constraints", Proceedings of the Offshore Mechanics and Arctic Engineering Symposium, June, 2001.

12. Kriebel, D.L., Berek, E.P., Chakrabarti, S.K. and Waters, J.K., "Wave-Current Loading on a Shallow Water Caisson," Proceedings of Offshore Technology Conference, Houston, TX, OTC 8067, 1996.

13. Huse, E., "Resonant Heave Damping of Tension Leg Platform," Proceedings of Offshore Technology Conference, Houston, TX, OTC 6317, 1990.

14. Huse, E., "Springing Damping of Tension Leg Platform," Proceedings of Offshore Technology Conference, Houston, TX, OTC 7446, 1994.

15. Huse, E., "Experimental Investigation of Deep Sea Riser Interaction," Proceedings of Offshore Technology Conference, Houston, TX, OTC 8070, 1996.

16. Huse, E., Kleiven, G., and Nielsen, F.G., "Large Scale Model Testing of Deep Sea Risers," Proceedings of Offshore Technology Conference, Houston, TX, OTC 8701, 1998.

17. SNAME, "Site Specific Assessment of Mobile Jack-Up Units", TR5-5A, First Edition, Revision 1, May, 1997.

18. Thuestad, T.C., Nielsen, F.G., "Submarine Impact with The Osberg Jacket", Proceedings of the Ninth Conference on Offshore Mechanics and Arctic Engineering, 1990, pp. 493-500.

19. Thuestad, T.C., Nielsen, F.G., "Damping of Resonant Oscillation of Jacket Structure Based on Full Scale Measurement", Proceedings of the Eleventh Conference on Offshore Mechanics and Arctic Engineering, 1992, pp. 223-230.

List of Symbols

Symbol	Description
a	cylinder radius or wave amplitude
A	projected area of the structure in water
B	ship beam or damping coefficient
$[C]$	stiffness matrix
c	damping coefficient or wave celerity
C_A	added mass coefficient
C_D	drag coefficient
C_f	correction factor or friction coefficient
C_L	lift coefficient
C_M	mass coefficient
C_p	pressure coefficient
C_s	surface effect coefficient
C_y	lateral force coefficient
D	structure (cylinder) diameter
d	water depth
E	Young's modulus
f	cyclic frequency or force per unit length
F	total force
f^*	frequency ratio
F_I	In-line force
F_L, f_l	lift force
f_n	structure natural frequency
Fr	Froude number
f_s	vortex shedding frequency
G	Green's function
g	gravitational acceleration
H	wave height
H_s	significant height
i	imaginary quantity (0,1)
I	moment of inertia
k	spring constant or wave number
K	mean surface roughness

Symbol	Description
KC	Keulegan-Carpenter number
K_s	Stability parameter
l	structure length
L	wave length
L_{pp}	length between perpendiculars
M	exciting moment
m_0	structure mass
m_k^i	mass or moment of inertia of module i in k^{th} mode
M_{lk}^{ji}	added mass matrix
N_{lk}^{ji}	damping matrix
N	number of modules
n	direction normal of the structure surface
p	dynamic pressure
$P(\cdot)$	cumulative probability function
$p(\cdot)$	probability density
Q	Bernoulli's constant
$Q(\cdot)$	probability of exceedance
R	restoring moment
R_A	aspect ratio
R_B	blockage ratio
R_E	elevation ratio
Re	Reynolds Number
r_i	radial coordinate of module i
R_k^i	radiation of module i due to unit displacement
$S(\cdot)$	energy density
S	immersed surface
St	Strouhal number
t	time
T	wave period
T_r	short-term record length of time
U	free stream velocity
u,v,w	velocity components in XYZ direction
$\dot{u},\dot{v},\dot{w}$	acceleration components in XYZ direction
U_{dev}	deviation of velocity from mean local velocity
U_M	mean current velocity
U_{MAX}	maximum current velocity
u_n	velocity of unit amplitude

List of Symbols

V	volume of structure
V_r	reduced velocity
W	chamber width
x_0	linear oscillation amplitude
y	vertical coordinate
y_1	distance from cylinder axis to free surface
y_2	distance from cylinder axis to bottom boundary
α	constant of proportionality
β	shear parameter, or beta parameter Re/KC
δ	logarithmic decrement
ε	phase angle
ϕ	fluid velocity potential
ϕ_i	incident wave velocity potential
ϕ_r	radiated wave velocity potential
ϕ_s	scattered wave velocity potential
γ	peakedness parameter
η	wave profile
λ	scale factor
μ	mean value
ν	kinematic viscosity of water
θ	ωt or polar coordinate
ρ	mass density of structure
ρ_s	mass density of structure
σ	rms value
ω	wave circular frequency
ω_n	natural frequency
ω_p	peak frequency
ψ	angular motion
ζ	damping factor

List of Acronyms

API	American Petroleum Institute
CFD	Computational Fluid Mechanics
CG	Center of Gravity
DAF	Dynamic Amplification Factor
DNV	Det Norske Veritas
ISSC	International Ship Structures Congress
JONSWAP	Joint North Sea Wave Project
MOB	Mobile Offshore Base
MWL	Mean Water Level
PM	Pierson-Moskowitz
SDOFS	Single Degree of Freedom System
SLIM	Single Leg Anchor Mooring
SPM	Single Point Mooring
SWL	Still Water Level
SWL	Still Water Level
TLP	Tension Leg Platform
TTR	Top Tensioned Riser

Conversion Factors

Item	English	Metric
Length	1 in	25.4 mm
Length	1 ft	0.3048 m
Volume	1 cu. ft	0.0283 m^3
Force	1 lbf	4.448 N
Force	1 kip	4448 N
Mass	1 lbm	0.4536 kg
Speed	1 knot	0.5144 m/s

Author Index

Abramowitz, M. 86, 118, 185, 206
Abramson, N. 83, 84, 118, 407, 421, 450
Achenbach, E. 323, 379
Ahilan, R.V. 191, 192, 207
Allen, D.W. 324, 325, 357, 373, 379
American Petroleum Institute 45, 47, 49, 273, 302, 316, 399, 401, 451
Arie, M. 317
Ascombe, G.T. 382
Bearman, P.W. 320, 325, 330, 331, 336, 338, 339, 340, 341, 342, 374, 380, 383
Bendat, J.S. 187, 197, 199, 200, 206, 219, 248
Berek, E.P. 451, 452
Biolley, F. 382
Bishop, R.E.D. 263, 316
Borgman, L.E. 233, 248
Brigham, E.O. 203, 206
Broch, J.T. 239, 248
Bruchi, R. 383
Bryndeum, M. 383
Chakrabarti, P. 439, 451
Chakrabarti, S.K. 79, 118, 163, 166, 173, 175, 206, 220, 229, 249, 266, 276, 316, 389, 399, 404, 411, 424, 432, 451, 452
Chantrel, J. 285, 316
Chaudhury, G. 359, 380
Cheng, Y. 318
Clough, R.W. 198, 202, 206
Coney, W.B. 173
Cotter, D.C. 79, 118, 229, 249, 316, 424, 451
Crede, C.E. 4, 7
Davies, M.E. 326, 380
Davis, J.T. 330, 380
DnV 359, 380
Doodson, A.T. 146, 173

Downie, M.J. 370, 371, 380
Durand, A. 382
Dyer, R.C. 191, 192, 207
Esperança, P.T.T. 380
Fernandes, A.C. 377, 380
Ferrari Jr., J.A. 318, 381, 383
Flatschart, R.B. 362, 363, 381
Fylling, I.V. 318
Fyrileiv, O. 383
Gerrard, J.H. 322, 381
Givoli, D. 281, 316
Goldstein, R.J. 157, 173
Gottlieb, O. 220, 223, 249
Graham, J.M.R. 380
Griffin, O.M. 261, 316
Gu, G. 399, 451
Gumbel, E.J. 185, 207
Hafen, B.E. 375, 376, 381
Hall, C. 380
Halse, K.H. 332, 350, 381
Hanna, S.Y. 389, 451
Harris, C.M. 4, 7
Hassan, Y. 263, 316
Heineche, E. 323, 379
Henning, D.L. 324, 325, 379
Himeno, Y. 317
Hoerner, S.F. 256, 263, 264, 316
Hover, F.S. 380
Humphries, J.A. 256, 258, 316
Huse, E. 373, 378, 381, 386, 388, 391, 392, 394, 395, 396, 397, 452
Ikeda, Y. 288, 293, 296, 317
Incecik, A. 380
Inoue, Y. 312, 317
Ishida, H. 382
ISSC 332, 381
Iwagaki, Y. 328, 382
Johnson, J.W. 49

Kaasen, K.E. 350, 382
Kallinderis, Y. 342, 343, 344, 345, 383
Katayama, M. 367, 368, 382
Kim, M.H. 280, 317
Kim, Y.B. 317
Kiya, M. 265, 266, 270, 317
Kleiven, G. 452
Kriebel, D.L. 405, 451, 452
Lamb, H. 2, 7
Landolt, A. 380
Larsen, C.M. 332, 382
Larsen, K. 318, 369
Le Cunff, C. 351, 352, 353, 382
LeMehaute, B. 254, 317
Li, Y. 364, 382
Libby, A.R. 168, 173
Lie, H. 355, 356, 382, 369
Lin, X.W. 380
Lin, Y.K. 183, 207
Littlebury, K.H. 136, 173
Liu, F.C. 381
Lyons, G.J. 357, 382
Madden, R. 173
Mair, W.A. 256, 261, 317
Marol, P. 285, 316
Martins, C.A. 384
Masch, F.D. 259, 260, 317
Maull, D.J. 261, 262, 317
McCarthy, D.J. 171, 172, 173
Meggett, D.J. 381
Meirovich, L. 310, 317
Meneghini, J.R. 318, 338, 339, 340, 380, 381, 383, 384
Milne-Thompson, L.M. 2, 7
Miya, E. 382
Moe, G. 310, 318
Moore, W.L. 259, 260, 317
Morgan, K. 349, 384
Morison, J.R. 42, 46, 49, 78, 233, 235, 315, 348, 355, 402, 439
Mørk, K.J. 333, 383
Morrison, D.G. 364, 382

Mukkamala, A. 451
Naess, A. 242, 245, 249
Newman, J.N. 276, 318
Nielsen, F.G. 445, 446, 447, 448, 449, 450, 451, 452
Noblesse, F. 276, 318
Nygaard, I. 380
O'Brien, M.P. 49
Ormberg, H. 314, 318
Palo, P. 316
Peltzer, R.D. 326, 327, 383
Penzien, J. 198, 202, 206
Piersol, A.G. 187, 197, 199, 200, 206, 219, 248
Price, S.J. 339, 383
Roberts, J.B. 93, 94, 118
Rooney, D.M. 327, 383
Saether, L.K. 378, 381
Saltara, F. 318, 381, 384
Sarpkaya, T. 30, 31, 32, 49, 339, 383
Schaaf, A.S. 49
Schulz, K. 342, 343, 344, 345, 383
Silva, R.M.C. 380
Siqueira, C.L.R. 271, 318, 361, 362, 381, 383
SNAME 440, 443, 452
Sodahl, N. 318
Solaas, F. 382
Sphaier, S.H. 380
Stansby, P.K. 257, 258, 261, 263, 264, 317, 318
Stegun, I.A. 86, 118, 185, 206
Sunderan, S. 312, 317
Tamura, H. 317
Tanaka, N. 317
Teigen, P. 242, 245, 249
Thompson, W.T. 111, 112, 114, 118
Thuestad, T.C. 445, 446, 447, 448, 449, 450, 451, 452
Timoshenko, S.P. 166, 173
Triantafyllou, M.S. 380
Tsahalis, D.T. 358, 384

Author Index

Tveit, Ø. 369
Unoki, K. 382
Vaicaitis, R. 335, 384
Vandiver, J.K. 318, 382
Verley, R. 383
Walker, D.H. 256, 258, 316
Waters, J.K. 452

Wung, C.C. 313, 318
Yamamoto, C.T. 349, 384
Yang, J.C.S. 225, 249
Young, R.A. 261, 262, 317
Zdravkovich, M.M. 360, 365, 377, 384
Zienkiewicz, O.C. 241, 249, 349, 384

Subject Index

Accelerometer 154, 156, 347, 378, 384, 385, 439
Bernoulli's Equations 5, 13, 14, 36, 403
Boundary Element Method 265, 267, 328
Cable 7, 149, 303, 311, 318, 339, 341–344, 348, 364
Cauchy 42, 43, 135, 140–142, 145, 146, 151, 170, 408
Continuity 5, 10–11
Continuous Systems 106–107, 230, 232, 233
Coupled Dynamic 305–307
Cross Spectra 193–194
Current Blockage Factor 272, 273, 399, 402
Damping Factor 58–62, 64, 70, 71, 77, 78, 82, 89, 116, 218, 219, 246, 282, 284, 287, 288, 301–304, 341, 379, 388–389, 391–402, 409, 414–415, 421–422, 443, 450
Drag Force 17–18, 20–23, 26–27, 41, 43, 48, 141–142, 154, 233–234, 248, 253, 271–272, 292–293, 320, 331, 333, 342, 355, 372–373, 393, 438–440
Duffings Equation 86–90, 92, 282
Eigenfrequency 333, 346, 350, 356, 415–416
Eigenvalues 237–239, 242, 348
Eigenvectors 237–239
Elastic Model 143, 145, 149, 417–422
Elastic Structures 111–115, 148–150, 346–347, 417, 423
Envelope Functions 189–193
Equation of Motion 55, 57, 61, 63, 69, 73–76, 80, 86–87, 90–92, 97, 106, 108–109, 113, 198, 202, 215, 219–222, 227, 235, 237, 240, 277, 283, 288, 306–310, 335, 340, 387, 406, 424–425
Extreme Values 177, 185, 186, 189–193, 241–244, 439, 442, 445
Field Experiment 349, 392–398
Finite Element Method 237, 239–241, 273, 280–281, 310, 355
Floating Structure 62, 66, 79, 90, 96–98, 107, 136, 141–144, 148, 148–149, 152, 190–194, 216, 224–225, 227–228, 233–235, 241–242, 252, 277–280, 288, 310–315, 369–371, 388, 422, 430–431
Free Surface 33–38, 42, 91, 108, 132, 146, 161, 256, 263, 268, 270, 273, 275, 289, 297, 410, 413–415, 439
Free Vibration 51–52, 57–63, 76–79, 82–83, 85, 106–107, 237–238, 254, 391
Frequency Domain 108–109, 194, 214–216, 219, 227–228, 247, 269, 333, 355–356, 365, 422, 428, 439, 441, 446
Froude Similitude 139–140, 144, 153, 155
Gaussian Distribution 181–183, 225, 241, 243–244, 441, 443
Hydrodynamic Coefficients 27, 41–42, 46–47, 227, 231, 233, 278, 234, 356, 428, 436
Hydroelasticity 144, 430–436
Impact Loading 175, 399–406
Instability 90, 170, 258, 262, 282, 284–286, 316, 364, 396, 423–428
Jack-Up Unit 438–445
Lift Force 17–24, 32–33, 43, 91, 253, 264, 289, 322, 329, 334–338, 349, 351, 353, 426, 428

Linear Diffraction 231, 274–278, 287–288, 298, 433
Mechanical Oscillation 27–29, 69–73, 130–131
Model Testing 120–121, 132, 143, 395, 400
Modes of Vibration 111–112, 148, 214, 237, 286, 325, 447, 450
Mooring Line 35, 108, 127, 150–153, 162–163, 192–193, 216, 220, 227–228, 230–231, 252, 287, 306, 312–316, 354
Nonlinear Systems 75–76, 215–218, 220–224, 232, 237
Numerical Applications 43, 73, 110, 193, 202, 220, 227–228, 236–241, 271, 273–282, 306, 310–314, 331, 335, 337–345, 349, 351–352, 361–363, 428–429, 433
Periodic Vibration 52–54
Potential Function 12, 273, 281, 338, 412
Probability Density Function 179–186, 242
Probability Distribution Function 178–184, 189–193, 242
Random Decrement Technique 224–227
Random Input 214, 232
Random Process 175–177, 185–186, 190, 225, 336
Random Variables 176–186, 440
Rayleigh Distribution 183–184, 241, 444
Regular Waves 42, 66, 109, 162, 202, 285, 402, 417, 427–429, 439–444
Response Spectra 193, 355
Restoring Force 54–55, 73, 82–94, 108, 218, 220–221, 282, 285, 386, 425, 427, 434
Rigid Body 94–100, 111, 236, 252, 277, 288, 346–348, 422, 430

Riser 113, 115, 123, 131–132, 151, 153–157, 163–164, 171, 252, 272, 305–310, 313–314, 319, 332–333, 348–364, 367, 372–378, 392–398
Roll 90–91, 128, 288–306, 369
Rotational Flow 11–12
Self-Excited Vibration 115–116
Shear Flow 127, 136, 256–257, 259–262, 254, 265–271, 325–327, 333, 350–351
Single Point Mooring 91, 227–228, 281
Spectral Density 187–189, 197, 220, 242, 364
Spoilers 264, 364–378
Steady Flow 15, 19, 20, 23, 51, 121, 252, 264, 271–272, 286, 320, 323, 331, 333–336, 340, 350, 365–366
Stochastic Dynamics 236–239
Stochastic Linearization 233–236, 355
Strouhal Similitude 143
Time Domain 218–224, 227–228, 232–233, 242, 314, 332, 336, 354–355, 425, 427, 438–440, 444–445
Transient Response 56–63
Transverse Force 33, 43
Uniform Flow 15, 19, 253–254, 256, 261, 262–269, 323, 328, 335, 350–351
Vibration Measurement 171
Vortex Formation 321, 324–330, 366–367, 377, 386
Vortex Induced Vibration 17, 258, 319, 350, 393
Vorticity 12, 24, 253, 261–262, 321–322, 325, 328, 337–339
Waves 33–35, 38, 43
Wave Equation 114